ADVANCING
EXCELLENCE

ADVANCING EXCELLENCE

The Making of Texas A&M University's Hagler Institute for Advanced Study

Tim Gregg

Texas A&M University Press
College Station

First edition

♾ This paper meets the requirements of ANSI/NISO Z39.48-1992 (Permanence of Paper).
Binding materials have been chosen for durability.

Library of Congress Cataloging-in-Publication Data

Names: Gregg, Tim, author. | Hutchison, Kay Bailey, 1943–writer of foreword. | Junkins, John L., writer of introduction, writer of afterword.
Title: Advancing excellence: the making of Texas A&M University's Hagler Institute for Advanced Study / Tim Gregg.
Description: First edition. | College Station: Texas A&M University Press, [2026] | Includes bibliographical references and index.
Identifiers: LCCN 2025031369 (print) | LCCN 2025031370 (ebook) | ISBN 9781648433771 (cloth) | ISBN 9781648433788 (ebook)
Subjects: LCSH: Hagler Institute for Advanced Study—History. | Hagler Institute for Advanced Study—Biography. | Texas A & M University—History. | Academic achievement—Texas—College Station—History—21st century. | Education, Higher—Aims and Objectives—Texas—College Station—History—21st century.
Classification: LCC LD5309 .G735 2026 (print) | LCC LD5309 (ebook)
LC record available at https://lccn.loc.gov/2025031369
LC ebook record available at https://lccn.loc.gov/2025031370

Unless otherwise noted, all photographs are from the collection of the Hagler Institute for Advanced Study.

Cover design by Dave Kessler Design.

Dedicated to Elouise Junkins, the First Lady of Excellence.

CONTENTS

A gallery of images follows page 78.

FOREWORD

Kay Bailey Hutchison

I embarked on a journey with John Junkins over twenty years ago. It started when I served on the Senate Appropriations Committee, and though we had Nobel laureates at Rice, UT Southwestern, UT Austin, and Texas A&M as well as many National Academy members, our Texas institutes were not winning their share of federal research grants. The science and medicine support was going to universities on the East and West Coasts. Our Texas research universities didn't have a sufficient understanding of the federal priorities, so they frequently weren't applying for the projects where they did have qualified experts.

I invited leaders of our research universities to Washington to hear from the federal research agencies. At those meetings over a three-year period, the academic leaders said they did not know what was being studied at the other universities and suggested there be a forum where the members of the National Academies would meet and have discussions and presentations on current research and share findings, hoping to increase awareness and collaboration. Peter O'Donnell, a philanthropist in Dallas, wanted to help this effort and agreed to have a meeting in his office with about ten leaders of Texas research universities to make a plan to achieve our goal: more academic research in Texas. John Junkins represented Texas A&M at that first meeting that included UT Southwestern Nobel laureates Drs. Michael Brown and Joe Goldstein, Rice University Pres. Malcolm Gillis, and MD Anderson Pres. John Mendelsohn.

Junkins immediately got it . . . and was on the ground floor of the creation of the Texas Academy of Medicine, Engineering, Science and Technology (TAMEST). The sole criterion for membership in the academy is that you have been elected by your peers to one of the prestigious National Academies: medicine, science, or engineering.

TAMEST celebrated its twentieth anniversary in 2024. TAMEST bestows an award for lifetime achievement in fostering science research. It is only given when there is a consensus of a deserving person or group. In 2022, it was given to John Junkins. His early support of TAMEST and establishment of the Hagler Institute for Advanced Study (HIAS) to bring America's best academic stars to Texas A&M have contributed significantly to our common goal.

This book describes John's leadership and commitment to bringing these thought leaders from around the world to Texas A&M. The institute gives them freedom to pursue their creativity without strings. The stories of Hagler Fellows are varied and fascinating in different ways. John's vision was supported by Chancellor John Sharp and the Texas A&M University System Board of Regents. Two milestones achieved by John cemented the future. First, Norm Augustine, the former Lockheed Martin CEO, was recruited to chair the advisory board; this gave prestige and credibility. Then, Jon Hagler, a successful former student, stabilized this project with an endowment to name the institute.

The book's interviews with Hagler Fellows and the impact the Institute has had on their productive careers are inspiring. Many of the Fellows have settled in College Station to become faculty members at Texas A&M. They are adding quality to the learning experience for students, many of whom have become academic leaders in universities throughout the United States. The student experience with the Fellows is covered with interviews in which the students discuss the impact engagement with Fellows had in shaping their own futures.

The Institute also has an association with the Lindau Nobel Laureate Meetings organization in Germany, where students can attend the annual German conference and hear lectures from Nobelists in their fields of accomplishment. These opportunities have given students experiences they might never have had access to without the A&M program.

This book chronicles how an idea becomes a dream and how the dream is molded into a plan. It is a how-to book for any entrepreneur who dreams big and has the determination to succeed, overcome obstacles, be willing to work harder than anyone else, and never give up.

John Junkins is an academic entrepreneur who has made an indelible imprint on Texas A&M University.

INTRODUCTION

John Junkins

Having spent roughly half my life as a member of the faculty at Texas A&M University—the former "Agricultural and Mechanical College of Texas"—I've developed a deep appreciation for this place. It's given a lot to me, as it has to countless others, not only for students, faculty, alumni, and friends, but also for the citizens of our state, our nation, and the world. This has been accomplished primarily through the extraordinary and meaningful research conducted here.

From its post–Civil War beginnings as a land-grant institution, Texas A&M has become a place where great minds are cultivated, nurtured, and given free rein. It is in that spirit that I had in mind the creation of an institute here to enhance excellence through the exchange of ideas and collaboration with the world's finest scholars, researchers, and industry leaders. That vision, cultivated in large part by the university's *Vision 2020* aspirational plan launched in 1997, eventually took shape as the Texas A&M University Institute for Advanced Study (TIAS). Today we're known as the Hagler Institute for Advanced Study, named for our largest donor and one of the university's great friends, Jon Hagler, Class of '58. In fact, it was Hagler and his former Texas A&M Corps of Cadets roommate, Ray Bowen, who masterminded the *Vision 2020* plan, Hagler as a successful and concerned alumnus and Bowen as president of the university.

Our Institute for Advanced Study owes much to both Hagler and Bowen, but there are two others with ties to Texas A&M whose aspirations and achievements established a model for what I believed the Institute could become.

Dr. Norman Borlaug arrived at Texas A&M as a distinguished professor of international agriculture in 1984, a year before I joined the university's aerospace engineering department. It would not be inaccurate to say that Borlaug serves as a model for the kinds of individuals who would eventually receive distinction as Fellows of the Hagler Institute for Advanced Study.

He was just a little ahead of our time.

Borlaug came to Texas A&M at the age of seventy. He brought with him the Nobel Peace Prize and a Presidential Medal of Freedom. He also had a well-deserved global reputation as the "Father of the Green Revolution." An Iowa farm boy turned microbiologist through his studies at the University of Minnesota, Borlaug accepted a position with the Rockefeller Foundation–sponsored Cooperative Mexican Agricultural Program in 1944. There, he was charged with leading efforts to improve wheat-production methods within the country. Borlaug, though, saw his mission as something larger than that. He was intent on leading the development of improved grain varieties to feed all the hungry people of the world.

Through his scientific expertise in plant pathology and genetics, Borlaug succeeded in creating multiple varieties of high-yield and disease-resistant wheat grain. In time, experts credited him with saving a billion people from starvation. It was for that work that he was given the Nobel Peace Prize in 1970 and the Presidential Medal of Freedom in 1977. He is arguably the most impactful Nobel Peace Prize winner in history, as measured by the direct impact of his work on saving human lives. After joining the faculty at Texas A&M, he also received the National Medal of Science in 2004 and the Congressional Gold Medal in 2006.

The first time I met Norman Borlaug, I was greatly impressed that Texas A&M had someone of his stature on our faculty. And given his age, I thought he was pretty robust for being an "old man." I was in my forties then. Now at eighty-two, I have a greater appreciation for Borlaug's contributions while here at Texas A&M.

As a longtime aerospace engineer, I've known of Frank Malina for many, many years. Malina was one of the pioneers whose work helped lay the foundation that ultimately enabled the United States to put men on the moon.

Collaborating with his mentor Theodore von Karman, Malina was the project engineer for the first US space vehicle, the WAC Corporal Rocket, funded by the US Army as the US response to the Werner von Braun–led German V2 program in the mid-1940s. In 1945, the WAC Corporal became the first US space vehicle when it flew to an altitude of over fifty miles. This was the founding project for the Jet Propulsion Laboratory—known as "JPL"—with von Karman as the first director.

In 1947, Malina published a landmark paper. In it, he became the first person to prove the theoretical possibility that a multistage rocket could escape the gravitational field of the earth and reach orbit. From that scientific foundation and within the framework of the organization that became JPL, Malina pointed humankind to the stars and in the process became JPL director himself, succeeding von Karman.

What I did not know until my arrival at Texas A&M was that Malina attended and graduated from the Agricultural and Mechanical College of Texas (AMC).

Malina was born in Brenham, Texas, just forty miles from the AMC campus he would attend. His parents were both immigrants from a part of Europe known as the "Czech lands." It was not until the end of World War I, after the Malinas had settled in Central Texas—a popular immigrant destination for their kind—that the nation of Czechoslovakia was created.

Malina received his degree in mechanical engineering from AMC in 1934. Based on his undergraduate performance, he went on to receive a graduate fellowship from the California Institute of Technology, where his career as a rocket scientist began in earnest.

In his pursuit of science, Malina harbored great misgivings about advanced technology enabling unimaginably destructive advances in warfare. His research in the fields of rocket propellants and rocket-engine design contributed substantially to the US war effort. He was especially impactful in helping to win the war in the Pacific. He invented jet-assisted takeoff rockets—labeled "JATO"—that enabled bombers to take off from the short runways of aircraft carriers to conduct bombing missions in Japan. However, after atomic bombs were dropped on Hiroshima and Nagasaki, bringing about the end of World War II, Malina became passionately opposed to both the advancement of nuclear weaponry and the "weaponization" of rocketry itself. In fact, following his recovery from a postwar nervous breakdown, one could go so far as to call Malina an antiwar pacifist of sorts.

During his time at Cal Tech in the war years, Frank Malina became friendly with counterculture individuals, including some members of the Communist Party. While he never joined the party himself, it was those sympathetic associations and friendships that led him into the crosshairs of federal investigations. He was targeted by US Senator Eugene McCarthy's infamous House Un-American Activities Committee as a "person of interest," much in the same manner as what more infamously happened to the "father of the atomic bomb," Robert Oppenheimer.

Malina never appeared before McCarthy's commission, but the suspicions implicit in the investigations of actions and associations likely prompted him to leave rocket science and pursue other interests. Thanks to his parents, he was fluent in several languages, including French. He moved to Paris in 1947, and from there, he time-shared his artistic hobbies while pursuing "science for peaceful purposes."

Soon after arriving in Paris, Malina went to work for the United Nations Educational, Scientific, and Cultural Organization, more commonly known as UNESCO. He began work on the Arid Zone Research Program, which, as its name implies, studied the scientific problems experienced in the desert regions of the world. Eventually, Malina became head of UNESCO's Division of Scientific Research. In 1953, he resigned from UNESCO to become a studio artist, pursuing his interest in kinetic sculpture that linked art, science, and technology. Although he did return to rocketry briefly in the late 1950s—his development of a hydrazine–nitric acid fuel was used to propel the Apollo Service and Lunar Excursion Modules on their journeys to the moon—Malina remained an artist for the rest of his life.

As a man of both science and the arts, Frank Malina exemplifies the diversity to be found among our more than 140 Hagler Fellows of the past dozen years. It is for that reason that the Hagler Institute for Advanced Study hands out the Frank Malina Renaissance Medallion for multidisciplinary contributions to technology, the arts, and critical thinking.

It is worth noting that I contributed a quarter of a million dollars to endow this award during the six months I spent as interim president of Texas A&M University from November 2020 to June 2021. The first recipient of the medallion was Paul Motheral, Texas A&M Class of '52, whose contributions as a civil engineer helped build the Interstate Highway System in Texas. Motheral is also a gifted musician elected to the Austin Jazz Society Hall of Fame for his saxophone artistry and his philanthropy.

In Frank Malina and Norman Borlaug we have examples of individuals with strong ties to Texas A&M whose talent, passion, and quest to know the unknown lifted them to great heights of national and international significance. In both instances, these extraordinarily impactful pioneers saw a duty to ensure that humanity was carried forward with them.

Both quite literally changed the world.

The Hagler Institute for Advanced Study seeks to affiliate an infinite sequence of superstars from many academic disciplines with the scholastic community at Texas A&M, both to further their work while in residence with us and to accelerate the advancement of our faculty and students. In the stories you will find in this book, I believe you will agree the individuals already brought to the Institute are succeeding in accomplishing that and more as their time and achievements here take root.

I assumed the position of founding director of this institute as a part-time job. I have continued in those duties while remaining active in my role as "Professor John Junkins," conducting research for external sponsors collaborating with our faculty and, most importantly, directing the research of a steady stream of PhD students and postdoctoral researchers. My intent in framing the design of the director position of the Hagler Institute is that an active scholar is best positioned to set the Institute's emphasis on academic excellence and collaboration and most able to attract these scholars as well as donors who underwrite the quality and stability of the organization.

But there is much more work that needs to be done, and I will discuss the path forward in the final chapter of this book.

The Hagler Institute for Advanced Study has become an important and permanent part of the Texas A&M landscape. The work we've done since 2011 has greatly enhanced the university's academic and research reputation. It has also enticed high-achieving faculty and students to make their academic homes here. Inspired by the iconic examples of Borlaug and Malina, the goals of this institute are not modest. We seek to—quite literally—transform academic excellence throughout every corner of our campus by allowing the ambitions of our current faculty to purposefully drive the selection of Nobel laureates—and National Academy inductees. We vet these nominations to invite only those with an earned reputation for world-class research excellence, innovation, mentoring, and effective collaboration. Many of our Fellows have joined our permanent faculty to help increase our National Academy membership from eleven in 2011 to sixty today—a rate of growth unrivaled at any other institution of higher learning in Texas.

In parallel, Texas A&M's funded research has surpassed the $1 billion mark and for the first time moved us into the top ten nationally in that category. In 2023, our College of Engineering research budget surpassed that of the Massachusetts Institute of Technology to achieve a number-one rank nationally. So many academic advancements are afoot that it is a challenge to summarize them all.

This book will allow you to feel the pulse of the process that led to the creation of the Hagler Institute for Advanced Study and, through the words of the Fellows themselves, give you a sense of the remarkable impact the Institute has made to date. As you will see, we are on an upwardly curving path to elevating the academic and research strengths of Texas A&M. All the while, we also purposefully seek to embody the core traditions and can-do spirit of our "Great American University."

We hope that what you'll learn about the Hagler Institute for Advanced Study within the pages of this book will inspire you. And, if you don't know us already, we look forward to a future opportunity to get better acquainted with you.

Ours is a close-knit group, and we welcome the opportunity to meet kindred spirits.

John Junkins
Founding Director
Hagler Institute for Advanced Study

ADVANCING EXCELLENCE

PART 1

The Road to Excellence

1

"Knowledge for Its Own Sake"

The Institute is pledged to assemble a group of scientists and scholars who with their pupils and assistants may devote themselves to the task of pushing beyond the present limits of human knowledge and to training those who may "carry on" in this sense.

—Mission statement of the Institute for Advanced Study by Founding Director Abraham Flexner, Organization Meeting, October 10, 1930, Princeton, New Jersey

It is commonly assumed that Albert Einstein, one of the greatest scientific thinkers of his time—or any time—was a member of the faculty at Princeton University. Einstein famously delivered a series of lectures on the Princeton campus in May 1921, part of a triumphant American tour recognizing the German native's accomplishments in the field of physics and in particular, his work on the reimagining of the mathematics of gravitation and the relative motion of celestial objects.

We know this today as Einstein's general theory of relativity.

Einstein's first lecture at Princeton was held in the school's largest lecture hall. Earlier that same day, he had been awarded an honorary degree from the university and hailed as a Christopher Columbus–like figure, "voyaging through strange seas of thought alone." By his fourth and final lecture during his stay on campus, Einstein was moved into a small classroom to better accommodate the shrinking audience able to comprehend his visionary geometrical and mathematical thinking.

Following his return to Berlin, Einstein published an essay on his experiences visiting the United States. "The American is friendly, self-confident, optimistic, and without envy," he wrote. On his return trips in the years that followed, Einstein was treated as a celebrity on par with the most famous American actors and authors of the day. It was on such a trip in February 1933 that Adolf Hitler rose to power as the new German chancellor. As a Jew, Einstein recognized that he would not be able to return to Germany, so he made permanent his posting to the Institute for Advanced Study in Princeton, New Jersey, where he settled and lived out the rest of his extraordinary life.

Although not directly affiliated with Princeton University, the Institute for Advanced Study benefited from its proximity to the renowned Ivy League school. How the Institute came to be located in Princeton township occurred, in large part, as a result of the success of a mercantile store in Newark, New Jersey.

Louis Bamberger was born into a German American family in Baltimore in 1855. His grandfather, Moses Hutzler, started the first Reform Jewish congregation in the United States. Prior to that, Moses had helped his sons open the Hutzler Brothers department store, and the family financially thrived as a result of the venture.

In 1892, Bamberger moved to Newark, where he bought a failing general-goods store named Hill & Craig, renaming it Louis Bamberger & Company. Following the Hutzler Brothers model, and with the help of his brother-in-law, Felix Fuld, Bamberger's quickly became a profitable venture. By 1912, its flagship location in downtown Newark covered an entire city block.

Bamberger and Fuld, along with Fuld's wife, Carrie (Bamberger's sister), became prominent members of Newark society and notable philanthropists. The trio contributed to numerous secular and Jewish causes, including the establishment of Newark's Beth Israel Hospital and the Newark Museum. Then in 1929, just weeks before that year's stock-market crash, which triggered the Great Depression, Bamberger's was sold to R. H. Macy and Company for an undisclosed sum believed to have been more than $25 million. With those proceeds, Louis and Carrie wished to make a major contribution to an important cause. Their initial thoughts were to build a medical school in Newark, both as a show of appreciation to the people of New Jersey for supporting their business venture and to populate their Beth Israel Hospital with Jewish doctors. To accomplish his objective, Bamberger asked two trusted advisers to identify an expert in the field of medical education. That man turned out to be Abraham Flexner, whose study of medical education became the well-known *Flexner Report*, published in 1910.

In his autobiography, Flexner wrote of his first meeting with Bamberger's agents:

> I informed them my competency was limited to the educational field and that in this field it seemed to me that the time was ripe for the creation in America of an institute in the field of general scholarship and science, resembling the Rockefeller Institute in the field of medicine—developed by my brother Simon—not a graduate school, training men in the known and to some extent in methods of research, but an institute where everyone—faculty and members—took for granted what was known and published, and in their individual ways endeavor to advance the frontiers of knowledge. I pointed out to them the part that Germany had played over a period of almost one hundred years, and that in consequence of the war [World War I] the German universities, academies, and institutes . . . upon which America had so greatly relied . . . had declined in importance and were likely to decline still further.

Flexner had been pondering this matter for some time. In 1921 he put together a memorandum on the subject for the General Education Board, which supported both higher education and medical schools in the United States. Ultimately, his ideas evolved into Flexner's famous treatise, "The Usefulness of Useless Knowledge," first published in *Harper's* magazine in 1939:

> Is it not a curious fact that in a world steeped in irrational hatreds which threaten civilization itself, men and women—old and young—detach themselves wholly or partly from the angry current of daily life to devote themselves to the cultivation of beauty, to the extension of knowledge, to the cure of disease, to the amelioration of suffering, just as though fanatics were not simultaneously engaged in spreading pain, ugliness, and suffering? The world has always been a sorry and confused sort of place—yet poets

> and artists and scientists have ignored the factors that would, if attended to, paralyze them. From a practical point of view, intellectual and spiritual life is, on the surface, a useless form of activity, in which men indulge because they procure for themselves greater satisfactions than are otherwise obtainable. In this paper I shall concern myself with the question of the extent to which the pursuit of these useless satisfactions proves unexpectedly the source from which undreamed-of utility is derived.

In other words, Flexner supported the concept of "knowledge for its own sake," a principle Einstein famously wrote about in his book *Mein Weltbild*, which was translated into English as *The World as I See It*. In it, he observed, "The pursuit of knowledge for its own sake, an almost fanatical love of justice, and the desire for personal independence—these are the features of the Jewish tradition which make me thank my stars that I belong to it."

The compelling case that Flexner made at his first meeting with Bamberger's advisers opened the doors to a subsequent dinner, during which Flexner met personally with Bamberger and his sister. After a series of subsequent meetings, Bamberger told Flexner that he and his sister had considered numerous other possibilities, but none had been as enticing or well-reasoned as the prospect of establishing an institute for advanced study.

In a letter to Flexner dated June 4, 1930, Bamberger confirmed his plans:

> We are asking you to serve with us as trustees of an institution of higher learning which we propose to endow with a substantial initial sum. . . . There is at present no lack of schools and colleges for the training of young men and women for the ordinary baccalaureate degrees. There is never likely to be an over-abundance of opportunities for men and women engaged in the pursuit of advanced study in the various fields of human knowledge. . . . There is no institution in the United States where scientists and scholars devote themselves at the same time to serious research and to the training of competent post-graduate students entirely independent of and spared from both the charms and the diversions inseparable

from an institution the major interest of which is the teaching of undergraduates.

In truth, Bamberger's letter summarized Flexner's own vision for his educational "Utopia." Flexner had also successfully convinced Louis and Carrie that Newark would not be the ideal location for such a project, given that it lacked an important university with a large library. The only place in New Jersey to fit that bill was Princeton.

Like both Bamberger and Einstein, Abraham Flexner possessed a German-Jewish ancestry. He was the first member of his family to complete high school and obtained his undergraduate degree in classics at Johns Hopkins University. He pursued graduate studies in psychology at both Harvard and the University of Berlin but failed to complete degree requirements at either school.

His first notable achievement came in his hometown of Louisville, Kentucky, where he established a college preparatory school. His interest in education ultimately led him to publish an assessment of the state of America's educational system, entitled *The American College: A Criticism*. His subsequent evaluation of medical schools throughout the United States and Canada—a work commissioned by the Carnegie Institute—led to his groundbreaking *Flexner Report*, which sparked major reforms in medical education. Thus, his was an inspired and fortuitous choice to collaborate with the Bambergers in the creation of the Institute for Advanced Study.

Launched in 1930, the Institute first opened its doors—in temporary offices on the Princeton University campus—in 1933. Notice of its first faculty appointments came in the autumn of 1932, when Flexner announced the establishment of a School of Mathematics and its first faculty: Oswald Veblen and Albert Einstein. Veblen had been a member of the faculty at Princeton, where his own work helped advance Einstein's theory of relativity.

Since that time, the Institute for Advanced Study has attracted some of the world's most influential scientific minds, including thirty-five Nobel laureates, forty-four of the sixty-two Fields medalists, and twenty-two of the twenty-five Abel Prize laureates. These visionaries include the likes of Hetty Goldman, George Kennan, Homer Thompson, and Robert Oppenheimer.

According to the Institute for Advanced Study website, "Research at IAS has provided foundations for progress in knowledge and applications across the sciences and humanities," including

- *the development of one of the first stored-program computers, whose structure (von Neumann architecture) formed the mathematical basis of computer software and influenced the development of modern computing;*
- *the foundations of game theory and much of the basis of modern theoretical meteorology;*
- *pioneering theories in the natural sciences, from string theory and astrophysics to systems biology, and their increasing interactions with mathematics;*
- *wide influence on the fields of global development, international relations, historical practice, and morality and ethics;* and
- *key texts in a range of historical disciplines, including essential contributions to the establishment of art history as a discipline in the United States.*

The first spinoff of the original Institute for Advanced Study was achieved in 1954 at Stanford University in California with the Center for Advanced Study in the Behavioral Sciences. In 1978, the National Humanities Center was established in Research Triangle Park near the campuses of Duke University, the University of North Carolina at Chapel Hill, and North Carolina State University. Today, these two entities along with the original IAS form what is called, somewhat unremarkably, "Some Institutes for Advanced Study," which also include the following:

- Radcliffe Institute for Advanced Study in Cambridge, Massachusetts
- Netherlands Institute for Advanced Study in Amsterdam, the Netherlands
- Swedish Collegium for Advanced Study in Uppsala, Sweden
- Berlin Institute for Advanced Study in Berlin, Germany
- Israel Institute for Advanced Studies in Jerusalem, Israel
- Nantes Institute for Advanced Study Foundation in Nantes, France

- Stellenbosch Institute for Advanced Study in Stellenbosch, South Africa

With no copyright or registered trademark on the phrase "institute *for* advanced study" or "institute *of* advanced study," numerous other academic/research pursuits around the world have chosen to use the terms, such as

- European Institute for Advanced Studies in Management in Brussels, Belgium
- Hong Kong Institute for Advanced Study
- *Institut des Hautes Études Scientifiques* in Bures-sur-Yvette near Paris, France
- Institute for Advanced Studies in Basic Sciences in Zanjan, Iran
- Adventist International Institute of Advanced Studies in Silang, Cavite, Philippines
- Institute of Advanced Legal Studies in London, England
- Krasnow Institute for Advanced Study in Fairfax, Virginia
- Institute for Advanced Study at University of Minnesota in Minneapolis, Minnesota
- Institute for Advanced Study of Human Sexuality in San Francisco, California
- Research Institute for Advanced Studies in Baltimore, Maryland
- Texas A&M University Institute for Advanced Study in College Station, Texas

As to why so much advanced study takes place in so many places, Philip Altbach, the founding director of the Boston College Center for International Higher Education, suggested in a piece written for *Inside Higher Ed* that for many institutions, the endeavor plays a role in "enhancing the image and burnishing the brand of the sponsoring university."

Altbach drove home this point in referencing a conference of advanced study institute leaders held at the Freiburg University Institute for Advanced Studies in Freiburg, Germany: "Brand building is part of contemporary higher education and the development of advanced study institutes signals that a university is attuned to the interdisciplinarity of the twenty-first century and is positioned to compete for the best scholars worldwide. All of the institutes represented at the conference are located at top research universities in their home country and tend to see themselves as elite pinnacles of inquiry in elite institutions. This type of institute indicates that a university has 'arrived' in the top echelons worldwide."

As the Texas A&M Institute for Advanced Study—now called the Hagler Institute—is the subject of this book, it is worth noting that the original Institute for Advanced Study was not the model for the concept created by Texas A&M Distinguished Professor John Junkins and launched under his direction in 2011.

"We turned the concept of an institute for advanced study on its head," Junkins says. At the time of this writing in 2025, Junkins remains director and chief recruiter for the Hagler Institute.

"Instead of seeking Einsteins—or people seeking to be of his caliber—to become full-time, Permanent Fellows," Junkins says, "I wanted primarily to bring in accomplished people on an interim basis whose presence here helps drive our faculty's ambitions. I wanted faculty nominations of extraordinarily attractive senior collaborators to initiate the selection process. If the occasional Hagler Fellow decides to take up full-time teaching and/or research residence here, all the better for Texas A&M's growing academic reputation."

Junkins, in fact, did model his Texas A&M program on a "Center for Advanced Studies in the Sciences" at the University of Virginia (UVA). Junkins's first teaching job after earning his PhD in aerospace engineering was at UVA from 1970 to 1978.

The UVA advanced studies program was the brainchild of Edgar Shannon Jr., the school's president. It was launched in 1965 seeking to attract eminent scholars and high-achieving students to the school. Frank Hereford, who succeeded Shannon as UVA president, chaired a campus committee charged with enhancing the university's academic distinction. "We fairly quickly came to the conclusion that the real key to excellence was the faculty," Hereford said later of the inspiration that led to the establishment of UVA's advanced studies program (now known as the Shannon Center). Esteemed scholars brought into the UVA fold through the initiative include English professor and poet

Rita Dove, Kenneth Thompson of the Political Science Department, Michael Fowler in physics, and chemist Robert Ireland.

John Junkins stored away that nugget of insight as his academic career progressed. Leaving UVA, he spent eight years at Virginia Tech before taking a position at Texas A&M in 1985. But not until 1999—long after he had been honored as the first endowed professor in A&M's College of Engineering, selected as a University Distinguished Professor, and named a member of the National Academy of Engineering—did the idea occur to Junkins to create an institute for advanced study of his own accord.

2

Inflection Points

In most lives, there comes a turning point, that moment of revelation, clarity, or understanding in which the course of a lifetime is set.

For John Junkins, that moment occurred one night during his senior year of high school in 1961 on a twisting two-lane Appalachian Mountain roadway just outside Ringgold, Georgia, not far from the Tennessee border. Junkins, along with his high school football coach, Crossland Clegg, had spent three days on the University of Chattanooga campus to get a feel for what college football would be like. A standout pole-vaulter as well as a two-way starter on his North Whitfield High School football team in Dalton, Georgia, Junkins had already garnered attention from Clemson University. While intent on studying engineering in college, Junkins believed his athletic ability was sufficient to enable him to compete at the college level. But what he saw on Coach Scrappy Moore's Moccasins practice field in Chattanooga changed his mind.

"What impressed me most about that experience," Junkins says today, "was that those college guys were bigger, faster, and meaner than me, and had a much more profane vocabulary than I did. It was apparent—or at least I thought so—that these young men were spending a lot more time in the weight room than they were in the classroom.

"On the way back to Dalton I knew that wasn't going to be the life for me."

"Coach," Junkins told Clegg, breaking a long silence during their nighttime drive home, "I'm not going to do this."

To know John Junkins is to know and admire his ability to tell a story. His oldest sister, Faye Gibbons, is an author of children's and young-adult books and a member of the Alabama Writers Hall of Fame. On the *Encyclopedia of Alabama* website, Gibbons's work is defined by its common themes: "broken homes, poverty, rural life, and the importance of family, nature, and the land." Her listing says, "Some themes in her stories—living with an alcoholic father, being a sometime ambivalent participant in fundamentalist churches, the longing to escape poverty—are autobiographical."

Junkins is proud of his sister's literary accomplishments. For those who want to know his life story in greater detail, he uses Gibbons's first published book, *Some Glad Morning*, as a primer into his own background. "The book was based on the life we lived when I was young," Junkins says. "It's about my mother finally leaving my father and moving her children deep into the mountain country where poverty was in abundance and education was lacking."

After a year, Junkins's parents reconciled and moved back to Dalton. His father, George, toiled as a machinist, while his mother, Lenell, spent time outside the home as a mill worker. The Junkins family constantly struggled to make ends meet, so from a young age, John worked to help pay bills. At school, athletics gave him an identity of which he could be proud. The camaraderie to be found with teammates also gave him a nickname, "Spats," for the spatula size of his hands.

"They started calling me that when I was a 6′2″, 140-pound freshman," Junkins says. "So the size of my hands really stood out. By the time I was a senior, I had added sixty pounds to my frame, so people were more inclined to call me, 'John.'"

After breaking the silence with his high school football coach on the way back from the trip to Chattanooga, Crossland Clegg was quick to respond.

"Hold that thought," Junkins remembers Clegg telling him. "Let me pull over here."

After stopping the car on the side of the darkened roadway, Clegg said to Junkins, "I want you to go out and stand in the headlights."

"Why, Coach?" Junkins asked.

"I want to shake your hand."

Sheepishly, Junkins did as he was told, and his coach was quick to follow.

"Approaching me," Junkins remembers, "Coach Clegg reached out and shook my hand. 'I believe that's the most mature thing I've heard come out of your big mouth since I've known you,' Clegg said. 'I don't want you to get your feelings hurt, but I strongly agree with you. I've never seen an athlete more tenacious than you are. You're long on desire, but you're short on the ability to play at the next level.' Clegg paused to let the words soak in, then added, 'I don't know what God's plan is for you, but I know it has to do with your mind.'"

As for the academic side of his high school experience, Junkins admits he was an "underachiever" and, in some respects, felt more comfortable tinkering with a car engine than trying to stay focused in a classroom. His father, thanks to his GI training, was a skilled mechanic and an industrious individual when not sidetracked by drink.

"I knew that I had a lot of gears in my academic transmission that weren't getting used in high school," Junkins says. "I hadn't taken many of the college preparatory classes that were offered at my high school. The math and science courses we did have weren't that strong. In hindsight, I guess you could say, I have no right to be sitting where I am today," which, in his duties as the head of the Hagler Institute for Advanced Study, is on the eighth floor of Rudder Tower, next to the Memorial Student Center on the Texas A&M University campus.

And a block removed from Kyle Field, the football home of the Texas A&M Aggies.

As an engineer, Junkins describes the new course he set for his life as a high school senior as an "inflection point," which anyone versed in the nomenclature of differential geometry knows is the point of a curve at which a change in the sign of curvature takes place.

In Junkins's case, the trajectory of his future academic career curved significantly toward the upside. He calls Clegg's influence "profound," and with each of his own students he pays forward Clegg's investment in him. Says Junkins, "Unlike many high school coaches who measure their worth by how many athletes they can export to play at the next level, Crossland Clegg was most interested in what was best for me as a person."

With his priorities shifted, Junkins chose to follow in the footsteps of his two older sisters and begin his pursuits in higher education by attending the nearby Berry College in Rome, Georgia. Founded by Martha Berry as the Boys Industrial School in 1902, the early guiding principle of the institution was to exchange instruction for construction of the campus, located on eighty-three acres that Berry had inherited from her father.

Today, Berry College rests on some twenty-seven thousand acres, making it the "largest" campus in the world. "And the most beautiful,"

Junkins adds. "If you've never been there, you should go see it. Automaker Henry Ford put a lot of money into that school, and it shows in the architecture of many of its buildings."

Junkins's year at Berry bolstered his academic credentials and his confidence. And in the process, he earned a conference championship in the pole vault for the Berry track and field team.

"I did it because I could," Junkins admits today.

Junkins's Berry College roommate was another athlete by the name of Spiros Pallas. A decathlete from Greece and his country's champion in the event at the age of seventeen, Pallas was pivotal in the creation of another personal inflection point in Junkins's life.

"As a result of my time at Berry, I knew I wanted to become an aerospace engineer," Junkins says. "While my worldview wasn't very wide back in Dalton, I was enthralled by the Soviets' launch of the Sputnik satellite in 1957 [when Junkins was fourteen years old] and a few years later with President Kennedy's speeches about America going to the moon." Junkins says the early years of the "Space Race" kindled a hope he could somehow be a part of launching humans toward the stars.

The obvious choice for a Georgia boy to pursue a future in engineering was the Georgia Institute of Technology in Atlanta. Junkins had contacted the school during his senior year of high school but concluded he would not be able to compete there academically given the shortcomings of the classroom curricula offered at his high school. After earning all A's in calculus, chemistry, and physics at Berry, Junkins was primed for academic success at the next level. Thus, he applied for admission to Georgia Tech and was accepted by the school.

Near the end of their final semester at Berry, Junkins and Pallas reflected on their future hopes and dreams while relaxing together in their dorm room. Pallas was curious as to the reason Junkins wanted to attend an all-male school. Being Greek and a bit of a ladies' man, that scenario was a nonstarter for Pallas, which is why he had enrolled at Auburn University.

While Georgia Tech had been an all-male academic enclave for most of its history to that point, the truth was that the school had begun admitting women in 1952. But to Pallas's point, even in 1962, women were still a small minority of the Georgia Tech student body.

The two friends stayed in contact over the summer, and over time, Pallas convinced Junkins to make a trip to Auburn with him so Junkins could get a closer look at the school. Junkins agreed, borrowing his father's car for the road trip to Alabama.

"Auburn was picturesque," Junkins says, remembering his first time on campus. "You had the rolling terrain, the huge oak trees . . . and lots of coeds!

"While Spiros was selecting his dorm room for the fall quarter, I borrowed his campus handbook and thumbed through its pages. Sure enough, Auburn also had an aerospace engineering tract. Better still, it had a co-op program that offered work-study opportunities. I was pretty keen on being a co-op to pay my way during my college years and it appeared Auburn encouraged that. I remember looking up from the handbook and telling Spiros, 'I think I'm going to go to school here!'"

Which leads to another John Junkins story.

"Looking through the handbook, I found the name of the college registrar: 'Pumphrey,' as best I can recall. The first thing I would need to do to attend Auburn was to submit an application to the school. So I found an Auburn telephone directory and thumbed through it. I discovered there was only one 'Pumphrey' listed, so I figured that was my target. The next day, on Sunday, I called his house.

"'Sir, I know this is highly unusual,' I said. 'I'm in town. I'm a prospective student, and I need some advice about how I can proceed to gain admittance.' This was just three weeks before classes began. After I had explained my situation, there was a long pause on the other end of the line. Finally, Mr. Pumphrey—at least I assumed it was Mr. Pumphrey—spoke.

"'Young man, where are you?' he asked. I told him.

"'I'll be there in five minutes,' he said.

"After he arrived, we spoke briefly and then he asked me to follow him back to his house. 'We can talk more there,' he said.

"Spiros joined me on the trip to the Pumphrey home. After we walked in, Pumphrey had us sit down at his kitchen table.

"'I just happened to have an application here,' he said. 'Why don't you fill it out. Your friend and I will get better acquainted in the parlor while you're doing that.'

"After I filled out the application, Pumphrey returned to the kitchen and took a look at the document. He was impressed by my test scores and my Berry grade-point average. He knew all about

Berry, which was why he had gone so far out of his way to help me. After looking over my application, he said, 'If your transcripts are in order, you get me a couple of references, and you pay your fees on time, you'll be admitted for the fall. I'll expect you to be back here in three weeks' time!'

"As a result," Junkins says, "I lost my registration fees at George Tech, but otherwise, I never looked back. I started classes at Auburn in the fall of 1962 and loved being there. During my first quarter, I applied for a work-study opportunity at NASA's Marshall Space Flight Center in Huntsville. I got the job, and it was there that I met Wernher von Braun . . . and I've been surfing on a space wave ever since."

Von Braun was a towering figure in the annals of human space flight. During World War II, he spearheaded Germany's quest to master rocket technology for military purposes. After the war, he surrendered to American intelligence forces and ultimately became one of the masterminds behind the US human spaceflight program. Von Braun and his expatriate German colleagues played important roles that led to the development of the Saturn V rocket that would ultimately carry Americans to the moon. They did much of their initial work in that capacity from the Marshall Space Flight Center located at the US Army's Redstone Arsenal near Huntsville, Alabama.

And yes, to get to the moon, NASA employed the cooperative educational services of many bright undergraduate engineering students like John Junkins. Such was the intensity of America's quest to beat the Soviets to humanity's closest celestial neighbor.

"My very first experience at Marshall blew me away," Junkins says, recounting yet another important milestone in his life. "My first boss was Helmut Bauer, who was a part of von Braun's cadre of German rocket scientists who contributed significantly to NASA's space aspirations. Bauer's specialty was aeroballistics. He was chief of the Dynamic Analysis Branch. The first time I entered his office, he stood up, shook my hand, offered me a seat, and then posed two questions with his heavy German accent.

"'You're from Auburn? You have good grades?'

"I assured him I was and I did.

"'Good!' he exclaimed. 'We're going to have some fun here!'

"I was on shaky ground as I attempted to answer his next question regarding my knowledge of Bessel functions. I knew sines and cosines, but I did not know Bessel functions.

"'OK, let me explain,' Bauer said. 'The first problem you're going to work on has to do with a rocket we are designing called the Saturn V. It's already on the drawing boards here and we are doing testing. This is the rocket that will launch a crew to the moon. It will weigh six and a half million pounds and 90 percent of that mass will be the liquid propellant.'

"Bauer paused and looked at me before asking another question, 'You know how to balance a broomstick on your finger?' We both smiled. 'That's what flying a rocket is all about. You have to balance this broomstick on your finger while you're pushing it through the sky.'"

Junkins's first project under Bauer's watch at Marshall was to lead an experiment on slosh. To keep the broomstick headed in the right direction, particularly when it contained millions of pounds of liquid fuel, one must damp out the propellant slosh, which can be created by a mere gust of wind. "Unless we control the slosh," Bauer told Junkins, "we might lose Florida. You will take data from experiments and compare that to our theoretical predictions."

Concluding their first meeting, Bauer presented Junkins with a handbook. "You have to know Bessel functions," he told his new protégé. "You read the first three chapters of this book and next week in this office, at this hour, you're going to give me a lecture. I want you to show me a graph of the Bessel functions. I want you to show me how to compute them. I want you to tell me what differential equation they satisfy."

"At that point of my life, I had never seen a differential equation," Junkins says, "but with a little help and an open mind, I was able to move myself up the learning curve sufficiently to make Bauer a believer in me."

For the rest of his time as an undergraduate student at Auburn, Junkins alternated quarters between classwork on the Auburn campus and his job at the Marshall Space Flight Center. While in Huntsville, he also took night classes toward his Auburn degree at the University of Alabama's Huntsville extension campus. Huntsville also proved to be fertile ground from a personal perspective in that Junkins met his future wife, Elouise, there. In addition, two of his Huntsville housemates, also Auburn students working at Marshall, began dating and later married Junkins's two sisters.

And a guest lecturer on the Huntsville campus provided yet another inflection point in Junkins's life.

"Samuel Herrick was the father of astrodynamics," Junkins says. "In fact, he invented the term around 1945, fusing astronomy with engineering. Not long after that, as a member of the astronomy faculty at UCLA teaching an engineering course in rocket navigation, Herrick devised a methodology to make a theoretical spaceflight to Mars. More than a few people in education thought he was a bit of a crackpot, but it turned out Herrick was a real visionary, and his work helped to pave the way for America's journeys into space and to the moon."

Thus, when it came time to pursue his graduate studies, Junkins's sights were firmly set on working with Herrick at UCLA. Upon completing his undergraduate education at Auburn, Junkins went to work full-time at Marshall while beginning a search for relevant employment to fund his graduate-school pursuits in the Los Angeles area. That job turned out to be with the Douglas Missile Systems Space Division, which was charged with building and launching the Delta rockets that have carried the majority of uninhabited scientific spacecraft into earth orbit as well as to the moon and Mars.

Upon moving to the Los Angeles area with Elouise—his new bride—Junkins started his new job and took a night class at UCLA just to get the hang of his new and dramatically different environs.

Keep in mind, this was also the heyday of "the '60s," and California's college campuses were the epicenter of the counterculture movement that exploded during the decade.

"There were several culture shocks when Elouise and I moved to the Los Angeles area," Junkins says. "It was 1966, a year before the 'Summer of Love,' and more than a few young people in California were 'hanging out of the trees.' Elouise and I did not adopt the 'tune-in, turn-on, and drop-out' mindset that became increasingly prevalent among college students. We had a plan: to get me through graduate school and then to move back east to raise a family.

"I remember walking to class on the UCLA campus one day when I came upon art students painting a live nude model. I remember thinking to myself, 'I'm in the wrong damn curriculum! What's wrong with this picture here?'"

To be sure, Junkins's life in the '60s could easily have been fodder for a song by the Mamas & the Papas.

Culture shocks, what can you say
Gridlock on the freeway
Grad-school strife
Uprooted wife
And a side job at a place
Launching rockets into space

Junkins confesses that his four years at UCLA were a challenge, but he was, in his own words, "young and full of vinegar."

Those who know Junkins today might well suggest that at the age of eighty-two he remains so.

On his way to obtaining both his master's and doctoral degrees from UCLA, Junkins was also carrying heavy work responsibilities and gaining firsthand, real-world experience as a "rocket scientist." During his time with Douglas, Junkins worked on the "trajectory optimization" for satellites the company was commissioned to launch into orbit for both NASA and the US military. "We launched eleven in one year," he recalls.

Later during Junkins's time with Douglas, the company would send rockets to both the moon and Mars.

"The level of rapid innovation and, shortly thereafter, high-stakes implementation on real missions was breathtaking and highly motivating," he says. "Collectively, I think of us back then as the Elon Musks of our day."

On the home front in Los Angeles, John and Elouise welcomed their first child, John Stephen, born in 1967. Both prior to and after having John Stephen, Elouise worked as an administrator in the personnel department of the Veterans Administration in Santa Monica. John says he and Elouise look back now at the California years and wonder how they could possibly have done all that they did and kept a smile on their faces.

The four years in California laid a foundation for all the years that followed.

3

Launching a Career

By the end of the 1960s, unrest and protests had become commonplace on college campuses coast to coast. Opposition to America's involvement in the war in Vietnam spurred outrage from an increasing number of students. Then on April 30, 1970, Pres. Richard Nixon made known to the American people that he had ordered US troops to invade the nation of Cambodia, thus escalating the war effort.

Protests, many of them violent, soon erupted on and off college campuses across the country. On May 2, students at Kent State University in Kent, Ohio, burned down the school's ROTC building. Two days later, on May 4, National Guard troops confronted protesters at the school and, in an ensuing melee, shot and killed four students. On May 6, at an evening rally held on the University of Virginia campus, activist Jerry Rubin—one of the so-called Chicago Seven who had instigated rioting during the 1968 Democratic National Convention in Chicago—called upon students nationwide to close down American universities across the country. Later that night, some two thousand UVA protesters assembled outside the residence of the school's president, Edgar Shannon, demanding that he sign a telegram of protest to send to Nixon.

On the front steps of his home, in the middle of the night, Shannon stood and listened. When it came his time to speak, he told the throng that he shared their concerns but could not condone their desire to shut down the school.

In an article written in 2018 for *Virginia*, the UVA alumni magazine, longtime newspaper editor Ernie Gates wrote about the episode:

> For one intense moment, the antiwar fervor of the 1960s converged on Carr's Hill at the University of Virginia. It was Wednesday, May 6, 1970, and a movement to shut down the University was

about to boil over. During that tumultuous week remembered as May Days, many classes were canceled, antiwar rallies swelled to the thousands, protesters occupied the Navy ROTC building, student marshals stood sentry against arson around the Academical Village, and billy club–wielding police stormed the Lawn and some fraternity houses, hauling dozens of fleeing students to jail. Through it all, Edgar Shannon walked a high wire: angering the governor, his board and many alumni by siding with the students against the war, but never bending in his determination to keep the University open.

A mile south and another mile west of the assemblage at the UVA president's mansion, twenty-six-year-old John Junkins, his wife, and their toddler son slept in peaceful tranquility in a newly constructed enclave of townhomes built for young members of the school's faculty. Junkins had arrived on campus at the outset of the 1970 spring semester to join the university's aerospace engineering department. "The first course I taught at Virginia was Analytical Dynamics," Junkins says today. "The class was made up of about fifteen PhD students, half of whom were older than me. I remember longing to look more distinguished."

At the outset of his postgraduate studies at UCLA, Junkins gave little thought to becoming an academic. While Sam Herrick's Huntsville lecture had inspired Junkins to attend UCLA and study under the acclaimed astronomer-turned-engineer, working with him proved a frustrating task. "He was painful," Junkins says. "In the entire four years I was at UCLA, I did not get my name as coauthor on a single paper with him, although I did make some significant contributions."

Fortunately, a young UCLA engineering assistant professor by the name of Peter Likins took Junkins under his wing, and the two grew close Likins became both a member of Junkins's graduate committee and an important mentor to him. Like Junkins, Likins had an athletic background. He wrestled collegially at Stanford University, going undefeated in dual-meet competition his entire four-year career. Likins joined the UCLA faculty in 1964, two years before Junkins's arrival at the school. A highly regarded teacher and researcher, Likins went on to become a Ford Fellow and an associate dean. He became dean of the School of Engineering and Applied Science at Columbia in 1976 and was named president of Lehigh University in 1982 at the age of forty-six. He then served as president at the University of Arizona from 1997 until his retirement in 2006.

Peter Likins is unique in that he is a member of both the National Academy of Engineering and the National Wrestling Hall of Fame in Stillwater, Oklahoma.

"Peter was the greatest classroom teacher I ever had," Junkins says. "He had the ability to go into a room and fill the center of gravity. He had a sixth sense that connected where the audience was and where he wanted to take them in his lectures. He spoke to his students in both an illuminating and motivating way, so it's not surprising his career took on the trajectory that it did.

"While we were both at UCLA, I talked to Peter about his life and how he worked as an academic. From what he told me, I concluded that when it came to his research interests, he invented his future rather than having projects assigned to him as I was experiencing working in industry. He decided what problems were important to pursue, then put together a proposal to explore that. He was very strong in fundamental mechanics and mathematics, and his vision and insight allowed him to 'see the future' and define meaningful research avenues. I also saw the admirable people-development aspect of how he worked, the bond that he had with his students.

"It didn't take long before I started thinking, 'That's who I want to be.'"

It was Likins's influence and recommendation that paved the way for Junkins to land his first full-time teaching position at the University of Virginia. "At that point, I was essentially unpublished," Junkins says. "I had significant industrial accomplishments having put together a lot of reports at NASA and Douglas Missiles and Space, but I didn't have any academic papers in print, although I had several in draft form. While accomplished as a young engineer, I was green as an academic. Peter Likins's recommendation got me in the door and I felt fortunate to land the job at Virginia."

In the weeks before Edgar Shannon's encounter with student war-protesters in the middle of the night outside his campus home, Junkins met the esteemed UVA president at a new-faculty event on campus.

"The University of Virginia campus is among the most historic in the country," Junkins says. "As most people know, it was established by Thomas Jefferson around his idea that a college should be an 'academical village.' The campus is renowned for its academic programs and its beautiful architecture. Jefferson himself designed many of the original central campus buildings, including the iconic library building called 'The Rotunda.'"

Although the university did not open its doors until March 7, 1825—about a year before Jefferson's death at the age of eighty-three—the cornerstone for the first building, "Pavilion VII," was laid in 1817. It is in that building that the Colonnade Club was located, where President Shannon held his new-faculty reception in the early months of 1970.

By that time, Shannon had been UVA president for more than a decade. He knew well the ravages of armed conflict, having served on the cruiser USS *Quincy* during World War II. On August 9, 1942, his ship was attacked and sunk by the Japanese off Savo Island in the Solomon Island area of the Pacific Ocean. Three hundred seventy American sailors on the *Quincy* died that day. Shannon survived.

Before the war, Shannon earned an English degree from Washington and Lee University in Lexington, Virginia. Then after the war, he continued his English studies, first obtaining a master's degree from Harvard, then, as a Rhodes Scholar, attended Oxford University, where he earned his PhD.

As president at UVA, Shannon quickly instituted two key programs to increase a perilously dwindling enrollment—particularly among top students—by bolstering the academic reputation of the school. First, he led the establishment of the Echols Scholars Program. Designed to entice Ivy League–caliber students to UVA, the initiative offered "ambitious academic privileges" to top-tier applicants to the school. Among the benefits of becoming an Echols Scholar were exclusive living-quarter arrangements, academic counseling services, and priority admission to select courses.

While the Echols program was a success, Shannon eventually realized enticements weren't enough to attract and retain the number of top students he wanted at the university. Elite undergraduates and graduate students alike most desired to learn under eminent scholars. This idea, perhaps somewhat obvious today, was brought to light for Shannon as his school sought to be included in a grant program sponsored by the National Science Foundation. The committee charged with submitting the grant proposal was chaired by Frank Hereford, a physics professor and dean of the Graduate School of Arts and Sciences. "We built the proposal around the concept of bringing to the University and retaining the very best people who could possibly be brought here," Hereford said. He ultimately succeeded Shannon as UVA president.

Upon meeting Edgar Shannon for the first time, John Junkins observed, "He had a casual yet impressive presence. He was a confident leader, completely at ease in his own skin. I liked him immediately."

And that feeling became mutual, as Junkins and Shannon shared a love of poetry. Shannon had studied poetry as a Rhodes Scholar and taught poetry courses in the English Department at Virginia before becoming president. "As we got acquainted," Junkins remembers, "he was surprised that an engineer knew of the works of Browning, Tennyson, and Wadsworth. Like Peter Likins, Edgar Shannon set an important example in my career."

UVA's Center for Advanced Studies in Science first came to Junkins's attention via a series of lectures given by faculty hired by the school through the center's advanced study initiatives. One presenter stands out in Junkins's mind half a century later.

"Doris Kuhlmann-Wilsdorf was a pioneer," Junkins says. "Women were rare in the fields of science and engineering back then, but Kuhlmann-Wilsdorf had a global reputation in materials science." Born in Bremen, Germany, Doris Kuhlmann attended the University of Göttingen during the years of World War II, receiving her doctorate in materials science in 1947. It was there she met her husband, Heinz Wilsdorf, also an accomplished materials scientist. Because of their German citizenship, the two were denied postwar visas to seek academic employment in both the United Kingdom and the United States. Thus, they wound up in South Africa, where Heinz joined the Council for Scientific and Industrial Research in Pretoria while Doris lectured at the University of the Witwatersrand in Johannesburg.

In 1956, the couple moved to the United States, where Heinz was offered and accepted a research-scientist position at the Franklin Institute in Philadelphia. Shortly after, Doris received an appointment as a research associate professor of metallurgical engineering within the University of Pennsylvania's School of Engineering and Applied Science, where her scholarly reputation blossomed. She might have remained at the Ivy League school, but in 1963, Heinz accepted the position of chairman of the newly established Department of Materials Science at the University of Virginia, and Doris followed as a lecturer. Three extraordinarily productive years later, she became the first woman named as a full professor at UVA outside the schools of Medicine and Nursing.

Throughout her career, Doris gained increasing international recognition for her pathbreaking work in plastic deformation, surface physics, and crystal defects. She wrote more than three hundred scientific papers and started two companies: HiPerCon, for high-performance contacts, and Kuhlmann-Wilsdorf Motors.

Wilsdorf Hall, erected in 2006 in honor of Heinz and Doris, is the University of Virginia's Materials Science Engineering and Nanotechnology building, encouraging collaborative and interdisciplinary study among departments in the School of Engineering.

"When I looked deeper into the Center for Advanced Studies while at UVA," Junkins says, "I was thoroughly impressed with the idea and the impact it was having."

During his time at UVA, Junkins earned tenure and was promoted to associate professor. But not until leaving the school did he achieve the title of full professor. That came in 1977 when Junkins departed the University of Virginia at the age of thirty-four for the Virginia Polytechnic Institute and State University in Blacksburg, Virginia, known more commonly as Virginia Tech.

"I joined the Department of Engineering Science and Mechanics at Virginia Tech with a much stronger critical mass of faculty overlapping my specialized technical interests," Junkins says. "Although I truly loved the University of Virginia, moving down the road to a stronger college of engineering was a tremendous move for me professionally. In addition to basic research, I did lots of applied things, grew a broader set of technical competencies, and produced some really good PhD students during those years.

"To say I reached 'technical maturity' at Virginia Tech might not quite be accurate, but my research and teaching prowess certainly took shape there."

In 1980 while at Virginia Tech and consulting with the Johns Hopkins University applied physics laboratory, Junkins developed a minimum-time spacecraft attitude maneuvering methodology utilizing the earth's magnetic field. This was implemented successfully in the pre-GPS Navy navigational satellite *NOVA-I*. For this work, he was honored with the 1983 American Institute of Aeronautics and Astronautics' Mechanics and Control of Flight Award. That was soon followed by the publication of Junkins's second book, entitled *Optimal Spacecraft Rotational Maneuvers*.

Eight years into his stay in Blacksburg, John Junkins was professionally content and personally happy. He and Elouise were raising two children: Stephen, born in Los Angeles, and Kathryn, born in Charlottesville. The couple had friends, both on and off campus, and as Junkins's forty-second birthday approached, Virginia felt like a place that could be home for the remainder of his days.

Then came a knock on his campus office door, initiating one of Junkins's all-time favorite stories.

"When I opened the door, there stood two men who had been friends and colleagues during my University of Virginia days: Skip Fletcher and Carl Erdman. They told me they just happened to be in the neighborhood and thought they'd drop by and say hello.

"'I'm sure that's not the case,' I said skeptically, 'but I'm glad to see you. Before you come in, though, please tell me the real reason you've come to town.'

"'We came to congratulate you,' Skip said.

"'Why am I deserving of your congratulations?'

"'You are,' Carl chimed in, 'going to hold the first endowed chair in engineering at Texas A&M University.'

"'That seems unlikely,' I said, gesturing for them to come into my office so we could get to the bottom of their flattering nonsense.

"After seating ourselves, I looked at both men, saying nothing but communicating my skepticism with one raised eyebrow.

"Skip and Carl reminded me that they both had joined the faculty at Texas A&M shortly after Junkins moved to Virginia Tech. They told me the engineering college there had just received its first endowment. After doing a blind search for an engineer who 'knew about space,' the dean of the college of engineering, Herbert Richardson, had decided that I fit that requirement and needed to be the first chair holder.

"'And where exactly is it that you moved to in your new jobs?' I asked. 'College Station,' they replied in unison. 'It's in Central Texas, about halfway between Houston and Austin.'

"'Well, you know,' I said, 'I have never been to College Station. Relocating there is not something I'm ambitious to do. So I'm really sorry you've gone to all this trouble, but I'm not interested.'"

Obviously, Junkins's story does not end there. Erdman and Fletcher told Junkins that Richardson was, at that very moment, waiting for their phone call. Junkins agreed to let them make the call, and after reaching their boss, they put Richardson on speakerphone.

By the way, Erdman and Fletcher had reached out to associates of Junkins to make sure their arrival coincided with his "office hours."

Junkins knew Herbert Richardson from the latter's time at the Massachusetts Institute of Technology and knew his *bona fides* were impressive. Fletcher had been chair of the department Junkins had left at the University of Virginia before Richardson recruited him to A&M. Erdman, a top-caliber nuclear engineer, had followed Fletcher to the Southwest Conference school.

On the phone, Richardson explained—according to Junkins's telling of the story—that he was looking to bolster the "space arm of aerospace engineering." Richardson told Junkins that Fletcher and Erdman had been advocating for their former UVA colleague and said he was impressed with the case they made.

"I very much appreciate hearing that," Junkins told Richardson on the phone, "but I'm just not interested in making a move at this time."

Let's let Junkins pick up the story from here.

"Richardson was quick with his reply. 'I hear you,' he said, 'and we will take no for your answer. But we're going to get our feelings hurt if you don't at least come visit, because we've gone through a lot of trouble to show you we're interested. We think you need to come down and talk with us.'

"Making our final approach into Houston Intercontinental Airport a few weeks later, I told Elouise as we looked out the window, 'Now, "little darlin"'—slipping into the Texas drawl of TV's J. R. Ewing—'I want you to understand something. There ain't no damned way we're moving to Texas.'"

For those too young to remember, J. R. Ewing was the rich and devious Texas oilman played by actor Larry Hagman on the popular television series *Dallas*, which aired for thirteen seasons on the CBS network from 1978 to 1991.

After landing in Houston, John and Elouise spent two days in College Station, during which time John met with Richardson.

"John, you like to talk, don't you?" Junkins laughs at the memory of Richardson's words spoken in his office at the end of his two-day visit. "So I want you to give me a lecture."

"On what?" Junkins asked.

Let's let the founding director of the Texas A&M Institute for Advanced Study describe the remainder of that conversation.

"Richardson said, 'Go to the whiteboard there. I want you to divide the board into three panels. In the left panel, I want you to describe—using bullet points if you'd like—your current professional state. How you would characterize where you are in your career.

"'Then in the middle panel, I would like for you to look five years into the future and, again using bullets there, tell me what you would like to achieve and what resources you need to get there.

"'And then, in the right panel, I want you to jump forward to the end of your career, and in a retrospective sense, put down some bullets that would characterize what you hope your career will have meant.'

"Herb knew my weak spot.

"After I finished—and it took a good while to assess my past, present, and future careers, I finally put the marker down and turned to Richardson. He was smiling. 'John, that's roughly what I thought you might be thinking. I tell you what I'm going to do right now. Today, I'm going to give you the resources—and then some—for your five-year plan. Let's get started building toward these lofty career goals you have set for your future. Let's build a top 10 department at Texas A&M.'

"'You're a really smart guy. I think you'd be crazy to turn that down.'

"I thanked Herb and told him that he had my attention but didn't give him an immediate answer. On the way to the airport, I discussed the matter with Elouise, and then as we were seated on our flight back to Virginia, I looked at my wife and again assumed the demeanor of J. R. Ewing. I told her, 'Little darling, I want you to take a good look around here, and if you see anything you want, you just let me know.'

"Elouise looked at me stunned. 'Oh my God! We're moving to Texas!'"

Herbert Richardson's approach left a lasting impression on Junkins, establishing a template that Junkins himself has used time and again in his own recruitment of Hagler Fellows.

4

The New Aggies

The April 1997 edition of *Texas Monthly* magazine posed a feature-story question to its readers: "Did You Hear the One About the New Aggies?"

The twenty-seven-year-old story's title phrasing evokes the long-standing University of Texas tradition of the "Aggie Joke," which itself is a spinoff of the unfortunate human tendency to make fun of people unlike themselves. Wikipedia defines an ethnic joke as a "remark aiming at humor relating to an ethnic, racial, or cultural group, often referring to an ethnic stereotype of the group in question for its punchline."

"In the state of Texas, it is no surprise that such a joke would be made with the Aggies as the subject. This is due to the rivalry with the University of Texas, because it has often been understood that it is harder to get into the University of Texas than Texas A&M." So says a post by "Patricja" in the University of Southern California's *Digital Folklore Archives* under the header "Aggie Joke."

In the aforementioned *Texas Monthly* article, writer Paul Burka summed up—in no joking manner—Texas A&M's backstory like this:

> It was born a stepchild in 1876, declared by the state constitution to be a "branch of the University of Texas," which would not even exist for another six years. For the first half century, the school faced repeated threats by the Legislature to shut it down. As recently as the 1950s it was more likely that the school would cease to exist than become a serious academic institution. Over the years, A&M has had to overcome politics, poverty, isolation, fire, and ridicule. Most of all, though, it has had to overcome Aggies. The lifetime love and loyalty that Aggies have for their school have been A&M's greatest asset—and its greatest liability.

It should be noted that as a "land-grant college," Texas A&M had an initial mission to focus on the teaching of practical agriculture and mechanical instruction—the "A" and the "M" contained within the school name—along with military science. In contrast, the historic practice of most colleges at that time centered on a liberal arts curriculum. Other American schools that were founded as land-grant institutions include Auburn, Virginia Tech, the University of California at Berkley, the University of Illinois, and the University of Georgia, as well as many colleges with "state" in their names, such as Penn State, Michigan State, and Ohio State.

At Texas A&M, the dichotomy between the mindset of "Old Ags" and more recent graduates of the university was perfectly captured in the opening line of Burka's *Texas Monthly* story, quoting then–University Pres. Ray Bowen.

"Texas A&M is not going to become a school of nerds."

According to Burka, Bowen's comment was an effort by the university's top administrator—himself a "Class of '58" graduate of "AMC," the Agricultural and Mechanical College of Texas (as Texas A&M was known until 1965, when "University" was incorporated into its official name)—to explain why "he [didn't] want to see the average Scholastic Assessment Test score of A&M students rise much higher than the then current 1174—even though scores at the University of Texas are higher."

"I don't ever want us to get to the point where test scores count more than leadership," said Bowen, but he *was* improving quality.

Dr. Ray Bowen served as Texas A&M president from 1994 to 2002. It was under his watch that the school became nationally recognized for its academic standing, the subject of the April 1997 *Texas Monthly* story. But before delving into his administrative achievements that elevated Texas A&M's scholastic brand, let's first take a look at Bowen the man . . . and Aggie.

"My father never got out of high school, but nonetheless he was a successful businessman in the Fort Worth area," Bowen said from the College Station home near campus where he still resides part-time. His father, Herbert, was a wholesale food broker, the "middleman" between grocers and manufacturers. "He gave us a good life," Bowen says, "and he knew how to enjoy life, too."

Herbert Bowen loved golf and was a member at Fort Worth's esteemed Colonial Country Club. It was there one evening that the Bowen family became AMC partisans.

"Dad came home from playing poker at the club that night," Bowen said, "and told my older brother, also named Herbert, that he was going to go to Texas A&M. Incredulous, my brother, who was eight years older than me, asked, 'Why???' Dad told him, 'One of the men at my poker table told us that if we wanted a boy to stay out of trouble and be a serious student, we needed to send him to Texas A&M, and not to send him anywhere else.'"

Thus did all three of the Bowen boys—Herbert, Jerry, and Ray—wind up attending Texas A&M. By the time it came Ray's turn to pack his footlocker and head to Aggieland, his father had the "off-to-college" routine down pat.

"Apparently Dad had read somewhere—or maybe his poker-playing friend at the club offered him this piece of advice—but when he took us to start classes at school, he didn't hang around once we got there. He just let us out of the car and left as quickly as possible.

"In my case, he pulled up in front of Dorm 12, got out, got my footlocker out of the trunk of the car, put it on the sidewalk, shook my hand, got back into his car, and left. I honestly don't remember him saying anything in the process.

"But over the course of his lifetime," Ray Bowen adds, "Dad was as much of an Aggie as any of his sons."

Bowen received his bachelor's degree in mechanical engineering in 1958. A year later, he earned a master's degree from the California Institute of Technology. He then returned to Texas A&M to complete his doctoral studies. Afterward, he began active military service as a first lieutenant in the United States Air Force, spending his entire three-year tour of duty as an instructor at the Air Force Institute of Technology at Wright-Patterson Air Force Base outside of Dayton, Ohio.

Texas A&M's requirement as a land-grant institution to provide military science instruction to its students inspired the establishment of the Corps of Cadets when the school was founded in 1876. For nearly a century, membership in the Corps was mandatory for all students, and all students at the college until 1963 were male . . . and white.

When Bowen first arrived on the A&M campus in 1954, Pres. David Morgan announced that membership in the Corps would be made optional after a student spent two years on campus. By the time Bowen graduated in 1958, that policy was revoked as a result of unrest between military and civilian students. Then in 1963, school Pres. James Earl Rudder, a World War II military hero and graduate of Texas A&M, opened enrollment to all students of all colors and to women as well as men. Two years later, he permanently ended mandatory involvement in the Corps, a pivotal moment in the transition of Texas A&M from a military-oriented college into a major university.

Today, the Corps of Cadets at Texas A&M University remains a prized element of the school's past and present, even though it comprises less than 10 percent of the more than 75,000 students now found on campus. Corps membership remains a qualifying criterion to be a part of one of Aggieland's finest traditions: the nationally acclaimed Fightin' Texas Aggie Band.

During his time in the Corps, Bowen was a Ross Volunteers platoon leader. In his final year at A&M, he served as deputy Corps commander to his roommate, Jon Hagler, the same Jon Hagler who serves now as the namesake for the Hagler Institute for Advanced Study.

From the common bond of their Corps experience, Bowen and Hagler went in dramatically different directions in embarking on what would turn out to be highly successful careers for both men. While Bowen ultimately became an academic administrator and president of both Oklahoma State University and Texas A&M, Hagler moved to the East Coast and became a wealthy investment broker. Nearly four decades after their shared college experience, the pair would reunite in an effort to better the academic reputation of their alma mater.

That endeavor became known as *Vision 2020*.

After leaving the Air Force and spending a year doing postdoctoral work at Johns Hopkins University, Bowen entered the teaching profession full-time as a member of the Mechanical Engineering Department and the Mathematical Sciences Department at Rice University in Houston. Bowen spent sixteen years at Rice before becoming dean of engineering at the University of Kentucky in 1983. Eight years later, he was named provost at Oklahoma State University, where he became interim president of the school in 1993.

Then a year later, Ray Bowen was named president at Texas A&M.

"Texas A&M had previously approached me about becoming their dean of engineering," Bowen says. "I was enamored by the prospect, but about that time, the president at Oklahoma State was forced to resign and I was offered the job as interim president. I took that position, but when they asked me to become president full-time, I said I wasn't interested in the job, primarily because I didn't like the way my predecessor, whose name was John Campbell, had been treated."

At Texas A&M, Bill Mobley stepped down as president on the flagship campus at the end of August 1993 to become chancellor of the Texas A&M University System. In the national search that followed, Ray Bowen was targeted as a possible candidate for Mobley's old job.

"When they called me up and asked if I had interest in becoming a candidate for the position, I asked just one question. I said, 'Do you have an internal candidate that's going to get the job?' I knew that in a lot of searches, the internal person is sort of the favorite, but the school still has to go through the process. The answer I got regarding the A&M search was, 'Absolutely not. It's a totally open search.' So I agreed to throw my name into the hat, thinking nothing was going to come out of it."

But Bowen got the job.

By the time Ray Bowen took over as Texas A&M president on June 1, 1994, John Junkins had become a fixture in the school's College of Engineering. Upon his arrival in 1985, the endowment intended for him was still being negotiated. To provide funding for his aerospace research initiatives, Junkins was made a TEES (Texas A&M Engineering Experiment Station) Distinguished Chair professor. A year later, he received the George Eppright Endowed Chair, named for the 1926 A&M graduate who served in the US Army Air Corps as a test pilot and then retired from the military as a colonel to become a successful and wealthy Texas rancher and farmer.

Upon his death in 1998, Eppright donated the majority of his estate—$4.4 million—to benefit Texas A&M athletics, the largest cash gift in the history of the 12th Man Foundation at the time. A residence hall on the A&M campus is also named for Eppright.

John Junkins himself racked up numerous awards and accolades in his early years at Texas A&M. He received the American Astronomical Society's Dirk Brouwer Award for contributions to spaceflight mechanics and astrodynamics, the John Leland Atwood Award bestowed jointly by the American Institute of Aeronautics and Astronautics and the American Society for Engineering Education, as well as multiple teaching awards for his work with A&M undergraduate and graduate students.

"My first five years in College Station were very busy," Junkins says today. "I brought four PhD students with me, and they helped me establish a strong research culture with my new Texas A&M students."

As for the hiring of Bowen, Junkins says, "He was a strong fellow engineer with a wonderful personality. I had instant respect for him."

And the feeling was mutual.

The two men finally met in the immediate aftermath of Junkins's election into the National Academy of Engineering in 1996. Junkins fondly recalls the occasion.

"When the announcement came out in the spring of '96, Ray Bowen planned an ambush," Junkins says. "He checked with my department head, Walter Haisler, and brought along about ten university vice presidents and barged into my office unannounced to congratulate me.

"He then asked me to explain in 'academic layman's terms' what I had done to deserve this election at the 'tender young age' of fifty-two. (The average age of election into the National Academy of Engineering is sixty-three.) I was borderline speechless, but we spent time conversing and he said he just wanted me to 'feel the love' and appreciation that the university leadership had for my accomplishments."

About a week later, Bowen hosted a reception in Junkins's honor in the Clayton Williams Building, which houses the executive offices of the Association of Former Students.

"It was a very nice gesture on his part and I greatly appreciated it," Junkins says. "At that time, I believe I was the fifth member of the faculty to become a member of one of the three congressionally mandated National Academies. This number increased over the next fourteen years to eleven by the fall of 2011."

Today, that number has grown more than fivefold thanks in large part to the work of the Hagler Institute.

In 1961, then–Texas A&M Pres. James Earl Rudder tasked a committee, called the Century Council, with developing goals for the next fifteen years, in time for the college's one hundredth anniversary in 1976. Included on the committee were former students as well as campus outsiders—including women, newspaper editors, and state office holders—all seeking to assist Rudder in undertaking a "self-evaluation" of the school.

"I hope you will look at A&M College with the future in mind," Rudder told the group at the first committee conference, "visualizing as best you can how all of us as a team can improve this institution and how we can better serve the people of Texas, especially the youth of this state."

Two years later, on April 29, 1963, Rudder addressed his all-male student body in the university's basketball arena, G. Rollie White Coliseum. The topic: the announcement that women would be given, albeit on a limited basis, admittance to the school leading into the upcoming fall semester.

"Face the facts of life," he told his students, according to his speech manuscript on file at A&M's Cushing Memorial Library. "There isn't much you can do. The board of directors has made the decision. You can pick up your marbles and leave and throw in with some other school. Or you can suck up your guts and work to make A&M great. Those that choose to defect should know they leave A&M in the hour of her greatest need."

Before the end of that summer, more change would come to campus as the Texas state legislature approved the renaming of the school from the "Agricultural and Mechanical College of Texas" to "Texas A&M University."

So it was in the spirit of Rudder's "Century Council" that Ray Bowen launched what would become known as *Vision 2020*. Credit for the idea that became *Vision 2020* falls squarely on the shoulders of yet another former member of the Corps of Cadets who rose to the top echelon of leadership at the school. Ed Davis, Class of '67, served his alma mater in a variety of important administrative positions over a span of nearly fifty years. Retired now, Davis was vice president for Fiscal Affairs of Texas A&M University, executive deputy chancellor for the Texas A&M University System, president of the Texas A&M Foundation, and from December 2006 until January 2008, interim president of Texas A&M's flagship campus.

As a cadet, Davis held the position of commander of the Corps during his senior year of school. In fact, his time as an undergraduate dovetailed with James Earl Rudder's rollout of his Century Council initiatives. After graduating, Davis spent four years in Vietnam on active duty in the US Army and then a year in Boston at Harvard Business School. In 1972, he, his wife, and their son returned to Texas A&M, where he continued working toward his MBA while holding down a full-time job in the university's business office. It was there Davis gained a deeper understanding of the challenges President Rudder had faced.

"My boss, Tom Cherry, had been vice president for finance at Trans World Airlines," Davis says. "He hated living in New York, so he took a job heading up the finance group here at Texas A&M. One day in the summer of '72, he called me into his office and said, 'Eddie Joe, we're getting ready to burst open with all these women coming to school here.'"

Rudder's initiative had finally taken hold at Texas A&M University.

"I've got an assignment for you," Davis remembers Cherry telling him. "I want you to go around and find out how many women's bathrooms there are on campus."

What Davis discovered was that there weren't nearly enough restroom facilities to accommodate the expected influx of coeds.

Just over a decade later, Davis had Cherry's job. After another decade had passed, Davis was named president of the Texas A&M Foundation, the school's money-raising enterprise: "The A&M Foundation was formed in 1953. I went there in 1993. Over the course of that first forty years, we had raised about $150 million. That wasn't a lot of money over that period of time. Back in the day, if you wore a maroon jacket and you drove a maroon Cadillac and your car horn honked the 'Aggie War Hymn' as you drove through town, you were a 'Big Aggie.' And in those days, most of the giving was in the form of an annual gift of a hundred dollars. It still only costs a hundred dollars to be a member of The Century Club.

"Our first campaign to seek what we now call 'major gifts' was back in the late eighties and early nineties. We set a campaign goal of $500 million. At the time, that was the largest campaign goal of any public university in the country. That was a big, big step for us, and sort of opened people's eyes that maybe there's more out there than just a bunch of hundred-dollar gifts."

When Davis came aboard, the foundation had reached the halfway mark toward its goal. By the time the campaign ended in 1996, the foundation had raised close to $750 million.

"I've always felt that fundraising is a product of both relationships and an understanding of what people want to connect to," Davis says. "It is important for donors to understand what they are giving to, and what we are trying to achieve."

To continue that sort of fundraising success, Davis knew the university needed academic programs worthy of major contributions.

"I remember the changes President Rudder brought to campus when I was a student here," Davis says. "Then when I started my job as chief financial officer of the university in 1983, we embarked on another intergenerational plan called *Target 2000*. I use the term 'intergenerational' because the idea is to look into the future as much as twenty years down the road and envision future success. That sort of thing excites donors and prospective donors and by about 1996, I told President Bowen I thought it was probably time to talk about launching another one of those strategic initiatives. It didn't take him long to 'study the horseshoe.'"

That's a Texas colloquialism you don't have to be a graduate of the A&M School of Veterinary Medicine and Biomedical Sciences to understand. In a nutshell, it means that the pros and cons of a plan turn out to be more closely aligned than would be thought . . . like the two ends of a curved horseshoe.

As for the matter of "Aggie jokes," Hagler Institute Founding Director John Junkins despises them, and for good reason. He attended a land-grant school—albeit without "State" in the name—at Auburn University and then taught at another, Virginia Tech—no "State" there, either—before arriving at Texas A&M. "Auburn jokes" populate many of the Alabama Crimson Tide fan pages found on the Internet, much as Aggie jokes continue to be a popular staple on chat sites frequented by Texas Longhorn supporters. In the Old Dominion State, University of Virginia supporters call them "Hokie Jokes" in reference to Virginia Tech's mascot, the "HokieBird," which is another name for a male turkey. Before becoming the Hokies, Virginia Tech athletic teams were called "the Fighting Gobblers."

Junkins takes tremendous pride in the school with which he's been affiliated for some four decades. And in testimony to his fervent belief that the "Aggie joke" must be relegated to the past, he offers this:

> Aggie jokes are recently falling flat in the face of a rapid evolution at Texas A&M where many of the school's academic and research metrics have surpassed those of many other universities and not just in the state of Texas. For example, Texas A&M's sponsored research budget of over $1.2 billion is now the highest in the Southwest. The College of Engineering's research budget recently surpassed MIT's to assume the top spot nationally in that prestigious category.
>
> And the number of National Academy Faculty at Texas A&M has increased more than five-fold from 11 to more than 60 over the past twelve years. All of these markers indicate pace-setting transformations that go beyond the borders of the Lone Star State.
>
> This explosive progress has, in large part, been catalyzed by the Hagler Institute for Advanced Study and the Chancellor's Research Initiative. A remarkable "metamorphosis" has taken place at Texas A&M over the past two decades that contrasts sharply with the traditional caricatures of the institution.
>
> Our efforts toward "advancing excellence" in Aggieland are young and still accelerating.
>
> "Tea-sips," be forewarned.

(Note from the Urban Dictionary website: "The 'tea-sip' nickname for University of Texas faithful refers to the World War II era, when UT students were alleged to have sat around campus in Austin 'sipping tea' while Aggies went off to fight the war.")

5

The Vision Evolves

Ray Bowen is more than willing to give credit where credit is due.

"Back in 2020, A&M put together a video," Bowen says, "about *Vision 2020*, the aspirational study that around 250 people helped me put together as president of the university. As I was being interviewed for the project, I told the woman asking the questions that the whole idea for *Vision 2020* belonged to Ed Davis. And she said, 'That's what he told us.'"

Bowen laughs while sharing that story. "After we unveiled the *Vision 2020* report in 1999, I had it out with Ed a few times about how he should take the credit. His reply was always, 'I know, but you had to say yes.'"

As for the fruits of the nearly two years' labor that went into crafting the *Vision 2020* plan, the website *Texas A&M Today* cites these major outcomes:

- Texas A&M's induction into the prestigious Association of American Universities (AAU)
- Research expenditures now exceeding $1.1 billion annually
- Enrollment growth by almost 45 percent to 77,000
- An escalation in hiring and retaining distinguished professors
- Achieving one of the nation's largest-ever fundraising campaigns
- Accommodating rapid physical growth with state-of-the-art facilities
- Accelerating national and global reach

At the core of *Vision 2020*'s success was the four-person group that spearheaded the initiative. Davis tagged that quartet "The Kitchen Cabinet."

"It was President Bowen; myself; Bob Harvey, a former A&M student body president and commander of the Corps who was a management consultant at McKinsey & Company; and Jon Hagler, a successful investment manager who, like Bob and myself, had also been Corps commander as a student."

Both Davis and Bowen knew Hagler well. Bowen had roomed with Hagler for their last two years as undergrads at A&M. Davis, meanwhile, had long courted Hagler professionally to provide financial advice and expertise to their school.

"Part of my job when I was at the Texas A&M University System was to be a liaison with the Texas legislature," Davis says. "I remember 1993 being a particularly contentious legislative session, and I wore out two sets of tires driving back and forth between College Station and the capitol in Austin trying to secure state funds for the system's endowment.

"During that time, I was approached about becoming president of the foundation, but I really didn't think of myself as a fundraiser, per se. Then one night, on my way home from Austin, I stopped in Elgin for a 'lobbyist's dinner' of a pack of crackers and a Coke, and it suddenly hit me: Maybe the foundation job wouldn't be so bad after all."

By late summer, Davis—who was commonly known around campus as "Eddie" or "Eddie Joe"—had assumed the helm of the Texas A&M Foundation. One of his first tasks was to assemble a group of advisers. One of his first calls to move forward on that front was to Jon Hagler.

"My first contact with Jon was when I was at the system. I had the same idea there, putting together an advisory group of investment and financial experts. I didn't know Jon at that time, other than by reputation. By then, he had been in the investment business for about thirty years and was well regarded by those I knew who were his clients. So, I called him—this would have been around 1987 or '88—and asked if he might be interested in becoming a pro bono adviser to help us look at how we should operate.

"Jon quickly turned me down that first time. He told me, 'Eddie, I've got a full-time business, I'm in Boston, and I just can't afford the time to help you right now.' After that, we had a very nice conversation, and I asked if I could call him again at some point in the future. He said, 'Eddie, you can call me any time.'"

Davis fast-forwards to 1993.

"After I got to the foundation, I gave Jon another call. I described to him the situation there, which was good but not the best, and asked, 'Any chance your circumstances have changed?' And he said, 'Absolutely! I just sold my business and I'm not quite sure what I'm going to do. I would be willing to come and sit down with you and look at your program and see if I can help you in a non-fiduciary way.'

"In other words, Jon wasn't looking for a seat on the board. He didn't want a hand in the decision-making process. He just wanted to help us as a fellow Aggie and a volunteer. And that began a wonderful and extraordinary postgraduate relationship between Jon Hagler and Texas A&M University."

That relationship first saw Hagler and his wife, Jo Ann, make a lead gift of $5 million in 1996 for the construction of a new foundation headquarters building on the Texas A&M campus. That building, which opened in 1999, is called the Jon L. Hagler Center. In addition, Hagler has also funded nearly a hundred Foundation Excellence Award scholarships to date . . .

. . . along with becoming the namesake for the Hagler Institute for Advanced Study.

His complete story follows in a chapter to come.

The Hagler Center opened its doors not long after the *Vision 2020* plan was formally adopted by the Texas A&M University System Board of Regents on May 28, 1999. At the core of the 197-page initiative were a set of overarching recommendations, known as *The Twelve Imperatives.*

- Elevate faculty and their teaching, research, and scholarship
- Strengthen graduate programs
- Enhance the undergraduate academic experience
- Build the letters, arts, and sciences core
- Build on the tradition of professional education
- Diversify and globalize the campus community
- Increase access to knowledge resources
- Enrich the campus's physical environment
- Build community and metropolitan connections
- Demand enlightened governance and leadership
- Attain resource parity with the best public universities
- Meet Texas A&M's land-grant commitment to the state

The on-campus response to the *Vision 2020* plan was exceedingly positive. Ellyn Perrone, who served A&M as vice president for governmental affairs at the time, recalled reading the final report.

"My first impression was that it was a very bold aspirational document," Perrone said. "It was about what we can be in twenty years and at the same time it was about preserving the heritage and traditions of the university.

"It was amazingly accepted without a lot of criticism or pushback," she continued. "It laid out imperatives and goals, with ideas on how to achieve those goals, but it was not a mandate. It left the opportunity down the road for stakeholders to again have input on how we actually navigate through those twenty years."

Another interested member of the Texas A&M community was John Junkins.

"I was intrigued by the scope of the project,'" he says, "and like many others on campus, I was curious to understand where the resources would come from to make the vision achievable."

Lurking in the back of Junkins's mind was a possible answer to that question. Soon, he submitted his idea to President Bowen's office.

An Institute for Advanced Study at Texas A&M University?

John L. Junkins
Distinguished Professor of Aerospace Engineering

September 24, 1999

During the 1970s, while on the faculty at the University of Virginia, I was in a position to observe operation of a unique advanced study institute (the Shannon Institute for Advanced Study) which brought many highly distinguished individuals to the University for appointments of up to one year. This institute was virtual in the sense that the members of the institute were housed in the academic units that nominated them, and had minimal facility requirements. The offices of the institute were the James Madison House which gave the institute an historical identity, provided a place for small meetings, and also apartments for short term stays. The Shannon Institute was highly successful for over a decade in bringing about 25 stars per year to the University and a good fraction of them ultimately accepted tenured appointments at UVA and contributed greatly to the advancement of the University. The Shannon Institute, however, had a fatal design flaw. It was operated as a tax on the University budget, and therefore proved to be politically vulnerable and ultimately the institute's budget was squeezed to the point that no further appointments could be made by the mid-1990s. The institute needed an endowment to make it permanent and immune to the challenges that led to the demise of the institute.

What would be the advantages of having an Advanced Study Institute at Texas A&M? First and foremost, in the context of *Vision 2020*, it would give us a powerful tool to elevate the faculty. We could attract stars in the particular areas that align with our areas of emphasis and form teams between these visiting distinguished faculty and our current faculty. The result, if we do it right, would be accelerating existing programs and occasionally attracting some of the stars to our faculty. It would project our image to the best and brightest people every year and the impact on peer perception would be immediate and highly significant. I think the case could be made that a permanent advanced study institute would rapidly transform both our quality and our image, and our students would be exposed to people and ideas that would likely be transformative and life changing. Who would decide who to recruit? I recommend that we use a committee of our current distinguished professors to review nominations solicited from across the University, and that very high standards be imposed to establish the "DNA" of the institute. A key cultural challenge would be to convince the President, the Provost, and the Deans that this much of the usual prerogatives of University administration should be turned over to a committee of distinguished professors. This would seem akin to letting the inmates run the asylum–but surely we can convince the powers that be that this is the way to proceed. How much would it cost? Well this depends on scale, but it would be significantly less than adding an average-sized department. Nonetheless, finding in excess of $1M/year of recurring funds is not easy and will require Presidential buy-in and leadership. This will likely require the idea to be attractive to every dean and the provost first, then a systematic effort made to develop the idea among the Distinguished Professors seems warranted.

That paper got Junkins into Ray Bowen's office.

"One of the things that's wonderful about John," Bowen says, "besides being an outstanding scientist and engineer, he's also a very strategic thinker. He thinks big. He thinks long term. And he had an idea, because of our *Vision 2020* aspirational study, of how to make our university among the top public universities in the country. We said it should be in the top ten and I still think that's a valid aspiration. John responded to that, and his proposal for an institute for advanced study characterized the kind of quality we knew we needed to look for."

Junkins told Bowen the Institute, as he envisioned it, would cost around $2 million per year.

"I remember the conversation I had with him," Bowen says. "I told him if I'd had the money, I would've funded the program instantly because I liked the idea. I had some knowledge of the Advanced Studies Institute at Princeton and how impactful it had been.

"But I had to tell John no."

Bowen says the most important achievement tied directly into the *Vision 2020* initiative was Texas A&M's admission into the Association of American Universities, also called the "AAU" (not to be confused with the "Amateur Athletic Union," whose position of prominence in the sporting world waned decades ago).

The Association of American Universities got its start in 1900 when six of the eight Ivy League schools were joined by the likes of Johns Hopkins University, the University of Chicago, the University of California, and Stanford University to strengthen and standardize doctoral programs on American campuses. Today, sixty-nine institutions of higher education in the United States are part of the consortium. Texas A&M and Stony Brook University joined the AAU ranks in 2001.

In a May 7, 2001, story that ran in the *Houston Chronicle* newspaper, President Bowen called the invitation to join the AAU "a great honor" for his school, adding that the designation "adds value to the degrees of all graduates, past and future."

A year and a half before Texas A&M received AAU recognition, on the night of November 18, 1999, Ray Bowen experienced his deepest despair as university president. It was that night the towering stack of timber erected on the north end of campus—to be ignited in advance of the school's annual football game against the University of Texas—collapsed, killing eleven students and a former student who were working on the Aggie Bonfire stack.

The aftermath of that catastrophic event defined Bowen's presidency far more than the AAU designation.

"There was a lot of sadness and a lot of crying that night," Bowen remembers.

Following the mishap, Bowen imposed a two-year moratorium on Bonfire. Then on February 4, 2002, he announced the tradition would be permanently discontinued as a sanctioned on-campus event. In 2008, the university reached a $2.1 million settlement with several families of those killed in the collapse.

"The Bonfire tragedy was ample reason to resign," Bowen said at a campus news conference on June 6, 2001, in which he announced his plans to step down at the conclusion of the ensuing academic year. "But in fact it was also ample reason to stay on and make sure the next Bonfire would be safe," he continued, as reported by the *Houston Chronicle* newspaper. "The right way to lean is the Bonfire caused me to stay longer than I might have otherwise."

The *Chronicle* story went on to say,

> Noting that he has already served as president for seven years, the fourth-longest tenure in the school's 125-year history, Bowen said he decided it was time to step down.
>
> "Now is simply the time," he said. "There was no other issue, no other agenda. No one offered me a job that's more attractive."
>
> Bowen, who earned bachelor's and doctoral degrees at A&M in mechanical engineering, said he plans to return to teaching after stepping down from his $250,000-a-year president's job.

He said "one of the miracles of higher education" has been A&M's transformation beginning in the 1960s from an all-male military academy to one of the nation's most respected research universities. With an enrollment of 44,000 students, A&M is the country's fourth-largest campus.

Bowen said he is proud that the university continued to enhance its academic reputation under his leadership, including being inducted into the highly selective Association of American Universities last month. In a self-effacing manner that was described as typical by associates, Bowen said he had little to do with the recognition, praising the faculty for its efforts.

His tenure also saw A&M open the George Bush Presidential Library and Museum in 1997.

A few long-term goals Bowen has proposed for the university include increasing diversity at the 85 percent white campus, strengthening its relatively weak liberal arts programs and attracting top-notch teachers.

6

Academic Master Plan

"I got into a political firestorm in chairing the search committee to find Ray Bowen's successor because, unbeknownst to me, Gov. Rick Perry had informally promised US Sen. Phil Gramm the presidency of the university."

By 2001, John Junkins's reputation as a distinguished professor on the Texas A&M University campus was such that he was asked to lead a twenty-four-member panel to find the school's next president. Insiders believed Gramm was a shoo-in. Georgia-born and University of Georgia–educated, Gramm had been a beloved and respected economics professor at Texas A&M from 1967 to 1978. He then left academia to serve Texas in the political arena, first as a five-term member of the US House of Representatives, then as a three-term US Senator.

Just days after being reelected to the House as a Democrat in 1982, Gramm was thrown off the House Budget Committee by its Republican leadership in a move they hoped would prompt Gramm to switch parties. In response, Gramm resigned his House seat, then ran as a Republican for his own vacancy in a February 12, 1983, special election, which he easily won.

Future Texas Gov. Rick Perry, a Texas A&M graduate, also changed party affiliations during his political career, becoming a Republican in 1989 on his way to an upset win over a Democratic incumbent to become Texas Agricultural Commissioner. Later, in 2000, Lieutenant Governor Perry became Texas governor when he replaced George W. Bush, who resigned from the office to prepare for his ascendancy to the US presidency.

In 2001, Governor Perry named Gramm's wife, Wendy, an economist who served in several positions in Ronald Reagan's presidential administration, to the Texas A&M University System Board of Regents. At the time, she was a member of the scandal-plagued Enron

Corporation's board of directors. By the end of 2002, Phil Gramm had stepped down from his Senate seat, a move that seemingly cleared the way for him to become A&M president.

The A&M System Regents ultimately had hiring authority for their flagship's new president. The chairman of the regents at the time was Erle Nye, a Texas A&M graduate and CEO of TXU, a burgeoning Texas-based energy company. Not surprisingly, given their lofty positions in business and government and their ties to Texas A&M, Nye and Gramm were close friends.

"When Erle Nye recruited me to lead the search committee," John Junkins says, "I asked for his word that the pick was not already a done deal. I told him, 'If you've already decided that Phil Gramm will be the next president, don't make me run a search effort because that process is a lot of work.'

"And he said, 'I give you my word. You give me at least three qualified candidates, well-qualified candidates, and we will choose the president from that group.'

"I said, 'OK, I'll do the search.'"

But from the outset, Junkins says, he realized the political pressures on both the selection committee and the board of regents were "intense."

"In narrowing the field, I tried to make sure everyone on the committee looked seriously at all of the people that came into the final consideration queue. After we had narrowed the search to around twenty candidates, I told the committee, 'OK, we're getting ready to get serious now. The first thing I want each of you to do is to prepare a five- to ten-sentence 'elevator pitch' on each of these twenty people.'"

Included in that group was Phil Gramm.

"By many measures, Gramm was the strongest candidate in the field," Junkins says today. "He was a former academic. He loved this university. And when he interviewed with us, he knew every one of the twenty-four search committee members. He knew where they went to high school and whom they were married to. And Phil offered a vision for the university that I thought was an attractive one."

Yet when it came time to put forth the three finalists for the job, Phil Gramm's name was not on the list.

"I didn't vote, but I can give you the sense of the committee," Junkins says today. "Given the political demography of the state of Texas at the time, the faculty here at Texas A&M was, shall we say, a little left of the conservative norm. So Gramm, in the search committee's collective thinking, would have been a polarizing figure for more than half of the faculty who were politically positioned to not be supportive of him based on his political track record with the Republican Party."

When it came time to announce the committee's choice of finalists—they were Robert Gates, a former director of the CIA who was the interim dean of the Bush School of Government and Public Service at Texas A&M; Richard Herman, provost and vice-chancellor for academic affairs at the University of Illinois at Urbana-Champaign; and Jon Whitmore, provost at the University of Iowa—Junkins took the list straight to General Howard Graves, the Texas A&M System chancellor.

"I knew because Gramm's name was not on the list," Junkins says, "Erle would not be pleased, despite his previous assurances he would support our recommended finalists. In meeting with Graves, I told him, 'You know what's happening regarding a push toward Gramm becoming president, don't you?' And he said, 'Yes, I know.' I said, 'If the regents reject this search and substitute their candidate, it may be legal, but it's not right. I have my own press release here and I'd like you to release it today, this afternoon.'"

And Graves did just that.

Junkins initially refused to answer reporters' questions about the finalists. By the time he got back to his office later that day, he had a message to call Nye.

"John, what in the hell do you think you're doing?" Junkins remembers the System Regents' chair asking him. "Erle," Junkins responded, "I'm doing my job. What are you doing?"

Junkins remembers being called onto the carpet . . . literally.

"I've called an emergency meeting of the Regents tomorrow," Nye told Junkins. "Be there. You're going to explain what you've done in this search." Junkins replied, "Erle, I'll be delighted to be there, I'll see you tomorrow."

A week later, the regents announced that Robert Gates had been named the twenty-second president of Texas A&M University. On the heels of that announcement, John Junkins received an email.

"Jon Hagler wrote me," Junkins says. "I don't remember exactly what his email said, but something along the lines of, 'The spinal calcium must run strong in your family!' I wrote him back, telling him that I didn't know my zip code was going to be at stake when I agreed to chair the search. I told Jon, 'I don't like when somebody takes a year out my life and then spits on my face and calls it "rain."' So that started a kind of long-distance friendship between Jon and me."

Well into the Gates presidency at Texas A&M, Erle Nye finally broke his silence on the search effort Junkins had led.

"John, I owe you an apology," Junkins remembers Nye telling him in a phone call between the two men. "We could not possibly have gotten a better president than Bob Gates. I was under a lot of political pressure."

"Tell me about it," Junkins replied to Nye. "The hot seat I was in chairing the committee was ridiculously hot. I only recently stopped receiving hate mail."

Robert Gates served as Texas A&M president for just over four years. He left the office on December 16, 2006. Two days later, Gates was sworn in as secretary of defense under Pres. George W. Bush. When Barack Obama succeeded Bush in the White House in 2009, Gates became the first secretary of defense in US history to serve under two presidents of different parties.

Gates's successor as Texas A&M University president also made history. Cuban-born Elsa Murano was not only the first woman to lead the school but the first Hispanic president as well. Previously, she had served in the George W. Bush White House as an undersecretary in the US Department of Agriculture before becoming dean of the College of Agriculture and Life Sciences and a vice-chancellor for the Texas A&M University System.

Joining Murano as the new provost at Texas A&M was Jeffrey Vitter, a highly regarded computer scientist with an impressive resume of his own. After receiving his PhD from Stanford University in 1980, Vitter spent the next twelve years on the faculty at Brown University, where he earned tenure at the age of twenty-nine. He then chaired the Department of Computer Science at Duke University for eight-and-a-half years before being named dean of the College of Science at Purdue University. He left Purdue to move to Texas A&M, and as the executive vice president of academics—a title he held in tandem with that of provost—he attempted to reignite Texas A&M's *Vision 2020* initiative by creating an "Academic Master Plan."

"As part of launching that plan," John Junkins says, "Vitter invited the submission of white paper proposals for projects and programs which would align with the 'Twelve Imperatives' contained within *Vision 2020*. This was about seven years since I'd put together my original one-page summary regarding an institute for advanced study at Texas A&M. Needless to say, that concept had 'evolved' in my mind in the ensuing years."

Junkins today believes more than 150 white papers were submitted to Vitter's office. Among those—in addition to Junkins's Institute for Advanced Study—was a proposal authored by a laser-physics pioneer to create an "Institute for Quantum Science" at Texas A&M.

Where a secretary might sit as a gatekeeper outside the door to Dr. Marlan Scully's office in Texas A&M's George P. Mitchell '40 Physics Building, visitors are greeted instead by the menacing figure of a towering grizzly bear. Fortunately, the animal is long dead and preserved through taxidermy, its presence a tribute to Scully's father.

"I grew up in Wyoming and my dad was a big-game hunter," Scully says with a warm and welcoming smile. "It was always his dream to take down a grizzly in Alaska—they call them brown bears or Kodiak bears there—but he never got a chance to do so. About ten years ago, after he passed, I made the trip and shot the bear for him. My wife and others were not terribly happy that I should do that, but it was a great hunt."

As for Scully's academic credentials, let's let the bio on his Physics and Astronomy Department "Distinguished Professor" page tell his story:

> Marlan O. Scully (Texas A&M and Princeton) is a laser physics pioneer. His work includes the first quantum theory of the laser with Lamb, the first demonstrations of lasing without inversion, the first demonstration of ultraslow light in hot gases, and the use of quantum coherence to detect anthrax in real time. Furthermore Scully's work on quantum coherence and correlation effects has shed new light on the foundations of quantum mechanics, e.g., the quantum eraser. He has been elected to the National Academy of Sciences, American Academy of Arts and Sciences, Academia Europaea, and Max Planck Society; has numerous awards including the APS Schawlow prize, OSA Townes Award, IEEE Quantum

Electronics Award, Franklin Institute's Elliott Cresson Medal, OSA Lomb Medal, and Humboldt Senior Faculty Prize. More recently he was named Harvard Loeb Lecturer, received an honorary doctorate from University of Ulm, and was awarded the OSA's DPG Hebert Walther Award.

Scully, John Junkins, and the late Jim Womack, a longtime member of the A&M Department of Veterinary Pathobiology and the first coordinator for the USDA Cattle Genome Research Program, all submitted Academic Master Plan white papers with each man signing off as a coauthor on the others' pet projects, providing each proposal with a National Academy "stamp of approval," according to Junkins.

Of the myriad of proposals submitted, just four were selected, with Scully and Junkins winners of what Junkins calls an "academic beauty contest." The half-million-dollars-a-year funding promised through the Academic Master Plan enabled Scully to launch the Institute for Quantum Science and Engineering (IQSE), which today includes the Center for Theoretical Physics and the Center for Biophotonics.

Defining biophotonics, the IQSE website reads, "Quantum spectroscopy combined with laser technology, can probe the vibrational states of molecules to generate specific spectral signatures that can be used to rapidly and reliably assess everything from the presence of foodborne pathogens and plant composition to the presence of pests in crop fields."

One reason Scully supported Junkins's institute for advanced study concept—although Junkins was unaware of this coincidence at the time—was that Scully had been tabbed to create an institute for advanced study at the University of New Mexico in the 1980s. That program never fully took root in Albuquerque, as Scully was also involved at the time with the Max Planck Institute for Quantum Physics in Munich, Germany.

Since 2005, Scully has had a joint appointment between Texas A&M and Princeton University, near which Abraham Flexner's original Institute for Advanced Study resides. The relationship with Princeton was spawned in the aftermath of the 2001 anthrax murders in which a series of letters containing the deadly spore were mailed to individuals across the country, killing five people. An army microbiologist, Dr. Bruce Ivins, was ultimately convicted of the crimes.

In 2007, Scully and his teams at Princeton and Texas A&M announced a laser technique to identify anthrax quickly and safely, a product of biophotonics.

"Our report shows how to use lasers to detect anthrax in real time as opposed to the cumbersome and unwieldy way it is done now," Scully said in the publication *Science Daily*. "We do our experiments 'on the fly,' so we can get a signature within a tiny fraction of a second. Our procedure can work for monitoring anthrax in mail, but it can also scan the whole atmosphere. And there are a lot of other potential applications."

In addition to Scully's IQSE, Texas A&M's Institute for Neuroscience and the Center of Digital Humanities Research were also spawned by Academic Master Plan funding.

An institute for advanced study was not.

John Junkins declined the half-million-dollars-a-year offer.

"I think I shocked a lot of people in the Texas A&M academic community when I turned down the money," Junkins says. "I was already fully employed as 'Professor John,' and I knew that if I undertook to start the Institute for Advanced Study too soon, with too few resources to make it matter to every dean on campus, I would have to be raising additional funds on a 24/7 basis, and I would probably fail."

Failing, too, was Elsa Murano's presidency, which never really took root at Texas A&M. Ultimately she was forced to resign less than halfway through her second year in charge.

Reporters Jeannie Kever and Mike Tolson of Hearst News Service wrote of her downfall,

> Elsa Murano's appointment as president of Texas A&M University looked like a watershed moment for a school trying to move beyond its conservative, small-town past to become one of the nation's top public research universities.
>
> The first woman and first Hispanic to head the campus, Murano symbolized the school's expansion into a modern, diverse world.
>
> But just 17 months after it began, a presidency that had seemed so filled with promise ended in an embarrassing display of acrimony and arguments over the independence of the office. More than the cultural clash between the old school and the new, Murano

> fell victim to a classic power struggle over who would control the university, especially its $600 million research enterprise.
>
> On the surface, it was a battle between Murano and Chancellor Mike McKinney. But insiders and influential alumni say the Murano affair is symptomatic of the damaging political influence of the state's most prominent Aggie, Gov. Rick Perry, which they fear threatens to derail A&M's drive into the top ranks of higher education.

Six weeks after Murano's departure, Vitter resigned as provost at Texas A&M. He moved on to become provost at the University of Kansas, then in 2015, he was named chancellor at the University of Mississippi, a post he held for three years.

"One positive outcome of Vitter's time at Texas A&M," says John Junkins, "at least for me, was that the university officially identified the Institute for Advanced Study as something it wanted to do."

7

The White Paper

Texas A&M University Institute for Advanced Study Proposal Draft

1. Rationale for the Creation of the Center or Institute

BACKGROUND

Texas A&M University has experienced significant development and growth over the past twenty years, especially in the first decade of the twenty-first century. Following the goals set forth in *Vision 2020*, the university hired over one thousand new faculty, including new appointments for over 450 newly created positions across the university. New faculty members were recruited at all levels, but most newly hired faculty members were at the junior ranks. Joining highly productive faculty already at Texas A&M, these newly appointed faculty will enhance research productivity and contribute additional value to the university's teaching mission.

With the creation of a Texas A&M University Institute for Advanced Study, the university can build on these developments to continue the momentum toward *Vision 2020*'s first imperative of "elevating the faculty and their teaching, research, and scholarship." This institute has been widely discussed on campus and was strongly endorsed by the Research Roadmap Committee for the Academic Master Plan. Attracting outstanding senior scholars to the Texas A&M campus on a temporary (e.g., one year) or more permanent basis, this institute would enhance research productivity among the faculty and enrich the educational experience for the students. Top scholars will contribute new research ideas, engage in collaborative research, provide mentoring for younger faculty (thereby accelerating their development), and contribute to the development of excellent graduate students.

MISSION

The Institute's mission is to provide a catalyst for high-quality, innovative research across the multiple disciplines represented in the university and to enhance the intellectual climate of the university. To fulfill this mission, the Institute will bring to campus top scholars who have distinguished themselves through professional accomplishments and have received important recognition for these contributions (e.g., appointed to the National Academies, selected to receive a Nobel Prize, or received other highly competitive and valued prizes and awards).

These scholars, freed from their normal obligations at their home institutions, will be invited and given the flexibility to engage in creative endeavors (e.g., conduct empirical research, write a book, engage in theoretical work) and to collaborate with Texas A&M faculty, students, and other TIAS Fellows in ways that will advance scholarship in one or more disciplines. As discussed below in section 4, each Fellow's appointment will be based on a rigorous screening process that includes a proposal written jointly by the candidate and a group of current Texas A&M faculty. The proposal will outline the activities planned during the scholar's appointment in the Institute. The Institute will provide an open and receptive environment for unrestricted academic inquiry while simultaneously encouraging the cross fertilization of ideas.

The Institute will use a variety of pedagogical approaches to promote high levels of academic interchange. While the TIAS Fellows will be given time and freedom to pursue solitary projects, they will also be encouraged to participate in collaborative activities with other TIAS Fellows, Texas A&M faculty, and Texas A&M students through periodic workshops and open forums (as well as informal meetings). In cases where there appear to be linkages among the TIAS Fellows' research, scholarly, or creative endeavors (either by design through the selection process or arising from formal and informal discussions during their visit), special workshops and forums will be developed for the Texas A&M University academic community.

GOALS

The Institute's advancement of *Vision 2020's* first imperative of "elevating the faculty and their teaching, research, and scholarship" will be achieved by focusing on four major goals:

- Enrich the intellectual climate of the university
- Enhance research productivity broadly throughout the university
- Recruit eminent scholars (TIAS Fellows) for an extended or regular (permanent) university faculty appointment
- Enrich the educational experience of graduate and undergraduate students

The pursuit and realization of these goals will enhance the visibility and reputation of Texas A&M University, thus leading to such outcomes as increased honors and recognitions for high-achieving faculty, increased external funding to support research and scholarship, and the ability to attract greater numbers of exceptional students (graduate and undergraduate) with extraordinary potential.

2. Impact on Education and Training of Students

Attracting senior scholars of national and international reputation, including members of national and international academies and Nobel laureates, to TIAS will directly strengthen our research programs, raise the level of scholarly productivity among our faculty and graduate students, and enhance the international reputation of Texas A&M. Equally important, the Institute will greatly benefit student learning and professional opportunities by bringing our students together with some of the world's most distinguished scholars. The interaction with these celebrated scholars will improve our research programs, benefit the intellectual life of the university community, and enrich the quality of life and educational experience of our graduate and undergraduate students. For example, students will have the opportunity to work collaboratively with TIAS Fellows on research projects, TIAS Fellows will participate in undergraduate and graduate research seminars, and graduate students will have the opportunity to consult with TIAS Fellows on their own research projects. The interactions and relationships formed may lead to future opportunities such as postdoc appointments at the TIAS Fellows' home institutions. All in all, the opportunity to know, consult, and work with TIAS Fellows will constitute a "high-impact" experience for Texas A&M students at the undergraduate and graduate levels. Such experiences are likely to have lasting impact on the students and redound to the benefit of everyone involved with the Institute.

The Institute envisioned will enhance the image and greatly increase the visibility for Texas A&M University and its faculty and students in the United States and the world. Unique programs fostered by these notable scholars will, in turn, be significant in retaining our best faculty and in recruiting the best faculty and students. Thus, TIAS will significantly enrich the quality of the university and the social and economic environments of the state and the region.

3. Sources and Future Expectations of Financial Support

The current start-up funds proposed for the Institute from within the Texas A&M University System for its initial five years are $500,000 per year, plus $400,000 per year for graduate fellowships from the Office of the President drawn from Heep Foundation funds, as well as $1 million per year for five years from the Office of the Chancellor. It is also anticipated that Texas A&M University colleges, direct beneficiaries of having these distinguished scholars on campus, will contribute to the funding of TIAS Fellows, and this would require close cooperation between the leaders of TIAS and of the university and system.

The creation of such an important center within Texas A&M University will, of course, require a sizable investment, and if it is to become a vehicle capable of raising our reputation to even greater international significance, it must have a high priority in future development campaigns of the university. An institute of international stature such as that proposed will eventually require an endowment of approaching $200 million. Several options should be pursued simultaneously, including seeking donors of endowed fellowships (analogous to the current system of chairs and professorships) linked directly to TIAS. In addition to general endowments to the Institute, individual significant donors should be sought whose gifts might endow programs or even the Institute itself. Importantly, such gifts would be recognized through prominent naming opportunities associated with TIAS. Endowment moneys also should be sought from research-related public sources at the private, federal, and state levels. Perhaps, during the start-up phase, current endowments from unfilled chairs (consistent with established gift agreements) could be used for temporary support of TIAS Fellows, allowing colleges and the university to leverage existing endowments. An important source of potential leverage in the support of some TIAS Fellows will come from their half-time salary support provided by their home institutions for sabbatical leave. Still, the actual growth rate for TIAS will depend on several factors, but an important one among them will be success in attracting permanent endowment.

Some of the core funding will be used for modest administrative support and to support travel and recruitment efforts. In steady state, it is anticipated that over 90 percent of the TIAS funds will be allocated to support the Faculty Fellows and student fellowships. During the five-year start-up phase, the endowments and the funds provided by the chancellor and president will be used specifically to support core TIAS activities and salaries (e.g., faculty fellowships, graduate fellowships). Some fractional level of cost-sharing from colleges will be needed to support successful TIAS Fellow nominations during the start-up phase. A longer-term goal of zero cost-sharing is intended for the Institute's steady state. The vice president for research and graduate studies—in consultation with the relevant college dean, TIAS director, and others as appropriate—will approve cost-sharing arrangements for support of a TIAS Fellow nominee. It is anticipated that the proportion of cost-sharing by colleges will decrease as endowment income increasingly supports the core TIAS programs.

4. Governance and Advisory Structure

The Texas A&M University Institute for Advanced Study will operate under the direction and guidance of the vice president for research and graduate studies. The reporting relationship and governance structure calls for the TIAS director to report to the vice president for research and graduate studies and to other important groups involved in the governance and operation of TIAS, an Administrative Council and an Advisory Board.

ADMINISTRATIVE COUNCIL

The primary responsibilities of the Administrative Council are to provide budgetary oversight and to recommend policies for guiding the activities of the Institute. As indicated in the figure, the Administrative Council provides guidance and counsel to the vice president for research and graduate studies in his oversight of the director and the Institute.

We expect the Administrative Council to have twelve to fifteen members composed of several Texas A&M deans, esteemed academy-level Texas A&M scholars (including National Academy members, Nobel laureates), vice presidents with a strong interest in the Institute, distinguished external scholars, and the Texas A&M Foundation president. The vice president for research and graduate studies will chair the council. Council members will be nominated by the provost and chosen by the president.

ADVISORY BOARD

The primary responsibilities of the Advisory Board are to solicit nominations for TIAS Fellows, working collaboratively with colleges and departments, and then to evaluate, select, recommend, issue invitations to, and help recruit TIAS Fellows. The TIAS director will serve as chair of the Advisory Board.

The Advisory Board will be composed of nine members, six who are elected and three who are appointed. Elected members will be nominated by (and be members of) a university electorate consisting of National Academy members, National Academy–level faculty, distinguished professors, and deans. The Advisory Board will develop the definition of "National Academy–level" faculty and make a recommendation to the vice president for research and graduate studies. Appointed members will be selected by the vice president for research and graduate studies, in consultation with the director of TIAS, after elected members are chosen. Chosen from the pool of members of the electorate mentioned above, the appointed members will be selected for the purpose of providing intellectual balance to the committee. Board members will serve staggered, three-year terms, with three members (two elected, one appointed) rotating off the board every year.

DIRECTOR

The director of the Texas A&M University Institute for Advanced Study will have several responsibilities. First, the director will have operational oversight of the Institute, including budget, staff, equipment and physical space. Second, the director will have primary responsibility for recruiting TIAS Fellows (in consultation with the Advisory Board). Third, the director will play an important role in supporting efforts to attract endowments for the Institute. Last, the director will build and maintain relationships with all relevant and important internal constituencies.

The director of TIAS will be nominated by the vice president for research and graduate studies and appointed by the provost and executive vice president for academic affairs. The search for the director will be conducted in a manner consistent with major academic administrative appointments at Texas A&M University. The recommended nominal term for the TIAS director is five years, subject to annual performance reviews. (See section 5 for discussion of the formal review in accordance with system and university policies.)

INTERIM APPOINTMENTS

To initiate the Institute, an interim director and the Advisory Board will be appointed for the first year of operation. The vice president for research and graduate studies will coordinate the search for the initial TIAS director, and the appointed Advisory Board will establish the procedures to initiate the elections for the first (elected) Advisory Board.

SELECTION PROCESS FOR TIAS FELLOWS

The selection process for TIAS Fellows will be highly important and must be managed with care. The selection of the Fellows will be based on the candidates' credentials, explained in the mission statement, and their ability to contribute to the four goals listed earlier. The selection process will include the following steps.

Prenomination process: initial identification of candidates—The first step in the process involves a "prenomination" to allow a screening of potential candidates. The process will involve a one-page description of the candidate and the value of having the candidate visit Texas A&M. Early in the process, there should be no direct contact with the nominee. The prenominations will be submitted to the Advisory Board through the director. The Advisory Board will review these prenominations, seeking additional information, if needed. The Advisory Board recommends which of the prescreened individuals should be recruited as TIAS Fellows. These nominations are recommended to the vice president for research and graduate studies by the TIAS director. The vice president for research and graduate studies takes the nomination to the Administrative Council for discussions on issues of space, cost-sharing, and so on. Resulting from this carefully triangulated vetting process will be a list of candidates receiving the final approval of the vice president for research and graduate studies to move forward in the process.

Initial contact with candidates—Candidates selected from the group of the prenomination and review process will receive initial contact only by the TIAS director and/or the vice president for research and graduate studies.

Full proposal prepared—When the contact provides a positive response, the director and Advisory Board member work with faculty members, department head(s), dean(s), and other relevant leaders to prepare a full proposal on the candidate. The full proposals are then reviewed and approved by the Advisory Board, and full recruitment of the candidate begins.

Discretion and confidentiality in the nomination process are absolutely vital. Contact with potential candidates will only be made after a careful screening and evaluation in the nomination process. The recruiting process for some candidates may require longer than one year. Thus, flexibility is required. For example, a candidate may be interested in coming to Texas A&M and serving as a TIAS Fellow but cannot visit in the year initially invited (e.g., because of the desire to coordinate the visit with sabbatical support from her/his home institution). The desired candidates should be cultivated for a subsequent year, depending on the particular circumstances for the delay, by an otherwise highly desirable candidate. Because of the very high standards and the target pool of well-qualified and internationally influential nominees, competition for invitations is expected to be strong. Care must be taken to ensure a continuing positive relationship with all candidates contacted and to protect the reputation of both the candidates and the university.

PHYSICAL SPACE

Space needs will grow as the Institute attracts greater numbers of TIAS Fellows, but initially TIAS will require modest physical facilities of only about 4,000 square feet. This initial space will provide five or six faculty offices for TIAS Faculty Fellows (approximately 1,500 square feet). It is anticipated that partnering colleges will provide some office space for the TIAS Fellows as well. In addition, space will be needed for the administrative offices for the director and three staff assistants (approximately 500 square feet). The remaining space is needed for a large conference room and two smaller meeting rooms (approximately 2,000 square feet). The cooperating colleges and departments are expected to provide laboratory facilities and any other research space and equipment for collaborative access for the TIAS Fellows during their one-year appointments.

5. Mechanisms for Periodic Review

An annual review of the Institute director's performance will be conducted by the vice president for research and graduate studies. Additionally, periodic reviews of the Institute's activities, development, and progress will be conducted in accordance with the Texas A&M University System Policies and the Texas A&M University Administrative Rule on Centers and Institutes. As such, a review will be conducted at least every five years as required by the policies mentioned above. The evaluation will include advice from the Institute's Administrative Council and the Advisory Board as well as other campus constituencies. The procedures used for the review and evaluation of the Institute will follow the guidelines and meet the specific requirements prescribed in the Texas A&M University System Policies and the Texas A&M University Administrative Rule on Centers and Institutes.

6. Potential Milestones for Evaluating Progress

During the first two years of operation, the following potential milestones could be used to facilitate the evaluation of progress in the development and implementation of the Institute:

- Appoint an interim director
- Create an Advisory Board and an Administrative Council
- Establish a space for the Institute on campus
- Search for and appoint the initial TIAS director
- Create and develop a fundraising/endowment plan
- Appoint the first five Fellows
- Establish links with the Honors Program to build the relationship between TIAS and the university's undergraduate program
- Appoint at least two of four Graduate (student) Fellows

During the next three years of operation, additional potential milestones include the following:

- Continue the growth and establishment of the center with evolving resources
- Continue plans for capital campaign with commitments of endowments
- Invite additional Fellows reflecting quality standards of the Institute and the university's scholarly heterogeneity and richness

7. Potential Internal Participants in the Institute

R. J. Q. **Adams**, Distinguished Professor of History
Kyle T. **Alfriend**, Professor of Aerospace Engineering and Member of the National Academy of Engineering
Fuller W. **Bazer**, Distinguished Professor of Animal Science
G. Kemble **Bennett**, Vice-Chancellor and Dean of Dwight Look College of Engineering
José Luis **Bermúdez**, Dean of Liberal Arts
Leonard L. **Berry**, Distinguished Professor of Marketing

Raymond J. **Carroll**, Distinguished Professor of Statistics
Abraham **Clearfield**, Distinguished Professor of Chemistry
C. Colleen **Cook**, Dean of Texas A&M University Libraries
Ryan C. **Crocker**, Dean of The Bush School of Government and Public Service
John L. **Crompton**, Distinguished Professor of Recreation, Park & Tourism Sciences
Ronald G. **Douglas**, Distinguished Professor of Mathematics
Kim R. **Dunbar**, Distinguished Professor of Chemistry
George C. **Edwards** III, Distinguished Professor of Political Science
Christine A. **Ehlig-Economides**, Professor of Petroleum Engineering and Member of the National Academy of Engineering
Margaret J. M. **Ezell**, Distinguished Professor of English
Ciprian I. **Foias**, Distinguished Professor of Mathematics
John A. **Gladysz**, Distinguished Professor of Chemistry
Richard J. **Golsan**, Distinguished Professor of European & Classical Languages & Cultures
D. Wayne **Goodman**, Distinguished Professor of Chemistry
Eleanor M. **Green**, Dean of Veterinary and Biomedical Sciences
Ricky W. **Griffin**, Distinguished Professor of Management
Rostislav **Grigorchuk**, Distinguished Professor of Mathematics
Timothy C. **Hall**, Distinguished Professor of Biology
Je-Chin **Han**, Distinguished Professor of Mechanical Engineering
Paul E. **Hardin**, Distinguished Professor of Biology
John C. **Hardy**, Distinguished Professor of Physics
Dudley R. **Herschbach**, Professor of Physics & Astronomy and Member of the National Academy of Science
Michael A. **Hitt**, Distinguished Professor of Management
Stephen A. **Holditch**, Professor of Petroleum Engineering and Member of the National Academy of Engineering
Mark A. **Hussey**, Vice-Chancellor and Dean of Agriculture and Life Sciences
R. Duane **Ireland**, Distinguished Professor of Management

Arthur E. **Johnson**, Distinguished Professor of Chemistry
William B. **Johnson**, Distinguished Professor of Mathematics
John L. **Junkins**, Distinguished Professor of Engineering and Member of the National Academy of Engineering
Howard B. **Kaplan**, Distinguished Professor of Sociology
Leonid V. **Keldysh**, Professor of Physics & Astronomy and Member of the National Academy of Science
Olga A. **Kocharovskaya**, Distinguished Professor of Physics
David M. **Lee**, Professor of Physics and Astronomy and Member of the National Academy of Science
W. John **Lee**, Professor of Petroleum Engineering and Member of the National Academy of Engineering
Yvonna S. **Lincoln**, Distinguished Professor of Educational Administration and Human Resource Development
Jerome M. **Loving**, Distinguished Professor of English
Joanne R. **Lupton**, Distinguished Professor of Nutrition and Food Science and Member of the Institute of Medicine
Ronald D. **Macfarlane**, Distinguished Professor of Chemistry
Bruce A. **McCarl**, Distinguished Professor of Ag Economics
John J. **McDermott**, Distinguished Professor of Philosophy & Humanities
Kenneth J. **Meier**, Distinguished Professor of Political Science
Kate C. **Miller**, Dean of Geosciences
Dimitri V. **Nanopoulos**, Distinguished Professor of Physics
Joseph B. **Natowitz**, Distinguished Professor of Chemistry
H. Joseph **Newton**, Dean of Science
Gerald R. **North**, Distinguished Professor of Atmospheric Sciences
Douglas J. **Palmer**, Dean of Education and Human Development
Gilles **Pisier**, Distinguished Professor of Mathematics
Valery **Pokrovsky**, Distinguished Professor of Physics
Christopher N. **Pope**, Distinguished Professor of Physics
Kumbakonam **Rajagopal**, Distinguished Professor of Mechanical Engineering
J. N. **Reddy**, Distinguished Professor of Mechanical Engineering

Kenneth F. **Reinschmidt**, Professor of Civil Engineering and Member of the National Academy of Engineering
Jose M. **Roesset**, Professor of Civil Engineering and Member of the National Academy of Engineering
B. Don **Russell** Jr., Distinguished Professor of Electrical & Computer Engineering and Member of the National Academy of Engineering
Stephen H. **Safe**, Distinguished Professor of Veterinary Physiology
William S. **Saric**, Distinguished Professor of Aerospace Engineering and Member of the National Academy of Engineering
Thomas R. **Saving**, Distinguished Professor of Economics
Marlan O. **Scully**, Distinguished Professor of Physics, Member of the National Academy of Science, and Member of the American Academy of Arts & Sciences
Clifford H. **Spiegelman**, Distinguished Professor of Statistics
Jerry R. **Strawser**, Dean of Mays Business School
Bjarne **Stroustrup**, Professor of Computer Science and Member of the National Academy of Engineering
Max D. **Summers**, Distinguished Professor of Entomology and Member of the National Academy of Science
Bruce **Thompson**, Distinguished Professor of Educational Psychology
Robert E. **Tribble**, Distinguished Professor of Physics
Jorge A. **Vanegas**, Dean of Architecture
P. Rajan **Varadarajan**, Distinguished Professor of Marketing
James E. **Womack**, Distinguished Professor of Veterinary Pathobiology and Member of the National Academy of Science

8

Chicken Oil Company

Although John Junkins initially passed on accepting a $500,000-a-year stipend to launch his Institute for Advanced Study at Texas A&M, the university still vetted his proposal through a committee formed by the school's acting provost, Karen Watson.

"I wanted to chair the committee and to start hunting funds before we opened the doors," Junkins says. "I wanted to have five years' worth of the money in hand. Because I had been the lead author, Karen didn't want me to lead the committee. She wanted someone at a greater arm's length from creating the concept—someone more objective about the idea than me—to head up the group because the committee's job would be to define how the Institute would function."

Michael Hitt with the Texas A&M Mays Business School was ultimately selected to chair the eight-person committee. The panel's work would not be to rubber-stamp Junkins's vision but to look critically at the model. The greatest adversary to that model turned out to be one of the school's most distinguished liberal arts professors.

Howard Kaplan joined the Texas A&M Sociology Department in 1988, three years after Junkins's own arrival in aerospace engineering. In his time as an academic, Kaplan became internationally known in the fields of medical sociology, social psychology, and deviant behavior. His book, *Deviant Behavior in Defense of Self*, published in 1980, offered insights based on a three-year study of nearly ten thousand seventh-grade students in Houston. In the book, Kaplan wrote, "The self-esteem motive is defined as the personal need to maximize the experience of positive self-attitudes and to minimize the experience of negative self-attitudes."

It can be argued that "deviant behavior in defense of self" may have arisen more than a few times during committee gatherings to discuss the Institute for Advanced Study concept.

"Howard felt that the Institute was going to be just another big win for engineering at Texas A&M and leave the humanities behind," Junkins recalls today. "Howard, representing many of his peers in liberal arts studies, wanted the Institute to be thematic, so from the outset, he and I were pretty much at polar opposites on how the Institute should be run."

Junkins says convincing Kaplan, who died in 2021, proved to be an arduous task.

"The rest of the committee was on board with the direction I felt the Institute should take," says Junkins today, "but Howard, God rest his soul, just wouldn't accept defeat."

Perhaps he was merely attempting to "maximize the experience of positive self-attitude."

By late 2010, Junkins's version of the Texas A&M Institute for Advanced Study had caught the attention of System Chancellor Mike McKinney, and a vote to move forward with the plan was placed on the agenda for the Texas A&M University System Board of Regents at their December 2010 meeting.

McKinney took over as A&M System chancellor in late 2006, appointed by Gov. Rick Perry. The two men had strong ties. McKinney had served as Perry's chief of staff when Perry first became governor in December 2000. They previously had become friends and political allies while serving together in the Texas State House of Representatives in the late 1980s.

McKinney grew up in Houston and received his undergraduate degree in biology from the University of Houston. He then enrolled in the University of Texas Medical Branch in Galveston, where he completed his training to become a doctor. He started his practice in 1976, moving to the small Texas town of Centerville, where he saw patients for the next sixteen years. While there, McKinney entered politics, becoming a city councilman and later town mayor. In 1984, he was elected to the Texas State House. A year later, he was appointed by Gov. George W. Bush to be commissioner of the Texas Department of Health and Human Services.

By 2003, McKinney had become CEO and senior vice president of the University of Texas Health Science Center in Houston. Three years later, at the behest of Texas A&M grad Perry, McKinney became chancellor of the Texas A&M University System. In attempting to do Perry's bidding while in that role, McKinney ran afoul of the faculty on the campus of the system's flagship.

Twice.

Not long after Elsa Murano was named Texas A&M's first woman president in 2008, McKinney announced that, in his opinion, faculty should be assigned either research or teaching duties and not do both. Rick Perry shared those sentiments, and McKinney's proclamation turned out to be the governor's first volley regarding proposed reforms in higher education. Murano strongly disagreed with McKinney's position, saying such a move would significantly damage the university's academic reputation.

McKinney then unexpectedly countered by suggesting the system's regents consider merging the positions of chancellor and flagship president to "save money." Ultimately, Murano was forced to resign less than a year and a half into her presidency. On the heels of her departure, the A&M Faculty Senate passed a historic "no-confidence" resolution regarding McKinney, the first system chancellor to be so admonished.

Perry's efforts to change higher education in Texas became the subject of another *Texas Monthly* story authored by Paul Burka. In it, Burka wrote, "Critics of higher education say that those who inhabit these contemporary temples are living off the fat of the land; that they answer to no one except one another; that, like the monks of old, they cultivate their vineyards, sniff heady wines, and raise lush crops on fecund soil while doing . . . what, exactly?"

At the core of Perry's desire to change the nature of higher education in Texas was the matter of faculty tenure reform, a subject for which a former University of Texas business professor—and Texas-born oilman—had become the standard bearer and Perry's most trusted counsel.

Jeff Sandefer's great-grandfather, Jefferson Davis Sandefer, was once president of Hardin-Simmons University in Abilene, Texas, the town where Jeff was born and raised. Jeff's father, Jakie, played football under the legendary coach Bud Wilkinson at the University of Oklahoma but made his name in the Texas oil patch. Jeff attended the University of Texas, majored in business, then followed his father into the oil business and became a millionaire. In 1990, Jeff returned to his alma mater as a part-time professor in the UT Graduate School of Business, a role he held for a dozen years.

Sandefer's departure from his college teaching job came, according to *Texas Monthly*'s Burka, "after a public dispute over the hiring of tenure-track professors to replace the type of instructors that Sandefer favored: part-timers with professional experience."

In the ensuing years, Sandefer authored a book called *The Innovative University: Changing the DNA of Higher Education from the Inside Out*. That work served as the basis of Rick Perry's call for reform. McKinney echoed those same sentiments from his place at the helm of the Texas A&M University System.

Among the changes the Texas Public Policy Foundation encouraged was the creation of an "Academic Financial Data Analysis." The proposal aimed to measure the financial worth of a faculty member and assigned a red or black number to each, resembling a spreadsheet.

Needless to say, that notion was not met favorably by the faculty on the A&M flagship campus, and McKinney became the target for blame.

"At that time, I think McKinney was looking to try to start a different conversation with the A&M faculty," John Junkins says today. "So when he learned about the Institute for Advanced Study, he embraced the idea immediately. The regents approved it and I was ready to become the first director. I met with McKinney to discuss funding. He committed to a funding package amounting to a half a million dollars a year if I could match that, and I felt certain that I could.

"But I never got his memorandum of understanding confirming our oral agreement," Junkins goes on to say. "A few days after we came to terms, he announced his 'retirement,' and our deal vanished into the wind."

McKinney left his post on July 1, 2011. Just six weeks later, his successor assumed control, another appointee with long-standing ties to Governor Perry. But this chancellor, the fourteenth to serve the Texas A&M University System, also had strong ties with Aggieland.

John Spencer Sharp grew up in the unincorporated Texas town of Placedo, not far from the Texas Gulf Coast. His father fought in World War II and then took a lifelong job in the oilfields. Sharp's mother was a schoolteacher. Both parents shaped their son in profound ways.

"My dad was a great guy," Sharp said in the book *Breaking Away: How the Texas A&M University System Changed the Game*. "He was like a lot of World War II vets of that time: quiet, loving, and he worked like a Trojan. He'd get up at five a.m. every morning, go off to work in the oilfields, then come home for supper in the evening. And then many times at night he would wake up with these nightmares. They were always the same bad dream about a Japanese soldier jumping into his foxhole and stabbing him.

"Probably the greatest lesson my dad taught me," Sharp adds, "was that you can't have too many friends."

The ability to network and recruit allies served Sharp well in his first career as a Texas politician and again as chancellor of the Texas A&M University System.

John Sharp graduated from Texas A&M as a member of the Class of 1972. Why he chose to attend the school boils down to his mother's insistence that he attend college somewhere, and his own random encounter, at the age of eleven, with a wounded pigeon.

"From the way it was flapping around, it seemed to have a broken wing," Sharp remembers today. "The bird let me pick it up and from that point, I saw it as my responsibility to try to nurse it back to good health. I fabricated a splint from a Popsicle stick and kept the bird in a cage for about a month, tending to its every need."

After that month, Sharp let the bird out of its cage and was disappointed when it flew away.

As a result of that experience, Sharp's mother, Venus, decided her son should become a veterinarian. Even though Mrs. Sharp attended the University of Texas, she had no aversion to pointing her offspring toward the best vet school around, the one at Texas A&M University.

By the time Sharp arrived on the A&M campus, he had given up on a pursuit of veterinary medicine and decided to major in political science. As significantly, he also became a member of the Corps of Cadets.

The Corps experience shaped Sharp in meaningful ways, as he discussed in *Breaking Away*:

> "Yeah, I'd have to say I probably was a bit of a smartass back then," Sharp says with a twinkle in his eye. "One day I was dragging my

footlocker down the hall of my barracks, and this guy started yelling at me, 'Hey, "fish Jones," you're scratching the floor!'"

[Freshmen cadets at A&M are called "fish" for the same reason first-year students at West Point are known as plebes: whittling young egos down to size. The word "fish" in the A&M Corps of Cadets is never capitalized.]

"Since my name wasn't Jones, I didn't pay any attention to him, so he kept yelling at me, kept calling me 'fish Jones.'"

"So finally, when I realized what was happening, I acknowledged the guy in the most flippant manner I could muster.

"'Hey, 'Fred,' how are you doing?'"

To which the response was, according to Sharp, "Blankety-blank, you blankety-blank-blank! My name's not Fred!"

"Well, my name's not 'Jones,' you asshole!"

Such was the occasion of Sharp's initial encounter with the first sergeant of his outfit.

"I learned a lot in those next five minutes," Sharp says, smiling.

"The Corps was the best training for life I could have ever had," Sharp says. "The philosophy is that in order to become a leader, you have to be torn down, become a follower, and work your way up."

Which Sharp did, all the way to student body president and a member of the Corps staff his senior year.

While Sharp's position as student body president was near the top of the leadership pyramid at Texas A&M, Rick Perry's role as an Aggie Yell Leader—the school's all-male sideline spirit squad at sporting events—usurped that, at least in the eyes of Old Ags and Aggie football faithful. For years, it was reported that Perry and Sharp were roommates within the Corps. They were not, but they were in the same "outfit"—Squadron 6—and lived on the same floor of their Corps barracks. From those halls, one of the great political rivalries of modern Texas times ultimately unfolded many years later.

After graduation, Perry flew refueling missions for three years as a US Air Force pilot. Sharp became a member of the US Army Reserve and took a full-time job as an analyst with the Texas Legislative Budget Board in Austin. While in that position, veteran Texas politico Sam Erwin asked Sharp if he would be interested in running for office himself. After thinking about the matter, Sharp realized he was interested, and upon Erwin's council, he moved in 1978 to Victoria, Texas, near his childhood home in Placedo. There, he set up a real-estate office as a precursor to his successful campaign later that same year for a spot in the Texas State House of Representatives.

He was twenty-eight years old.

Thus began a political career that saw Sharp elected to the Texas Senate in 1982, the position of Texas railroad commissioner in 1986, and the post of comptroller of public accounts for the State of Texas in 1990, in which Sharp was the last Democrat to win a statewide race.

As comptroller—pronounced "controller"—Sharp made his name and earned a national reputation for austerity and forward thinking. He created the Texas Performance Review, a statewide audit system that, during difficult financial times, saved Texas taxpayers billions of dollars and enabled the state to avoid the indignity of implementing a state income tax.

The "wildcatter" mindset runs deep in the hearts of Texans.

Nearing the end of his second term as comptroller, Sharp upped the ante by throwing his hat into the ring at an even loftier level. "Nothing much was at stake in the 1998 Texas lieutenant governor's race," *Breaking Away* offers, "except the very heart and soul of the Lone Star State":

> With Gov. George W. Bush a shoo-in for reelection and expected to become the next president of the United States, his second-in-command in Austin would become his successor as the state's chief executive. If Republican Rick Perry won, Bush could be assured of leaving Texas in capable GOP hands. If John Sharp emerged victorious, Democrats would regain the governor's mansion. Before Bush, Bill Clements had been the only Republican governor of Texas dating back to 1874.
>
> Sharp versus Perry was the political equivalent of Frazier versus Ali. Both men entered the fray with undefeated records at the polls. Their title fight was a brutal one, with both men inflicting punishment on their rival.
>
> A few blows may have been below the belt, but both political pugilists stood standing at the final bell.
>
> Unlike Joe Frazier's victory over Muhammad Ali that March night in Madison Square Garden so many years ago, Sharp v. Perry

was anything but a unanimous decision. Perry, the Texas agriculture commissioner, bested Sharp by a margin of less than 2 percent of the more than 3.5 million votes cast.

After the defeated Sharp finished out his term as comptroller, he announced his departure from politics. He took a job in the executive ranks of the Dallas-based global tax-consulting firm Ryan, LLC. Perry indeed became governor after George W. Bush left for the White House and served four terms.

Perry and Sharp eventually put their contentious race for lieutenant governor behind them. As Sharp recalls, atonement came about six years later through a chance encounter at a College Station western apparel store. Then in 2005, Perry appointed Sharp to the Texas Tax Reform Commission. In 2007, the two men worked together for the passage of Texas Proposition 15, an effort to beef up the state's support for cancer research.

Sharp considered runs for governor in 2006 and the US Senate in 2008, but by then, Texas had become a Republican stronghold. Instead, he continued working with Ryan. In Sharp's words, "I was making way more money there than I was worth."

Perry had contacted Sharp in 2006 to gauge his interest in becoming chancellor of the Texas A&M University System. Sharp told Perry that it would be difficult for him to leave Ryan.

"I had stock options with the company," Sharp says, "and if I would have left, it would have cost me an insane amount of money."

Five years later, though, the time was right for the sixty-one-year-old Sharp to assume the helm of the Texas A&M University System and return to Aggieland.

Before the end of Sharp's second day in his new job, John Junkins walked into the chancellor's office intent on having a discussion with the system's new leader about the Texas A&M Institute for Advanced Study. Junkins had come close to securing a commitment from the previous chancellor and was hoping to get the attention of the new man in charge.

"Before I first met Dr. Junkins," said Sharp, "someone had told me Texas A&M was way behind in National Academy membership." In his interview for this book, Sharp actually called Junkins by his first name, but since both men involved in this important milestone of

Hagler Institute history share the same given name, to avoid confusion, we'll formalize references to each.

"We had something like eleven faculty who were in the academy," Sharp continued, "while the University of Texas had more than sixty. I came into the chancellor's role intent on setting up a funding mechanism to recruit those people and bolster our numbers.

"When Dr. Junkins told me what his Institute for Advanced Study was about, it seemed to perfectly dovetail into what we were trying to do. It seemed, given the folks he was after who were members of a National Academy or better [Nobel Prize winners], his institute would be a good way to try to convince that level of academic that they needed to come here and consider staying. And both Dr. Junkins and I thought we could do that by waiving some resources in front of 'em.

"I just thought immediately that the two things matched up real good."

By the end of that same week, in August 2011, Sharp and Junkins met again. Junkins suggested a lunch meeting, and Sharp recommended the Chicken Oil Company, a spinoff of the nationally famous College Station bar called the Dixie Chicken located across the street from Texas A&M's main campus. And it was there, at the Chicken Oil Company restaurant along historic South College Avenue north of campus in the city of Bryan, that Sharp agreed to help fund the Texas A&M Institute for Advanced Study by contributing $5 million over five years to sustain the endeavor.

And yeah, after John Junkins had spent some fifteen years championing the cause, it was that easy.

"I didn't know Chancellor Sharp had already been mulling over a concern that my vision for the Institute would help address," says Junkins. "At the time, I thought it was my overwhelming charisma that won him over.

"After our first meeting, I knew he liked the idea a lot, but he said he would have to figure out how to do the accounting, move money, and all that stuff, so he would need a little time. I told him we could talk about money whenever he was ready to do so. It turned out he was ready to have that conversation when we met at the Chicken Oil Company."

"He's not a cheap date," Sharp laughs today about that fateful meeting at an Aggieland landmark severely damaged by fire in 2022. Three years later, the restaurant reopened for business after "rising from the ashes."

As for a definitive view of his collaborator, Junkins offers, "John Sharp puts the 'swash' in 'swashbuckler.' He makes bold decisions faster than any academic leader I have ever met."

The eighth floor of Rudder Tower is home to Texas A&M University's Hagler Institute for Advanced Study. The building is named for James Earl Rudder, an A&M graduate, World War II hero, major general in the United States Army, and from 1959 to 1970, president of Texas A&M.

Mae Jemison, a physician and former NASA astronaut, strides under the saber arch formed by members of the Texas A&M Corps of Cadets Ross Volunteers at the Texas A&M Institute for Advanced Study's Induction Gala in 2024.

Founding Director, Dr. John Junkins (left in photo), introduces signature donor and the namesake for the Texas A&M Institute for Advanced Study, Jon L. Hagler '58, at the 2016 Induction Gala.

Dr. Christodoulos Floudas, a renowned Princeton University chemical engineering professor and a 2013–14 Institute for Advanced Study Fellow, waits to present a seminar. Floudas served as director of the Texas A&M Energy Institute until his death in 2016.

Two icons in their field: Founding Director John Junkins and Norm Augustine, the inaugural chair of the Institute for Advanced Study External Advisory Board.

John Sharp, former Chancellor of the Texas A&M University System, at the announcement of a new class of Hagler Fellows. A former Texas state comptroller, Sharp put together the funding package that launched the Texas A&M Institute for Advanced Study in 2011.

Dr. Jack Dongarra (above) and Dr. Rakesh Agrawal (below), both members of the 2014–2015 Institute for Advanced Study Class of Fellows, engage in meaningful dialogue with students and peers.

Dr. Clifford Fry, associate director of the Hagler Institute, joined the staff in 2013.

Amanda Scott, associate director of the Hagler Institute, joined the staff in 2016.

Dr. H. Vincent Poor (right, facing camera), a 2018–19 Hagler Fellow and Eminent Scholar.

Founding Director John Junkins with 2020–21 Hagler Fellow Rachel Moran, now a professor of law and director of the Education Law Program at the Texas A&M School of Law in Fort Worth.

Dr. Ed Fry, former deputy director of the Institute for Advanced Study, assists 2020–2021 Hagler Fellow Dr. Paula Hammond as she signs the induction book at the annual gala. Dr. Hammond is a member of three national academies: Engineering, Medicine, and Sciences.

Dr. Wolfgang Schleich came to the Institute for Advanced Study in 2013 from Ulm University in Ulm, Germany. Specializing in quantum optics, he is a fellow of the American Physical Society, the Institute of Physics, the European Optical Society, and the Optical Society of America.

Dr. William Unruh was a member of the Hagler Institute's 2018–19 Class of Fellows. His research interests include contributions to general relativity. Working with the late Stephen Hawking, Unruh helped refine the foundations of quantum mechanics in relation to black holes.

Dr. James Hubbard, a 2017–2018 Hagler Institute Fellow, received his PhD in mechanical engineering from the Massachusetts Institute of Technology, where he later served as a member of the faculty. He is now a University Distinguished Professor of mechanical engineering at Texas A&M.

Dr. Donna Strickland, a Hagler Fellow in 2021–2022, won the Nobel Prize in Physics in 2018 for work as a PhD student in the high-energy laser field at the University of Rochester in New York. She became only the third woman in history, along with Marie Curie and Maria Goeppert Mayer, to receive the Nobel Prize in Physics.

The late Harold Adams '62 (center), was a world-renowned architect and a 2014–2015 Hagler fellow. He is the only Texas A&M graduate to be honored as an Institute for Advanced Study Fellow.

John Junkins with Ray Rothrock '77 at the 2023 annual gala. Rothrock received the Frank J. Malina Renaissance Medallion for outstanding achievements in a technical field and indelible contributions to the arts.

The late Robert Skelton, 2014–2015 Hagler Fellow and former eminent professor in Texas A&M's Department of Aerospace Engineering.

John Junkins and his wife Elouise flank former U.S. Senator Kay Bailey Hutchison, who played a key role in launching the Texas Academy of Medicine, Engineering, Science & Technology (TAMEST). Junkins is a recipient of TAMEST's Kay Bailey Hutchison Distinguished Service Award.

9

"Perennial Blue-Ribbon Panelist"

Norm Augustine

Of Norman Ralph Augustine, the Texas A&M Institute for Advanced Study's first big "get," no less an authority than *Defense News* offers this:

> When Norm Augustine's name is mentioned, most think about his role in the merger of Lockheed Corp. and Martin Marietta to create the biggest defense company on the globe. And certainly that was a huge factor in Augustine landing on our list of the 10 top defense influencers for the last 30 years. But his imprint on industry beyond Lockheed Martin was very profound, as was his influence during his two stints in the Pentagon. And then, of course, there are his laws—those gems of wisdom with a heavy dose of cynicism that spur both a chuckle and knowing snicker among anyone in defense, corporate America or both.

But wait! There's more . . . from the Smithsonian Institution's *Air & Space* magician in a story dated May 8, 2009, with the headline, "NASA Needs Direction? Call Norm Augustine!":

> Norman Augustine, that perennial blue-ribbon panelist, just accepted the easiest gig of his career—or the hardest. He's been asked by the White House to review NASA's plans for human space exploration, to help chart a "vigorous and sustainable" path forward. If Augustine wanted, he could just dust off the report from another panel he chaired in 1990, slap on a new cover, and be finished by Monday—because the findings and "concerns"

identified by the Advisory Committee on the Future of the U.S. Space Program are as relevant today as they were two decades ago.

In fact, it wouldn't take too much of a stretch to assume the first two letters of the acronym representing the US space program—NASA—stand for . . . "Norm Augustine."

Augustine's curriculum vitae could easily run as long as this book. Just look at his list of awards as posted to Wikipedia:

- Eagle Scout, 1952
- Member of the National Academy of Engineering, 1983
- National Space Club Goddard Award, 1991
- Fellow of the American Academy of Arts and Sciences, 1992
- Rotary National Award for Space Achievement National Space Trophy, 1992
- Silver Buffalo Award, 1994
- Electronic Industries Association Medal of Honor, 1994
- Golden Plate Award of the American Academy of Achievement, 1995
- The Washingtonian's Business Leader of the Year, 1997
- National Medal of Technology and Innovation, 1997, "For visionary leadership of the aerospace industry, for championing technical and managerial solutions to the many challenges in civil and defense systems, and for contributions to the United States world preeminence in aerospace"
- The NASA Distinguished Public Service Medal, 1997
- IEEE-HKN Eta Kappa Nu Eminent Member, 2001
- Space Foundation's General James E. Hill Lifetime Space Achievement Award in 2002. The highest honor bestowed by the Space Foundation, the award recognizes outstanding individuals who have distinguished themselves through lifetime contributions to the welfare of betterment of humankind through the exploration, development and use of space, or the use of space technology, information, themes or resources in academic, cultural, industrial or other pursuits of broad benefit to humanity. Augustine was the first recipient.

- Public Welfare Medal from the National Academy of Sciences, 2006
- USO's Freedom's Finest Award, 2004
- The Harold W. McGraw Hill Jr. Prize in Education, 2006
- The 2006 BENS Eisenhower Award [Business Executives for National Security]
- The 2007 Bower Award for Business Leadership, from The Franklin Institute
- NAA Wright Brothers Memorial Trophy, 2008
- National Science Board Vannevar Bush Award, 2008
- IRI Medal from the Industrial Research Institute, 2009
- The American Chemical Society Public Service Award, 2009
- B. Kenneth West Lifetime Achievement Award, 2009
- NAS Award in Aeronautical Engineering from the National Academy of Sciences, 2010
- Drexel University Engineering Leader of the year, 2011
- The Wings Club Distinguished Achievement Award, 2011
- Character Education Partnership's American Patriot of Character Award, 2012
- Montgomery County Business Hall of Fame, 2012
- Industry Week Manufacturing Hall of Fame, 2012
- Smithsonian Air & Space Museum, Lifetime Achievement Trophy Award, 2014
- Arthur C. Clarke Lifetime Achievement Award, 2014
- American Astronautical Society Space Flight Award, 2014
- Advisory Board, Journal of Science Policy & Governance, 2015
- Tech Council of Maryland Lifetime Achievement Award, 2015
- International Von Karman Wings Award, 2015 (For his visionary leadership, contributions to the aerospace industry and distinguished service to the nation's defense, security and space programs)

No doubt Augustine's list of honors and recognition goes on past 2015, but Wikipedia editors can be forgiven for moving on to other things. The point—Augustine's contributions to the scientific and engineering worlds, let alone the world in general—is well taken.

John Junkins frequently refers to Augustine as a "national treasure."

One of the first people Junkins contacted after receiving the green light from Chancellor John Sharp to establish the Texas A&M Institute for Advanced Study was Norm Augustine. Junkins and Augustine were acquainted through their mutual membership in the National Academy of Engineering. In addition, Junkins had also teamed with Augustine's Lockheed Corporation as an academic and Augustine had sat in on review meetings in which Junkins had made presentations.

Since creating his initial white paper mapping out his plan to establish the Institute, Junkins had altered his view of the role an "advisory board" would play. Rather than using that term to describe the body which would initiate the process of nominating Fellows, Junkins also wanted a strong external advisory group resembling a board of directors. And at the top of his wish list to head up that ensemble was Norm Augustine.

"After getting John Sharp's buy-in in mid-August 2011," Junkins says, "the National Academy was set to meet the first week of October. Not long after I realized the Institute for Advanced Study was finally going to become a reality, I sent Norm Augustine an email. I told him I had a really important matter we needed to discuss. 'I want your advice,' I wrote, and I suggested we could get together during that year's National Academy meeting. I added in the email, 'I'm not going to ask you for money, so don't worry about that.'

"He wrote back saying he would be happy to meet and we chose a time during one of the more mundane parts on the National Academy agenda."

Junkins arranged for their meeting to be held in one of the many small breakout rooms used during the annual gathering, but when the time came, the previously scheduled session in the room that had been reserved had run long. And so, as a precursor to bigger things to come, Junkins and Augustine were placed in the executive boardroom of the National Academies building located in Washington, DC.

Junkins recalls, "As we sat down at one end of this giant mahogany boardroom table, Norm joked, 'This already feels like a very expensive meeting!'"

Norm Augustine says one of the first defining moments of his young life came while a student at East Denver High.

"There was a teacher there whom I never had in class," Augustine recalls. "He was a tough old bird. Everybody knew about him. One day, out of the blue, he sent me a note and told me I needed to report to his office immediately. He wanted to talk to me.

"After I was handed the note, I dashed down the hall, got to his office, sat down, and he listened as he said, point blank, 'What are you going to do when you get out of here? I've heard you're a pretty bright student.' I thought about his question for a moment, and then answered, 'I think I'd like to be a forest ranger.'

"With that remark, he gave me a very stern look and said, 'That's not what you're going to do. If you do, you'll be wasting my time . . . and your life!'"

"I was stunned by his reaction," Augustine says. "I was just seventeen years old and I hadn't asked to meet with him in the first place. But he continued.

"'Young man,' he said. 'I have two envelopes here.' As he said those words, he handed me what turned out to be application forms to a pair of colleges. One was Williams College, a small private liberal arts school in Williamstown, Massachusetts, that I had never heard of. The other envelope contained an application to Princeton University. At that point, I'd never been east of Wichita, Kansas, and I knew my parents couldn't afford to send me to either school. He said, 'Look, if you can get into either one of these places, they'll pay your way through. But first you have to fill out the application forms.'

"He handed the envelopes to me and told me to bring them back to him when I was done."

Augustine's application to Princeton landed him a trip to New Jersey and an interview at the school.

"During my Princeton interview they asked me what I wanted to study," Augustine remembers. "Not being familiar with the concept of an 'Ivy League school,' I replied, honestly, that I wanted to study forestry. Everyone in the room, except me, burst out laughing. They thought that was the funniest thing they had ever heard. They told me the closest thing they had to forestry was geological engineering.

"To make a long story short," Augustine recounts, "I attended Princeton and as a freshman declared my major to be geological engineering."

Ultimately, Augustine graduated from Princeton magna cum laude with a BSE in aeronautical engineering. After graduation in 1958, he joined the Douglas Aircraft Company in California, and worked as a research engineer, program manager, and chief engineer. Seven years later in 1965—at the age of thirty—Augustine moved to Washington and a job in the Office of the Secretary of Defense as assistant director of defense research and engineering.

Ten years later, Augustine became undersecretary of the US Army, serving during the administrations of Pres. Gerald Ford and Pres. Jimmy Carter.

When Augustine huddled with John Junkins about the Texas A&M Institute for Advanced Study in 2011, he did so willingly and eagerly despite his lofty credentials and many important responsibilities.

"One of the things I admired most about John back then was that he was not only very competent, but boy, he was also determined," Augustine says. "He was one of those people that when he sunk his teeth into a task, you knew he was going to make something special happen.

"Having attended Princeton myself, where the original Institute for Advanced Study was located, I thought I knew well the concept behind John's idea. And I knew a lot about Texas A&M because my son had attended school there as a member of the Class of '85. He studied engineering and loved it."

But as Augustine would soon learn in that first meeting, Junkins's vision for an institute for advanced study ran counter to the Princeton concept. Rather than having a fixed set of thirty permanent residents who each annually invited others to collaborate with them, Junkins wanted the current Texas A&M faculty to nominate elite candidates to become visiting Fellows at their school.

In that Washington meeting, Junkins explained to Augustine that he had many problems to solve and needed well-regarded "subject-matter experts" to move the initiative forward. To do that, Junkins's plan was to put together an advisory board, and to do so he needed a board chair. Was Augustine interested in the job?

"I asked John, 'Who's already on the board?'" Augustine chuckles at the recollection. "John replied, 'Well, there's no one on the board yet.' In response to that bit of information, I thought, 'Well, that's interesting. I'm going to be chairman of a board that doesn't yet exist of an institute that doesn't yet exist.' But I knew John well enough to know that if anybody could make this work, he could. I also felt like this would be a great thing for Texas A&M and, potentially, for the country as a whole. If John could create a success with his institute, it would be a great thing for other universities to copy."

As the two men concluded that first meeting, Junkins told his new advisory board chair, "Over the next five years, I want to establish beyond an absolute shadow of doubt that the operational plan we've laid out is an incredibly efficient and inexpensive way to accomplish our goals of boosting the academic integrity and reputation of Texas A&M."

Greg Augustine would have been proud and appreciative of his father's involvement with his alma mater.

When Greg was trying to figure out where to go to college, he asked his father to put together a list of the best engineers he knew. After Norm obliged his son, the two discovered that a disproportionate number of the names on that list had attended Texas A&M.

"After Greg and I visited the Texas A&M campus," Norm says today, "he knew it was the place for him. My son was an Aggie through and through, and just loved Texas A&M."

Sadly, Greg Augustine would take his own life.

"We lost him in 1999. But he was very successful during his time in the engineering world. He became a systems engineer and did some really exciting stuff and kept getting promoted. But everywhere he went, and in everything he did, my son was always an Aggie first."

And in becoming chairman of the Texas A&M Institute for Advanced Study's External Advisory Board, Norm Augustine would soon help reimagine the "Spirit of Aggieland" in ways both diverse and profound.

PART 2

Profiles

10

Ashley's Dad

Clifford Fry

Since September 2013, Dr. Clifford Fry, with a PhD in economics from Texas A&M University, has served as the associate director of the Hagler Institute for Advanced Study. His work with Founding Director John Junkins has been instrumental in growing the enterprise, both in number of Fellows and donations to support the mission of the Institute.

Fry's duties at the Institute include creating and analyzing financial reports, writing most of the agreements for Fellows, and often negotiating the terms of those agreements. He helps Fellows with their working and living arrangements while on campus, and he frequently spends personal time with them to enhance their connection to the Institute and integrate them into the College Station community. Fry also managed the Hagler Institute during the time Junkins served as interim university president, from January through May of 2021, overseeing the organization's day-to-day activities.

Clifford Fry's story serves as the perfect pivot point in the history of the Institute leading to the profiles of select Fellows, which follow. Not only has Fry played a central role in the growth of the Institute, but he has also framed the Institute story for more than a decade, providing written content for newsletters, press releases, annual reports, and its website. Thus, it is right and reasonable to give him the opportunity to tell his own story here.

• • •

I first met John Junkins when I moved my family onto Shady Drive in College Station back in the 1990s. I was leaving academia after spending more than twenty years as a finance professor at the University of Houston to enter the private sector full-time near the Texas A&M campus where I had attended college in the 1960s.

After graduating from Henrietta King High School in Kingsville, Texas, I enrolled at the A&M College of Texas—as it was known then—in the summer of 1963, a member of the Aggie Class of '67. My intent was to become a pilot in the US Air Force, but those ambitions failed to take flight—both figuratively and literally—because my vision failed to meet the standard of the times. My instructors within the Corps of Cadets then wanted me to set my sights—pun intended—on becoming a navigator, but my young cadet heart was set on being a pilot, and navigator, though important, just did not sound glorious enough.

It is funny how seemingly small choices can affect your whole life.

When I was in the Corps, a navigator's visual acuity was measured by a process in which instructors would give students a picture of a hypothetical bombing target. It could be anything: a dam, a factory, a city, and so on. Then an array of photos was put before the student navigator, and he would be told to identify the hypothetical target. For reasons I cannot explain, I could not successfully do that. My instructors and I eventually concluded the world would be a safer place if I wasn't in it dropping bombs.

From there, my interest turned to economics.

My undergraduate experience at Texas A&M—the school officially became Texas A&M "University" in the fall of 1963, not long after I began classes my first year—benefited significantly from a development grant the economics department received at the time. It was a million dollars, which was a lot of money back then. The funds were earmarked to hire top faculty in economics. Several of those people ultimately came from Michigan State, where the economics department was in a state of turmoil.

Over what, I can't recall.

Unquestionably, the most notable member of that group—and a man who would make a big difference in my life—was Thomas Saving. He was young, enthusiastic, and already highly regarded in his field for his breakthroughs in research.

In 2000, more than thirty years after I studied under Dr. Saving, Democratic Pres. Bill Clinton appointed him as a public trustee of the Social Security and Medicare Trust Funds. In 2001, Republican Pres. George W. Bush named Saving to the bipartisan President's Commission to Strengthen Social Security. That gives you some idea of Dr. Saving's stature as an economist. As an educator, his presence at Texas A&M persuaded me to seek my PhD in economics and to do so from my alma mater, something most people do not do.

After receiving my doctorate, I went to Washington State University to teach. After two years away from home, I moved back to Texas and took a position with the Federal Reserve Bank in Dallas. In 1975, I joined the faculty at the University of Houston, where I spent the next twenty years. Throughout that time and no matter where I was, I kept in touch with Tom Saving.

I always thought Dr. Saving's surname foretold his career choice.

It was Saving who encouraged me to become a consultant. In addition to his work at Texas A&M, Saving helped start the nationally focused economics consulting firm RRC Inc., in 1979. For ten years, Saving was president of the company. He continued to serve as chairman of the board until his death at the age of ninety in 2024.

While at the University of Houston, I had success consulting as a side gig to my teaching and administrative duties. For eight years I was head of the U of H Finance Department. I enjoyed academia, and I enjoyed teaching very much, but the lure of entering the private sector full-time was enticing. From Houston, I frequently referred business to RRC, and in 1992, Saving invited me to join the firm and I accepted his offer, moving back to College Station with my wife and family. Two years later, Saving told me he was leaving his role as RRC president to become director of Texas A&M's Private Enterprise Research Center. He asked if I was ready to buy out his shares in the company. It didn't take me long to accept that offer, and for the next almost twenty years I helped run RRC and made the kind of money I wanted to make.

Which was a lot more than most college professors make.

Much of the work I did at RRC was as an expert witness in large corporate lawsuits. One case, in which Proctor & Gamble sued Amway International for continuously besmirching their reputation and product lines, ended up being three separate lawsuits lasting a total of seven years. In time, I will admit to you here, I got tired of working with lawyers.

In 2013, I decided it was time to "retire," at least from the business-consulting world.

When I first moved onto Shady Drive, I lived across the street from John Junkins. After I joined RRC, I upgraded my family's living accommodations and moved to the end of our block to a house next door to Junkins that had previously been owned by liquor tycoon Forest "Bert" Wheeler. For sixty years, the Wheeler family had owned a chain of liquor stores in the Houston area. Bert wound up investing in land throughout Texas, including College Station, and ultimately established his residence in Aggieland. Bert was handicapped during the last decade of his life, and to accommodate his physical limitations, an elevator was installed into his immense two-story home on Shady Drive.

That elevator appealed to my wife, Judy, and me because our youngest daughter, Ashley, is afflicted with Rett syndrome, a rare neurological disorder that affects the way the brain develops, causing a progressive loss of motor skills and language. Rett syndrome primarily affects females, and Ashley's case was one of the first to be diagnosed in the United States after the condition was made known in the English-speaking world.

That's made her almost as much of a big deal in the scientific and medical communities as she is around the Fry house.

When I first became acquainted with John Junkins, I had no idea of his stature at Texas A&M. Unquestionably, he is one of the brightest people I've ever met, but I had no idea of his accomplishments as an academic in the field of aerospace engineering. From early on in our friendship, John shared with me his vision for an "institute for advanced study" at Texas A&M, which, he thought, could bring top-tier academic achievers to the university, thus bolstering the school's reputation and enhancing academics across the entire university.

Thinking "small" is not John's strong suit.

The Wheeler home had an apartment built into the detached-garage complex adjacent to the back of the house, and John and I spent hours there discussing ideas for his proposed institute. John did most of the talking because his vision was remarkably detailed and surprisingly clear. I listened, and the more I heard the more I encouraged him to keep the dream alive. As this book suggests, he obviously did just that . . . and so much more.

After the Texas A&M Institute for Advanced Study was started in 2011, John and his administrative assistant, Lisa Jordan, spent most of their time getting the ball rolling. That forced John to step back from his beloved pursuits in the aerospace engineering department. After getting the first class of world-class scholars put together—a group of six—John realized he needed help.

Given that I was looking for a way to make use of my increasingly free time, John suggested we have lunch together to discuss how I could "give something back to Texas A&M." That opportunity, as it turned out, was for me to come aboard and join him at the Institute.

I started in September of 2013 and viewed one of my key roles with the organization—and still do—as helping John spend more time working on his aerospace engineering research projects and less time worrying about the day-to-day operations of the Institute.

Back in my consulting days in Houston, one of my clients made me take a personality test before they would hire me. At the time, I thought that was a little extreme, but later I realized the move was highly astute. I took the test, and to my surprise it revealed that I had the aptitude to be a highly paid salesman.

I don't think John knew about that when he hired me.

Because of my background in economics, John did want me to take a closer look at the financial reports he was both receiving and generating. He also knew that having been both a department chair at the University of Houston and an owner of a successful business enterprise, I had dealt with people a lot in my working life. After he gave me a rundown of how the Institute was set up and how it functioned, I thought, "OK, what kind of a contribution can I make here?"

One thing that caught my attention early was that the Texas A&M Foundation had assigned a group of donors to the Institute to serve as "advocates" for the work that we were intending to do. These were prominent and successful people who supported the university with sizable financial gifts, but I knew those people couldn't possibly know as much about the Institute as they needed to know to be real advocates. So I decided it would be my job to educate our advocates. I began traveling to meet them—individually and in groups—to enlighten them on the value the Texas A&M Institute for Advanced Study would bring to Texas A&M.

One of my first trips was back to Houston to meet with Jerry Cox.

In 2012, Cox and his wife, Kay, received the E. King Gill Award from the 12th Man Foundation, the fundraising arm of the Texas A&M athletic department. The award recognizes those "who have shown the commitment as a volunteer advocate of the athletic program through exemplary service and leadership." More times than not, that means they've given a lot of money.

Jerry is five years younger than me, a member of the A&M Class of '72. When we met at the offices of his company, Cox & Perkins Exploration, he was gracious and seemed eager to listen to what I had come to tell him. After I finished, I was shocked when he said, "I'll give you $500,000 to help with the endowment." I thought I was just explaining the program to him so he would know what to tell others about it. Turned out, I *was* a pretty good salesman.

Cox's gift wasn't the first outside donation to come to the Institute from an Aggie alum. That distinction belongs to Dr. Yong Xu, who earned his PhD in biology from Texas A&M in 1993 as a Rockefeller Foundation fellow. Xu—whom his Aggie friends call Eric—is best known as the cofounder of Baidu, the world's largest Chinese Internet search engine. After becoming fabulously rich, Xu wanted to honor one of his A&M biology professors, Tim Hall, by establishing a chair in the biology department in Hall's name. Instead, John Junkins convinced Xu to fund a TIAS chair in Hall's honor, and he did so to the tune of $1.5 million in matching money. The ensuing $3 million endowment established the first major chair in support of the Institute's Fellows initiative.

Xu's contribution came early in my time at the Institute. I didn't know Jerry Cox before I met him. Likewise, I more or less cold-called Tom Powell.

Another Aggie alum with business pursuits in Houston, Powell was a member of the Class of '62. He majored in mathematics, went to work for Westinghouse Electric right out of college, then started two companies which he merged to become Powell Industries. He had made the Texas A&M Foundation's list of advocates as a result of his ten years on the board of visitors for Texas A&M University–Galveston and his funding of the Powell Marine Engineering Complex there.

Powell and I hit it off quickly. From my perspective, we shared a bond, but apparently not the kind where close friends divulge all their secrets to one another. In the months after I first met Tom, he donated $1.5 million to the Institute . . . without telling me! I later learned that when the science department development people went to see him, he told them he would give to science but that a million dollars of his donation—later raised to a million and a half—would need to be used specifically to create a science chair for Fellows at the Texas A&M Institute for Advanced Study!

That wasn't exactly what the science department had come to ask for, but he was insistent on how he wanted the deal structured. Meanwhile I had no idea that Tom was intending to support the Institute in such a generous manner. Needless to say, when I finally heard about his gift—which the Institute was able to match—it was a very nice surprise.

In my early travels on behalf of the Institute, I went to Houston, I went to Dallas, I went to Austin, The Woodlands, Corpus Christi, and almost anywhere else there was an advocate or a Texas A&M alumni chapter. I thought Aggies needed to know about their new Institute for Advanced Study.

The donations described here were early financial milestones for what today is known as the Hagler Institute for Advanced Study. Another of the organization's milestones in its early years took place when the first class of Fellows was inducted at the Institute's first black-tie gala—which has become a proud tradition for us—in the spring of 2013. That was before I joined John, but as his neighbor and friend, I was honored when he invited Judy and me to the event.

The "Gala"—now always capitalized—is currently administered by our outstanding program coordinator and assistant director, Amanda Scott. She does a great job creating a top-shelf event, from the food and entertainment to the red carpet and march of inductees under the Ross Volunteers' saber archway. The Ross Volunteers are a special unit of the Texas A&M Corps of Cadets that also serves as the official honor guard for the governor of Texas.

Another of my contributions to the Institute came in the form of *Cornerstone*, the organization's annual-report magazine, first published in February 2014. Through the years, I've done much of the writing for that publication, as well as our newsletter and reports for our administrative council and external advisory board.

Writing comes naturally to me in that I've been a songwriter for most of my adult life. You might not know it to look at me now, but I am the founding father, rhythm guitarist, and one of the vocalists for the musical group, Dr. Fry's Texas Medicine Band. We were a pretty big deal back in our heyday—the late '80s and early '90s—in places like Europe, Australia, and New Zealand. Our songs remained popular on radio and relevant in those countries into the twenty-first century.

Our stage equipment, from speakers to lights to drum set, stand at the ready in my aforementioned garage apartment, where I have jam sessions with some of my old bandmates and new friends even to this day.

Need a band for your next party?

In 2015, Texas A&M Pres. Michael Young—not to be confused with the 2022–23 Nobel Prize–winning Hagler Fellow of the same name—relocated the TIAS offices from the English Department in the Liberal Arts and Humanities Building to the Jack K. Williams Administration Building, where Young's own office was located. That move added immensely to the reputation and prestige of the Institute. Also in 2015, after learning about the success of the Institute and the Chancellor's Research Initiative, which helps turn visiting Fellows into Permanent Fellows of the Institute, Texas Gov. Greg Abbott established the Governor's University Research Initiative aimed at "helping Texas public institutions of higher education recruit distinguished researchers from around the world." As it has turned out, a large percentage of those funds have come to Texas A&M to turn Institute for Advanced Study Fellows into permanent and full-time members of the university's faculty community.

Without question, our organization's biggest milestone to date was the $20 million gift alum Jon Hagler gave to the Institute in 2017, thus making him the namesake for our venture. You'll read about that donation and the man behind it in the pages to come.

Now let me backtrack a little and tell you more about our daughter, Ashley.

Ashley was born in 1980. At about eighteen months, she began losing use of her hands, her face started to become pale, and gradually, she quit vocalizing. Eventually, she just sat on the floor rocking in her own world and would not make eye contact with anyone. What finally got our full attention and frightened us about her condition was a screaming episode where she grabbed her head in a manner somewhere between intense pain and heartbreaking frustration. My wife was so unnerved by the episode—she says she still hears Ashley's screams in her head—that we soon ramped up our already significant efforts to determine the nature of her condition and find a health-care provider who could help her.

We took Ashley to a specialist in Houston who, like the others before him, did not know what was wrong. What he thought he knew was that she would never talk and would require total care for the rest of her life. He suggested we institutionalize her, this child we cherished, and get on with our lives.

It is hard to describe how I felt hearing those words: "shock," certainly, but also a mix of sadness and fear. As Ashley's parents, Judy and I were in a world of confusion that seemed to permeate the infinite.

We had told the doctor that Ashley had been mostly normal for the first year and a half of her life. But when we read his report, we saw that he did not believe or had not listened carefully to what we had said. He wrote that she never had use of her hands and never "vocalized." His diagnosis was disappointing, but Judy and I did not give up our quest to find answers.

Rett syndrome was not known in America at the time of Ashley's regression. Andreas Rett, a Viennese pediatrician, first described the condition in 1966, but as his writings were in German, his findings did not become widely known in the English-speaking world. A Swedish doctor, Bengt Hagberg, published the first English article about the condition in 1983 and named it after Dr. Rett.

Before we ever heard the term "Rett syndrome," Judy and I thought we had a narrow window of time to help Ashley regain her function. We spent many a night studying everything we could get our hands on (I didn't use the Internet back then), from academic journals on child development to the writings on health of famed psychic Edgar Cayce. This focus on Ashley, though, had a price in our family. At this time our daughter Kristin was thirteen years old and healthy, and while she was not totally ignored, she was not getting the attention she also needed.

The most likely possible diagnosis we could come up with was that Ashley was severely autistic, and for three years we headed in that direction. We engaged with a well-known New York couple, Barry and Susie Kaufman, who were purported to have cured their autistic son. We enrolled in their autism-treatment program, and while we never "cured" Ashley, we learned much from the experience.

Most importantly, perhaps, the Kaufmans helped us understand that "happiness is a choice."

Finally, in 1984, Ashley was examined at Houston's Texas Children's Hospital by a small group of doctors that concluded she was suffering from Rett syndrome. That group included Dr. Huda Zoghbi, who was then an intern at the Baylor College of Medicine studying advanced genetics. Sixteen years later, Dr. Zoghbi led a team of researchers that found the mutations in the MECP2 gene that cause Rett syndrome. In the Baylor College of Medicine press conference announcing that discovery, Ashley was in attendance as the "guest of honor." Dr. Zoghbi brought her to the podium and told the assembled group that Ashley had inspired her to search for the cause of Rett syndrome. Dr. Zoghbi used Ashley's picture in several journal articles and videos and pictures of Ashley at many conferences, thus making Ashley the most famous member of the Fry family.

Needless to say, Dr. Zoghbi has been a friend to our family for many years. In 2019, she became a friend to the Hagler Institute for Advanced Study as well, when she was named a Distinguished Lecturer. She gave that lecture in 2020. Subsequently, the Hagler Institute helped support a major conference organized by Dr. Zoghbi.

• • •

Partly due to Dr. Zoghbi's discovery, Texas now has one of the leading Rett treatment centers in the world, the Blue Bird Circle Rett Center at Texas Children's Hospital. The center's recently retired medical director Dr. Daniel G. Glaze estimates that the group in Houston has seen between eight hundred and one thousand patients with Rett syndrome from all over the world.

And Judy and Clifford Fry's beloved youngest daughter, Ashley, was the first.

11

First Fellow

Alan Needleman, 2012–2013

Growing up in Philadelphia, Alan Needleman spent a lot of time on the streets . . .

"We played stickball in the streets," Needleman recalled from his home today along University Avenue in College Station, just a few miles from the Texas A&M campus where he is a full-time faculty member of the Department of Materials Science and Engineering. "We played tackle football in the streets. I loved to play all kinds of sports when I was a kid, although none of it was what you would call 'organized.'

"It turned out I wasn't very athletic, so that dispelled any hopes I might have had of becoming a 'big-leaguer' someday."

Instead, Needleman focused on his schoolwork, and those efforts have paid off in making him an "all-star" in the world of scientific research. Think of his place in the National Academy of Engineering as being akin to a spot in Cooperstown, New York, and the National Baseball Hall of Fame located there.

"The family culture I grew up in," says Needleman, "was sort of an Eastern European Jewish culture. It dictated that you kept going to school. And it never occurred to me not to do that. My father had a small shoe store, and I decided early on I definitely didn't want to follow in his footsteps for the rest of my life. I worked days, weekends, and some summers in his store, but I didn't want that life for me when I was grown."

When it came time for college, Needleman attended the University of Pennsylvania, thus becoming an "Ivy Leaguer" despite his father's limited financial resources.

"That was easy!" Needleman says of his academic upgrade to Penn. "In Philadelphia, the high school I wound up going to,

Central High School, was actually the second institution of learning in Pennsylvania—after the University of Pennsylvania—to offer a bachelor's degree. Back in colonial times, you got a bachelor's degree when you graduated from high school."

Of course, not that many people in America sought to advance their education in the 1700s, perhaps the Founding Fathers and a few more. Most folks were like Paul Revere, who left school at the age of thirteen to take up an apprenticeship silversmithing with his father.

"At Central," Needleman says, "there was a program called the Mayor's Scholarship, which basically meant if you had good enough grades, you got a free scholarship to Penn. My future father-in-law went that route through Central back in the 1930s. It was by far the cheapest way for me to go to college, and explains how I wound up attending an Ivy League school."

For many years, John Junkins has described an Institute for Advanced Study Fellowship at Texas A&M as the "mother of all sabbaticals." And if there is a shining example of what a Faculty Fellow of the Institute can be, it is Alan Needleman.

Needleman came to Texas A&M from the University of North Texas, but he is best known for his thirty-four years as a professor of engineering at Brown University, another Ivy League school located in Providence, Rhode Island. His many accolades include membership in the National Academy of Engineering and receiving both the Drucker and Timoshenko Medals from the American Society of Mechanical Engineering. It was during his acceptance speech for the Timoshenko Medal in 2011 that Needleman talked about what inspired his lifetime academic and research interests:

> My life in mechanics began my senior year at the University of Pennsylvania. I took a course in continuum mechanics from Hsuan Yeh who was Dean of the Towne School. The course was so interesting that I decided that was what I wanted to study in graduate school. I went to graduate school at Harvard and was very lucky that a young faculty member named John Hutchinson agreed to be my thesis adviser. My PhD thesis involved the finite strain, finite element analysis of a two dimensional periodic array of circular holes (motivated by the pioneering ductile fracture

> studies of Frank McClintock and Jim Rice). This initiated me into two emerging developments in solid mechanics: finite element methods and materials mechanics. As John Hutchinson remarked in his Timoshenko Medal address, we did not realize we were participating in a revolution.
>
> At that time finite element methods were not well regarded by much of the solid mechanics community. Too many years passed before computational mechanics was accepted and integrated into mainstream solid mechanics. I am pleased that the situation is very different now as evidenced by the Timoshenko Medal having been awarded to Oleg Zienkiewicz, Tinsley Oden, Ted Belytschko and Tom Hughes. Materials mechanics was more readily accepted and became a source of wonderfully challenging problems. It has stimulated the development of new mechanics theories and experimental methods, particularly those aimed at understanding small size scale phenomena. Conversely, it has led to solid mechanics having a significant impact on materials science and engineering.

The process by which prospective Faculty Fellows are nominated for inclusion in the Texas A&M Institute for Advanced Study has gone basically unchanged since John Junkins first outlined the methodology in his initial white paper proposal, although the nomination form has been expanded from one page to two. Submissions are coordinated between the college or school dean and the department heads under whom the fellow will conduct her or his work. Many nominees are of a "National Academy level of distinction as appropriate for their particular discipline." Individuals with a strong mentorship record are also desired.

Needleman's nomination as a TIAS Faculty Fellow was submitted on January 26, 2012, by Ramesh Talreja, a professor of aerospace engineering within the Texas A&M College of Engineering. Talreja wrote that Needleman was best known for his "pioneering work in the mechanics and physics of deformation, fractures, and instability of complicated mechanics" as well as "computational materials science."

As for how Needleman's presence as a Faculty Fellow would benefit Texas A&M University, Talreja offered, "Dr. Needleman has a strong record of collaboration with researchers in different materials fields.

He has pioneered many areas, which he has advanced by working with young researchers who have been inspired by him. It is anticipated that he will bring his vast knowledge and experience to [Texas A&M] and his guidance will help the engineering community here in making leap-frog advances in emerging areas such as atomistic modeling for self-assembly, self-healing, nonstructural materials, etc."

Needleman retired from the faculty at Brown in 2009, as he and his wife, Wanda—noted in her own right for her work as a psychiatrist and psychoanalyst—made the decision to leave the East Coast to be closer to their grandchildren living in Texas. Wanda's parents had made the same decision and moved to Providence, Rhode Island, to be closer to their two grandchildren after Alan had joined the faculty at Brown. The Needlemans' daughter, Deborah, and her husband, Marshall Armintor, were teaching in the Department of English at the University of North Texas, so Wanda and Alan relocated to Denton, where the UNT campus is located. Alan had hoped to continue working, and in 2010 he joined the UNT College of Engineering faculty, where he helped establish the Texas Materials Modeling Network.

In building that network, which included both faculty and students from several Texas schools, Needleman was an occasional visitor to the Texas A&M campus. In addition to Talreja, Needleman knew Dimitris Lagoudas and Amine Benzerga, also members of the College of Engineering's aerospace department. In fact, Benzerga had studied under Needleman as a postdoc at Brown.

When the Needleman nomination form to become a Faculty Fellow reached the director's desk at the Texas A&M Institute for Advanced Study, John Junkins instantly recognized the name.

"I knew Alan," says Junkins. "We had met at a National Academy gathering in Washington. I also had attended one of the seminars he put on here with Lagoudas, Benzerga, and company. I knew he enjoyed coming here to engage with more distinguished colleagues than those to be found at UNT. So on one of his trips to College Station, I ambushed him, took him to lunch, and told him about the Institute. When I was done, I asked him if he might be interested in becoming one of our first Fellows. He told me he thought he would be interested and that he would talk to his wife about the opportunity."

As for Needleman being offered "the mother of all sabbaticals," let's first explore the history of these academic interludes.

The dictionary definition of sabbatical is “a period of paid leave granted to a university teacher or other worker for study or travel, traditionally one year for every seven years worked”: The term is a derivative of the Hebrew word *shmita*, which is defined in the Book of Leviticus: “God spoke to Moses at Mount Sinai, telling him to speak to the Israelites and say to them: When you come to the land that I am giving you, the land must be given a rest period, a sabbath to God. For six years you may plant your fields, prune your vineyards, and harvest your crops, but the seventh year is a sabbath of sabbaths for the land” (Leviticus 25:20).

In the *Journal of Higher Education Policy and Management*, Bruce McFarlane of the Education University of Hong Kong offers details regarding the history of sabbaticals in higher education:

> Harvard University first formally instituted sabbatical leave in 1880 under the leadership of its President Charles Eliot. Other elite private US universities including Cornell, Brown, and Wellesley College followed Harvard’s lead in the late nineteenth and early twentieth century, along with prestigious state institutions such as California and Illinois, by establishing their own sabbatical schemes.
>
> Harvard’s policy was intended as a means of attracting new academic appointees and to provide existing faculty “with a respite for career renewal.” Before 1880 the policy on leaves of absence was ad hoc but from this date Harvard decided to introduce an approach that was “more liberal” in the mutual best interests of both the institution and the professoriate. The new policy would provide leave for a year on half pay for the purpose of “health, rest, study, or the prosecution of original work in literature or science” and was deemed to occur no more often than once every seven years.
>
> Hence, the historical purpose of the sabbatical was to facilitate rejuvenation and restoration of academic faculty connecting closely with the etymological roots of the word.

Alan Needleman’s first sabbatical-like experience came early in his academic career. While at the Massachusetts Institute of Technology in Cambridge, Needleman accepted an offer to do research work at the Technical University of Denmark (DTU), located in the northern suburbs of Copenhagen. The year was 1973. Needleman hadn't been at MIT seven years, but the opportunity overseas was enticing, and he would be paid for his service.

It was at DTU that Needleman first met Viggo Tvergaard, a young professor of solid mechanics at the school. Their work together there began a collaboration that has lasted a half century. One of their most recent collaborations, done after Needleman became a Permanent Fellow of the Texas A&M Institute for Advance Study, resulted in a paper published in 2021 and entitled "A Mechanical Model of Blastocyst Hatching."

In fact, there are two Needlemans cited on the work with Tvergaard: Alan and his son, Daniel, a professor of applied physics and molecular and cellular biology at Harvard University. The Needleman Lab there is named for him and not his father, who received both his master's and PhD from the school.

Of his children, Needleman says with a smile, "Daniel was born shortly after we arrived in Providence, and it's been no secret what a proud papa I was then and remain today, of both our children.

"The way I describe Daniel is that he's just like me, except for three things. He's taller than I am. He's better looking than me. And he's smarter than I am. Aside from that, he's also a lot nicer than me."

The Needlemans' other child, Deborah—whose job at the University of North Texas brought Alan and his wife to Texas—was born in Denmark while Alan was at DTU.

As for the matter of "blastocyst hatching," the abstract of the Tvergaard/Needlemans paper offers clarification: "The blastocyst and the *zona pellucida* are modeled as concentric thick-walled initially spherical shells embedded in a viscous medium. Each shell is characterized by a nonlinear elastic—viscous—constitutive relation. The stiffer outer shell (the *zona pellucida*) contains an opening. The softer inner shell (the blastocyst) is subject to a continually increasing pressure, which can eventually drive the escape of the inner shell from the outer shell ('hatching')."

Translating that into layman's terms, Alan Needleman says, "It turns out people hatch from eggs. After conception, an egg forms with a hard outer shell protecting the soft inner part. A chemical process then triggers an opening in the hard shell which enables the part that becomes us to escape and attach itself to the wall of the uterus.

"This happens in all mammals, thus, Dr. Seuss was spot on when he wrote *Horton Hatches the Egg.*"

That book, published in 1940 by Theodor Geisel under his internationally famous pen name, introduced the Horton the Elephant character to readers. The improbable egg that Horton hatches turns out to be half-bird . . . and half-elephant.

Who knew? Well, Dr. Seuss knew, that's "Who."

Forgive the attempted word play regarding the subject of Geisel's sequel involving his beloved elephant character. You may know that book, too, from your own childhood or that of your children.

Horton Hears a Who.

As for Alan Needleman, another important relationship in his life also began during his initial time in Denmark. Ramesh Talreja, who submitted Needleman as a Texas A&M Institute for Advanced Study nominee in 2013, earned his PhD in solid mechanics from the Technical University of Denmark. Today, Talreja is a joint professor in both aerospace engineering and materials science and engineering at Texas A&M.

"I recall my first meeting with Alan when he came to the Department of Solid Mechanics at DTU as a postdoctoral fellow," says Talreja. "Alan joined us at the urging of his PhD adviser, Prof. John Hutchinson at Harvard University. Hutchinson had close ties with our department at DTU. My work was in experimental mechanics there while Alan was involved with computational mechanics, so we did not have much interaction, but we often had lunch together as it was a tradition to embrace newcomers at the school.

"In the years that followed, Alan came frequently to DTU. And now, since he came to Texas A&M as a Permanent Fellow of the Hagler Institute, I still have the opportunity to discuss matters involving the mechanics of materials with him. Over the years, Alan and I have found many common interests, especially as I moved increasingly into modeling of failure in composite materials."

Thanks to Talreja's strong recommendation and an endorsement by Dimitris Lagoudas, at the time the senior associate dean for research, associate vice-chancellor for engineering research, and deputy director of the Texas A&M Engineering Experiment Station, John Junkins judged Needleman a perfect fit to become one of the first Faculty Fellows.

"Based on what I had experienced in my visits to the school, Texas A&M just seemed like an interesting place to be," Needleman says of becoming a Faculty Fellow. "John Junkins can be very persuasive, but what I discovered here was a place with an upbeat atmosphere and an expanding set of great opportunities."

After his initial six-month TIAS Fellowship ended in 2013, Needleman continued an association with the school. An on-campus friendship forged with Marlan Scully led to an agreement in which Needleman would maintain his connections at Texas A&M by being paid a month of summer salary to become a part of Scully's Institute of Quantum Science and Engineering.

Within a year, Needleman was offered a permanent faculty position under the Chancellor's Research Initiative and became a Permanent Fellow in 2015.

"The offer was a good one," Needleman says today. "So good that my wife was willing to move away from the grandchildren."

But not so far away as to preclude regular visits to and from Denton.

Now settled into his work as a University Distinguished Professor of Materials Science and Mechanical Engineering, Needleman is in a unique position to gauge the quality of the academic environment at Texas A&M versus that of an Ivy League school like Brown University, where he spent more than three decades. After all, A&M is courting academics like Needleman to put itself in the rarefied academic air of institutions like Brown.

"The most obvious difference," Needleman says, "is that the school here is a lot bigger than Brown, both in size and available resources. In fact, you could move the whole Brown campus here and it would take you a month to find it. But there are surprising similarities between the two places.

"The students here are very loyal to the university like they were at Brown. There's also an emphasis on both campuses—at least in my scope of understanding—to let people do the things they're interested in. That's important. It's liberating and makes for successful scholars and researchers both today and tomorrow."

Not long after Needleman's arrival as a TIAS Fellow, Texas A&M established its Materials Science and Engineering Department. Prior to that, according to Talreja, MSEN had been a graduate degree program in the Colleges of Engineering and Science. "After Alan became a Permanent Fellow," Talreja says, "he brought a much-needed visibility to a new department and has been instrumental in its growth and enhanced reputation ever since. His congeniality and friendly interaction with students and faculty in the department has brought him the respect and admiration he enjoys."

In helping build a brand within the Materials Science and Engineering Department at Texas A&M, Needleman has worked hard to create a sense of community.

"One of the things I got involved in with the Institute for Advanced Study was hosting a dinner every semester along with fellow Fellows Harold Adams and Bob Skelton. What the three of us did was invite faculty from a wide array of disciplines outside the Institute to come to our dinner. We encouraged each of them to bring along a younger associate with whom they regularly interacted. And if there was room, we also wanted students to attend.

"I think it's important to develop a sense of community with people in different areas. These dinners enabled us to spread the tentacles of the Institute to people who were not necessarily associated with it."

Needleman adds there was one other benefit to this kind of community engagement.

"A great example of that involved my friend Marlan Scully. One of his associates in the Quantum Physics Institute was David Lee, who won the Nobel Prize in Physics in 1996. Just imagine that you're a student who comes to that dinner and you get to meet a Nobel Prize winner.

"I think that makes an impression."

As for what impressions Needleman has made on the Texas A&M academic landscape, rest assured they are profound, although he may be the wrong person to ask.

"I don't know that I can tell you what effect I've had on others," Needleman says, "but I can tell you what effect my time here at Texas A&M as a Permanent Hagler Fellow has had on me. It has given me an opportunity to get involved here and to see things that I just wouldn't otherwise have seen. I don't know if this makes sense, but it's like kinetic theory.

"I can't tell you what the fine points of kinetic theory are, but it's the simplest theory of gases where you have a bunch of the little ping pong balls and they bump into each other and a small percentage of them stick together. If you don't have the action going on, then none of them will stick. So the idea of the Hagler Institute is to get human interactions started. Maybe only a small percentage of those will stick, but those that do can lead to very significant accomplishments."

12

Probing the Genetic Code

Leif Andersson, 2013–2014

It's not uncommon to hear mention of the Wolf Prize around the Texas A&M University campus. In certain circles—and according to the all-knowing authority Wikipedia.com—the Wolf Prize is known as the "Nobel Prize of Agriculture." Given that the "A" in "Texas A&M" stands for "Agricultural," Wolf Prize winners dot the school's academic history.

Perry L. Adkisson served as chancellor of the Texas A&M University System from 1986 to 1990. During his tenure, he brought four new regional universities into the system fold; significantly increased funding for the system's research, teaching, and extension programs; and led the effort to locate the George H. W. Bush Library at the flagship campus in College Station.

But before assuming his wide-ranging administrative responsibilities, Adkisson was a renowned Texas A&M entomologist. For his work, he received the three most prestigious international awards in agriculture: the Alexander Von Humboldt Award, the World Food Prize, and the Wolf Prize. His research on integrated pest-management methods played a key role in reducing insecticide use on US crops by 50 percent.

Other Texas A&M Wolf Prize winners include Albert Cotton, Jim Womack, Fuller Bazer, Allen Bard, and most recently in 2014, Hagler Institute Permanent Fellow Leif Andersson.

"Prof. Andersson has led the path in the development of genomic and marker assisted selection as a means to identify superior breeding stock," reads the Wolf Foundation's website of Andersson's award. "These advances in livestock selection have replace[d] the more classical phenotypic selection methods and are an essential contribution to sustainable feeding of a growing world population."

This work matters from the perspective of the Wolf Foundation, which has awarded the Wolf Prize since 1978, because it "benefits humankind."

"The Wolf Prize is awarded annually," says the foundation's website, "to prominent scientists and artists from around the world for their achievements and contributions in the interest of humanity and friendly relations among peoples, regardless of race, gender, religion or political view."

Other Wolf Prize winners include musicians Paul McCartney, Stevie Wonder, and Placido Domingo, as well as artists Jasper Johns and Marc Chagall.

The Wolf Foundation's headquarters are in Herzliya, Israel. Its founder, Ricardo Wolf, was a German-born Jew who emigrated to Cuba as a young man prior to World War I. He made his fortune developing techniques to recover iron residue from the smelting process. At the outset of the Cuban Revolution in 1953, Wolf became a moral and financial supporter of Fidel Castro. In return for that support, Castro named Wolf as Cuban ambassador to Israel in 1961. Wolf then permanently relocated to the Jewish homeland after Cuba cut diplomatic ties with Israel in 1973.

Leif Andersson's background and his interest in his career calling are a bit less politically charged than Wolf's.

"Something which is beautiful with Scandinavia is that we have a lot of nature," Andersson said from his office on the campus of Uppsala University in Uppsala, Sweden. Andersson splits his time between Texas A&M, the oldest public university in Texas founded in 1876, and Uppsala, which is the oldest continually operating university not only in Andersson's native country of Sweden but also the entirety of Scandinavia.

Uppsala University was established in 1477.

"I started birdwatching when I was ten, maybe eleven years old," Andersson said. "We have similar birds in Sweden to what you find in Texas, but because they evolved separately, there are distinct differences."

Because his childhood hobby kept him in nature much of the time, Andersson became influenced by society's increasing concern about the environment. That pathway was paved, he says, by American conservationist and author Rachel Carson and her book *Silent Spring.*

"That book awakened people's understanding of what we were doing to the environment," Andersson says today. "We were releasing a lot of chemicals into the environment, polluting our lakes and so on. That seemed to really matter to me so I joined a biology club for kids when I was growing up. We did various field excursions and eventually I got involved in nature protection.

"So basically, I went to the university to work toward environmental conservation."

Unlike Ricardo Wolf, who saw Cuban society diminished in the years preceding Castro's revolution, Andersson was the beneficiary of a positive change within his motherland.

"Neither of my parents had an academic background," he says. "They only had six years of school. Both my father and my mother started to work when they were thirteen. Sweden, in the early 1900s, was a poor country, but in time we had a very positive change in our economic development. And with that change, what Sweden did very well was to initiate a welfare system where major investments were made in education for young people. When I was growing up in Stockholm, any kid could, for free, get into high school and university. So a university education was open for everyone.

"That is, I think, a beautiful system," he says.

While at the University of Stockholm, Andersson studied both chemistry and biology. It was there that he first became intrigued by genetics. His undergraduate project involved the study of genetics in Atlantic herring.

"This was in the late '70s before DNA technology was developed," he says. "We used protein polymorphisms, and that had a very limited resolution. My idea was to study the genetic differences between various populations of herring, but they all looked the same. There was no difference whatsoever. It looked like all herring in the Atlantic Ocean and in the Baltic Sea belonged to the same panmictic population."

Despite this failure, Andersson was convinced that species are influenced by and adapt to their environment through genetic mutations. No less an authority than Charles Darwin said so. After graduating from college, Andersson received a job offer from the Swedish University of Agricultural Sciences. Researchers there were looking for a geneticist to analyze horse data.

"They had been doing paternity testing of horses for more than ten years and had data on about forty thousand horses," Andersson says. "What I'd done in college fit well with this area of research. I applied and got the job."

The website VisitSweden.com suggests one think of Uppsala as the "Cambridge of Sweden." As Cambridge, Massachusetts, is home to both Harvard and the Massachusetts Institute of Technology, both the Swedish University of Agricultural Sciences and Uppsala University are also located within close proximity.

"I was still living in Stockholm at the time," Andersson says. "Uppsala was a forty-five-minute train ride away. This was long before laptops and the Internet, so I spent most of my time on my daily commute reading. In fact, one of the books I read during those trips was Charles Darwin's *The Origin of Species*."

Andersson says he was inspired by Darwin's book and his descriptions of how he had used domestic animals as models for evolutionary change.

"If you look at the first chapter of that book," Andersson points out, "Darwin explains that humans have selected domestic animals and plants and created a lot of changes in their phenotype. Darwin points out that this 'artificial selection' was the same thing that happens in nature through the evolutionary process of natural selection. Those individual organisms that best fit into the environment and its lifestyle survive better and have more progeny, thus shifting the genetics of the population. So, as I grew to understand Darwin's views, I realized that domesticated animals are a model for evolutionary change."

And with the advent of DNA technology in the early 1980s, Leif Andersson became one of the pioneers in the field of genetics and genomics of domesticated animals.

"You could say that, inspired by Darwin, I took the next step to use modern technologies to understand what changes have happened when we transform wild species into domesticated species."

On the other side of the Atlantic Ocean and halfway across the North American continent, Jim Womack was doing work at Texas A&M similar to Andersson's. Womack was the son of a basketball coach who himself became a standout basketball player at Abilene Christian College. It was there that Womack took a course in genetics that altered the course of his life. Rather than become a coach, like his father, or a dentist—he enrolled in the genetics class needing a science prerequisite to qualify for dental school—Womack pursued a PhD in genetics at Oregon State University. A decade later, he joined the Department of Veterinary Pathobiology at Texas A&M, where he pivoted away from his study of mouse genetics to conduct research in a field more conducive to the Texas culture. Thus, he developed the school's comparative genetics program for cattle.

Womack, who died in 2023, won his Wolf Prize in 2001 "for the use of recombinant DNA technology to revolutionize animal sciences."

"My introduction to Texas A&M came back in the early 1980s when I first met Jim Womack," Andersson says today. "Jim was one of the pioneers of animal genomics. He was working with mice for a number of years at the Jackson Laboratory in Maine. But when he took the position at A&M, he decided to do cattle genomics. In the early '80s, he visited Uppsala, where I was doing my PhD at the time. That was the first time I met him. Several years later, Jim invited me to come to Texas A&M and give a lecture. And from that point forward, our paths crossed at a scientific meeting somewhere almost every year. Jim was president of the International Society for Animal Genomics and it was through that organization that we became friends."

Jim Womack's name was on the list of the Texas A&M College of Veterinary Medicine and Biomedical Sciences (CVMBS) faculty supporting Andersson's 2012 nomination to become a Fellow of the Texas A&M University Institute for Advanced Study. According to the form, Andersson was "scheduled to visit TAMU in May 2013 to present a lecture and discuss collaborative research ongoing with several faculty members in the CVMBS. During this time, he will be presented with information related to the Whole System Genome Initiative and One Health concept and asked to discuss the opportunities, advantages, and potential pitfalls of such interdisciplinary, multi-institutional collaborations. The expectation is that he will become a TIAS Fellow to assist the CVMBS in strategic planning to help implement the above initiatives."

Also supporting Andersson's nomination was Bhanu Chowdhary, at the time the associate dean for research and graduate studies at Texas A&M.

Chowdhary had collaborated with Andersson while earning his doctoral degree at the Swedish University of Agricultural Sciences in Uppsala. "I was already an established researcher," Andersson says. "Bhanu and I published together." The title of their 1994 paper was "Assignment of the Bovine Immunoglobulin Gamma Heavy Chain (IGHG) Gene to Chromosome 21q24 by in situ Hybridization."

News of Andersson's selection as a TIAS Fellow came as somewhat of a surprise to the man himself.

"They didn't contact me to let me know I was under consideration," Andersson says today, laughing. "Suddenly, one day, I get a call in Uppsala telling me I had become a 'TIAS Fellow.' I asked them if they could let me think about it and see how it fit with my other commitments. I asked for a couple of days to think about it, maybe a week or something like that, and then I decided it could be an interesting opportunity."

Shortly after his arrival in College Station, in January of 2014, Andersson received another call, notifying him he had been selected to receive the Wolf Prize.

"Prominent Swedish Animal Scientist Currently Serving at Texas A&M Wins Prestigious Wolf Prize," the headline read in the *Texas A&M Today* newsletter.

"Dr. Andersson is highly deserving of the Wolf Prize in Agriculture," said David Threadgill in the *Today* article. Threadgill then served as director of the Whole Systems Genomic Initiative (WSGI) at Texas A&M. "[Andersson] is the leading geneticist using the latest genomic tools to reveal the genetic control of many important production traits in agricultural animals. There is no other scientist who has been as successful over the last ten years as Dr. Andersson has been in studying many different species and traits."

Texas A&M Interim Pres. Mark Hussey was also quoted in the story. "I congratulate Dr. Andersson on being awarded the prestigious Wolf Prize," Hussey said. "His breakthrough work in genomic technologies is an example of the positive impact that he and our Texas A&M University faculty are having on Texas, the nation, and the world."

Another member of the Texas A&M administration offering congratulations to Andersson in the *Today* piece was Eleanor M. Green, the Carl B. King Dean of Veterinary Medicine:

> We are excited about Dr. Andersson's recognition as a recipient of the Wolf Prize in Agriculture. His international reputation and expertise in functional genomics, combined with the world-class genomics faculty already in place at the CVM, will be integral in fostering innovative One Health collaborations and leading-edge discovery. Special thanks to TIAS and to our university leadership, who provided the opportunity to bring this world-renowned scholar to our campus. This effort, led by Texas A&M University Chancellor John Sharp and directed by Dr. John Junkins, has provided a wealth of opportunity to Texas A&M and to the communities we serve.

That story went on to describe how Andersson's work might someday benefit humans in addition to the animals humans have domesticated. His work on the genetic basis of muscle physiology and motor coordination in horses, the story offered, may have important implications for human diseases such as cerebral palsy.

Which is why, when Andersson returns home to Sweden and his research work at Uppsala University, he is now a member of the medical faculty.

"Since my program was very successful," Anderson says, "I was offered a position at the KTH Royal Institute of Technology in Stockholm, which is a medical research school. The medical faculty at Uppsala also wanted to keep my research at Upsala. Because of the competitive nature that exists between the two schools, I received a good offer to move to the medical faculty in Uppsala, and that's where I've been for the past twenty-nine years, in the Department of Medical Biochemistry and Microbiology.

"When I agreed to become a Permanent Fellow within the Hagler Institute," Andersson says, "I chose to remain with the veterinary faculty at Texas A&M. There are a number of people there that are doing good research work, and I'm pleased to have the opportunity to frequently collaborate with them."

In the spring 2020 issue of *CVM Today* magazine, then–Texas A&M graduate student Margaret Preigh profiled Andersson's continuing fascination with "things he can't explain."

> Andersson is among the first to have applied molecular genetic techniques to domestic animals. He has worked with pigs,

> chickens, horses, finches, dogs, herring, cattle, and more. . . . The broad scope and open mind Andersson maintains in his research efforts have paid off, resulting in a diverse array of findings and prestigious recognitions such as membership in the US National Academy of Sciences and the 2014 Wolf Prize in Agriculture. . . .
>
> At the root of Andersson's success is not only a brilliant mind, but also curiosity and the trust that his research, no matter how abstract it may seem at times, brings important contributions and applications to the real world.
>
> "I have this keen interest to understand how things function," Andersson said. "Research is trying to understand the world, I would say. When you get insight, new knowledge that no one has had before, that is really rewarding. . . . These new genomic tools that we have give us basic new knowledge about biology," he said. "That improved basic knowledge can open up the possibilities for practical applications, so you can breed healthier animals, you can also maintain biodiversity in natural populations."

In addition to his Wolf Prize and membership in the National Academy of Science, Andersson is also affiliated with another prestigious assembly you may not have heard of.

Benjamin Franklin founded the American Philosophical Society in 1743 for the purpose of—in Franklin's own words—"promoting useful knowledge." Members represent all fields of science, humanities, and the arts, and have included George Washington, Louis Pasteur, Thomas Edison, Albert Einstein, Charles Darwin, and Stephen Hawking. Also elected to membership in Andersson's 2017 class: Pres. Barack Obama.

"Unfortunately, he didn't make it to the ceremony that year," Andersson says. "But I had heard him give a speech before."

That was at Andersson's 2013 induction into the National Academy of Science, that group's sesquicentennial year. In his remarks, Obama reflected on the founding of the organization:

> One hundred and fifty years ago, the nation, as all of you know, was in the midst of the Civil War, and the Union had recently suffered a devastating defeat at Fredericksburg. The road ahead seemed

> long and uncertain. Confederate advances in weapons technology cast a dark shadow on the Union.
>
> The previous spring, in the waters outside of Hampton Roads, the ironclad Confederate battleship *Virginia* had sunk two wooden Union ships and advanced on a third, and this endangered the Union blockade of [the state of] Virginia, threatening Union forces along the Potomac River. And then, overnight, the USS *Monitor*, an ironclad herself, arrived and fought the *Virginia* to a draw in the world's first battle between iron-sided ships.
>
> There was no victor, but the era of ironclad warfare had begun. And it brought unexpected challenges for President Lincoln and his Navy as they expanded this fleet in early 1863, because aboard their new iron-side battleships, sailors found that the iron siding made the ships' compasses unpredictable, so it skewed navigation, and they were bumping into things and going the wrong way. So the basic physics of magnetism undermined the usefulness of the ironclad vessels, even as the Confederates were stocking up on them.
>
> And that's where your predecessors came in. Because in March of 1863, President Lincoln and Congress established the National Academy of Sciences as an independent and nonprofit institution charged with the mission to provide the government with the scientific advice that it needed. And this was advice that was particularly useful in the thick of battle.

As for Leif Andersson's reflection on his own life's work, he offers this.

"I think the domestication of plants and animals is perhaps the most important invention humans have done. It really changed our lifestyle when we took control of food production. Today, I praise all the domestic animals that we have access to, and I think it's really important that we honor them and take good care of them.

"One of the many challenges for our future, in addition to my lifelong concern for the environment, is to have food-production systems where you take care of animal welfare. I think it's good that we should have more plant-based food, thus reducing how much animal protein we need to produce.

"But at the same time, we also must maintain high ethical standards in the treatment of domesticated animals. I would hate to see our domestic animals go away, because I think they are our 'fellows' here on earth.

"We should treat them well."

13

"Sense of Community"

Robert Levine, 2013–2014

Since 2005, Robert Levine has served as general editor of *The Norton Anthology of American Literature*. Published by W. W. Norton & Company, *The Norton Anthology* is a collection of written works by acclaimed American authors. Included are pamphlets, poems, entire novels, and works of nonfiction. The book today—in its tenth edition—is primarily a teaching tool, with "innovative resources to tackle today's teaching challenges," according to the W. W. Norton website.

Levine is the author or editor of more than twenty books on a variety of topics, from literary studies of Frederick Douglass, Herman Melville, and Nathaniel Hawthorne to matters as far-reaching as race, transnationalism, conspiracy, and romance.

As a "man of letters," it is fitting to give Levine a chance to describe his association with Texas A&M University. When asked about his time with the Institute for Advanced Study, Levine offered this reply in an email response:

> From 2013 through 2014 I was a Visiting Faculty Fellow at what is now called the Hagler Institute for Advanced Study. At the time I was working on a book about Frederick Douglass. Instead of staying in residence at A&M, I made three trips a semester for three semesters, three to five days at a time, to share my ongoing work on the book.
>
> When I visited, I gave a lecture from the book in progress, visited undergraduate classrooms, and led a late afternoon workshop for graduate students in English on recent critical work in the field. Then we would all go out for a cheap (and good) Indian dinner. I was treated extraordinarily well by people at the Hagler

> Institute and by people in the English Department. The book that emerged from these visits, *The Lives of Frederick Douglass,* was published in 2016 by Harvard University Press. Over my many visits, I had presented every chapter of the book in talks or in writing workshops, and the book was much better for the critical attention it got from people at A&M. Overall, it was a great experience, introducing me to a great university I had known little about. My thanks to you, John, and Clifford for being perfect hosts.
>
> The book I produced while at the Institute led to my next book: *The Failed Promise: Reconstruction, Frederick Douglass, and the Impeachment of Andrew Johnson,* published by W. W. Norton in 2021. Clifford and John invited me back to A&M to participate in Hagler activities, and I gave a lecture on the book in September 2022 (and also visited some classes). Once again, people at the Institute were incredibly friendly and supportive. It was a great trip. I should probably also say, as a left of center person, that I had heard A&M was "conservative." It may well be conservative in comparison to the University of Texas-Austin, where I've given four lectures over the years, but I consistently had more fun and had a better sense of supportive community when visiting Aggieland. I remain grateful for the support and the friendships I've made there. And I'd encourage you to bring in more people from the humanities as Hagler Fellows!

Levine's Institute for Advanced Study nomination was put forth by Larry Reynolds, a member of the English Department at Texas A&M. Like Levine, Reynolds focused his scholarly pursuits on nineteenth-century American literature. Nathaniel Hawthorne's work is a favorite subject of both men.

As an interesting aside, Reynolds took a somewhat circuitous route to becoming a humanities professor. His undergraduate degree from the University of Cincinnati was in aerospace engineering. He then received his Master's in English from The Ohio State University before claiming a PhD in American Literature from Duke University.

Levine's nomination was also supported by José Bermúdez, then dean of the Texas A&M College of Liberal Arts. When Levine was offered his TIAS Fellowship, the plan was for him to spend a semester on campus. However, Levine had also recently received a Guggenheim

Fellowship, so he parsed his duties at Texas A&M into smaller chunks of three to five days at a time.

"I had the best of both worlds," Levine says today of his time as both a Guggenheim and TIAS Fellow in 2013–14. As a result of the Guggenheim award, he was able to take time off from his full-time teaching pursuits at the University of Maryland to focus on writing a new book from home.

"Then I flew to Texas and spent time at Texas A&M," Levine says. "The Institute covered all my expenses and gave me a good honorarium for each visit.

"Maybe because I was visiting, I had a different relation to the department than I was accustomed to at my own school. It was my understanding and my experience that TIAS Fellows could choose to work with students or not. I sort of wanted to do both every time I visited the campus because I wanted to justify going there."

Levine laughs as he hears himself say those words. On most of his week-long visits, he would either give a lecture, meet with graduate students, or attend undergraduate classes. "I always stayed really busy," he says. The topic of conversation was frequently about matters relating to his book.

But not always.

"I felt like my presence was making some sort of impact on both the faculty and students with whom I met," he says, "but I got a lot out of the TIAS experience myself.

"The majority of times I came in for my visits, I would bring in a chapter of the book that I was working on and share it with the group." Levine says, "and I got really good feedback. During my fellowship period at Texas A&M, I covered the entirety of the book with the people there. The book got better because these were smart people who gave me great input."

Levine published *The Lives of Frederick Douglass* through Harvard University Press in 2016. On the website *Open Letters Monthly*, reviewer Steve Donoghue wrote of the book and its author, "This is a sustained performance of first-rate literary analysis on Levine's part. And he also performs that most crucial of all the services of literary criticism. He compels his readers back to the sources."

Levine cites one example of how his colleagues at Texas A&M improved the book through their feedback: "At one lecture, many of those in attendance made it clear that they thought my chapter didn't really work. And from my point of view as a writer, that was great. I mean, why would I want to publish a chapter that doesn't work?

"The chapter was an important one, about the relationship Douglass had with Abraham Lincoln. The feedback was basically that the chapter came off a little flat. So I reconceived the chapter. I added John Brown to the discussion, creating a triangulation in the story line and it made the chapter a lot better."

Levine says the reason he shared his work with his associates at A&M was that he wanted honest opinions.

"I'd been visiting for a while when I gave the lecture about the chapter the group thought was flat," Levine says. "If it had been my first lecture, I think people would've been too polite to tell me truthfully what they thought. But because they knew me, I think they realized it would be helpful for me to hear that what I had written wasn't working."

Really? Academics would be "too polite" to offer commentary of someone else's work because they're "a stranger"? One of the core principles of the academic world is to analyze and deconstruct anything and everything that comes across one's path.

"I've given several lectures at the University of Texas in Austin," Levine says, "and one thing I can say with certainty is that the people at Texas A&M are a lot nicer on the whole. I think I was seen as a 'special visitor' to Texas A&M who was part of this new and important program on campus, and they just wanted to be polite.

"Honest, but gracious."

Which, Levine adds, isn't typical when academics congregate.

"If I had been attending an academic conference somewhere else, that would have been a very different story. It's my experience that in those settings if you're not on top of things or they disagree with your fundamental premise, they'll let you know emphatically."

Levine's interest in Frederick Douglass as a subject of literary analysis dates back to his arrival at the University of Maryland in 1983.

"I grew up in Massachusetts, a pretty liberal and very segregated state at the time," Levine says. "Both my parents were good liberal Democrats who never said anything bad or racist around the house. I went to Columbia as an undergraduate. I majored in English and did very little with race there, maybe nothing. I then attended Stanford

University to get my master's and PhD, and there were no Black figures in my dissertation. When I got my first job out of school at the University of Maryland in 1983 and had Black colleagues and Black students, I wanted to be able to talk to them, and, as a result, I got really interested in matters of race."

Levine's page on the Hagler Institute website calls him "an influential force in American and African American literature."

"Seeing that my early approach to nineteenth-century American literature was overly focused on white writers, I decided I wanted to broaden and expand my scope after I settled in at Maryland," Levine says. "My second book, but my first book that dealt with issues of race, came out in 1997. I started working on that around 1990 or 1991."

The title of that book was *Martin Delany, Frederick Douglass, and the Politics of Representative Identity*. Of that book, the University of North Carolina Press offers this:

> The differences between Frederick Douglass and Martin Delany have historically been reduced to a simple binary pronouncement: assimilationist versus separatist. Now Robert S. Levine restores the relationship of these two important nineteenth-century African American writers to its original complexity. He explores their debates over issues like abolitionism, emigration, and nationalism, illuminating each man's influence on the other's political vision. He also examines Delany and Douglass's debates in relation to their own writings and to the work of Harriet Beecher Stowe. Though each saw himself as the single best representative of his race, Douglass has been accorded that role by history.

In the introduction to his book, *The Lives of Frederick Douglass*, which took much of its shape while Levine was a TIAS Fellow at Texas A&M, he wrote,

> This is a book about autobiography that is also a literary biography. In *The Lives of Frederick Douglass*, I examine a wide range of Douglass's autobiographical writings from the 1840s to the 1890s, focusing on his efforts over time, and often in very different historical and rhetorical contexts, to describe his relationships with a number of nineteenth-century figures, not only Lincoln and Thomas Auld,

> but also John Brown, William Lloyd Garrison, Gerrit Smith, and others. I explore Douglass's evolving and sometimes contradictory positions on race, violence, nation, and black diasporic community. By emphasizing the complex seriality and performative nature of Douglass's autobiographical writings, I present a new and less familiar Douglass who is surprisingly canny and elusive, capable of depicting himself as both a black Benjamin Franklin and a Toussaint L'Ouverture, while ever aware of what his autobiographical writings and lectures might accomplish (as political and cultural interventions) at specific historical moments.

"My Black colleagues in the English Department at Maryland were more interested in women writers," Levine says. "I found Douglass appealing because he was so influential across the breadth of the latter half of the nineteenth century. While being both eloquent and precise with his words, his opinions about matters and people sometimes changed from writing to writing. He also had serious literary interests himself—which I think speaks highly of him—ranging from Shakespeare to Dickens.

"But the big thing I like most about Frederick Douglass is that he was a native Marylander!"

Levine says one of the great privileges of his life was getting involved with the late University of Maryland historian Ira Berlin in establishing Douglass Square on the University of Maryland campus.

Another of Levine's favorite scholarly subjects is the writer Herman Melville, author of the classic novel, *Moby Dick*. When asked to compare Douglass and Melville, Levine revels in the topic.

"They were very different in the sense that Douglass was a public figure and Melville wasn't," Levine says. "But they had interesting overlaps in the 1840s and 1850s. There was a point when they both were in New Bedford, Massachusetts, at the same time.

"There is a scene in *Moby Dick*, in which Ishmael, the story's narrator, walks into a Black church in New Bedford and kind of freaks out. Melville wrote very powerfully against racism. So that makes him different from one of his most notable contemporaries, Nathaniel Hawthorne, whom I've also studied for years. Hawthorne wasn't fully engaged with matters of race. Melville, on the other hand, wrote a novella about a slave revolt. It's very sympathetic to Black people rebelling against slavery. He wrote that two years after Frederick Douglass wrote a novella about *his* slave revolt. So I find that very interesting."

And the subject of a book he published in 2008.

"Back in 2005, I co-chaired a conference in New Bedford on Herman Melville and Frederick Douglass. It led to me writing a book called *Frederick Douglass and Herman Melville: Essays in Relation*. There are eighteen essays about the two men in that book. It's become a standard study of how we might think about Douglass and Melville together. There are important things they have in common.

"I like the way they write," Levine goes on to say. "As a literary person, that's certainly one thing I admire about them. They're also very honest, very socially engaged, even though Melville kept his distance as a social activist who spent much of his most productive years thinking about how society and culture work."

Levine's interests in the accomplishments of others lie beyond writers who were born and died within the nineteenth century. For instance, a couple of Texas A&M students he grew to know during his time as a TIAS Fellow impressed him immensely.

"While I was there, Larry Reynolds introduced me to his best undergraduate student, a young woman by the name of Hillary Roegelein. If I remember right, she graduated with a 4.0. She was really, really smart. When she applied to graduate school in English, I recruited her. She was accepted at both North Carolina and Maryland, but she wound up coming to Maryland, and I directed her dissertation. I thought that was very neat, that I met her while I was at Texas A&M and then had her as a doctoral student.

"I also worked with a PhD student at Texas A&M by the name of Rene Treviño. While he was there, he decided to pursue a graduate certificate in African studies. As a result, we got to know each other quite well. When he was having some trouble writing an essay in his pursuit of the African studies certification, I helped him with it. His essay was eventually published, and I think it helped him get his first full-time teaching job at Cal State University–Long Beach. It's hard to get a job as an English professor these days, and I was pleased when he got the job at Long Beach."

Levine also maintains ties with some of his former colleagues at Texas A&M.

"I'm still friends with and a strong supporter of Ira Dworkin, who is an associate professor in the English Department at Texas A&M. Among the areas he specializes in are race and ethnicity studies and African American and African diaspora literature, so we have a lot in common.

"Also, it was an honor to know and continue to know Distinguished Professors Larry Reynolds and Jerry [Jerome] Loving."

Levine has also kept in touch with Lucia Hodgson, a member of the A&M English Department during his TIAS Fellowship.

"She set up all my meetings with the graduate students while I was there." Levine says. "After many of those meetings, we would go out for dinner at a local Indian restaurant, which Lucia arranged. During her nine years at Texas A&M, she founded a Critical Childhood Studies working group funded by the Melbern G. Glasscock Center for Humanities Research. Unfortunately, she was denied tenure at Texas A&M, after which she returned to Sweden, her country of origin."

Today, Hodgson works as a guest researcher at Linnaeus University in Sweden, where she was recently awarded a three-year project grant from the Swedish Research Council to produce a "cultural history of nineteenth-century Swedish Texans and their inhabitation of Texas land as a form of settler colonialism interrelated with the enslavement of African-descended peoples."

The State of Texas sided with the Confederacy during the American Civil War—as, apparently, did many of its Swedish immigrants of the day—and the state's political leanings as a whole remain strongly conservative. On January 1, 2024, Texas Senate Bill 17 went into effect on state-owned college campuses—including Texas A&M—banning diversity, equity, and inclusion (DEI) offices, staff, training, mandatory statements, programs, and activities based on race, color, ethnicity, gender identity, or sexual orientation.

That new law postdates Levine's time as a TIAS Fellow by a decade, or even his most recent visit to Aggieland in 2022 to promote his *Failed Promise* book.

There are striking similarities between the current political climate in much of America and places like the state of Texas with those years that immediately followed the Civil War. Levine is an expert on the latter.

His thoughts on today?

"There's a whole bunch of complicated issues that are entangled here," Levine begins. "As far as DEI goes, I'm not involved with that. I'm not an administrator. I don't set hiring guidelines. I and others I've gotten to know at Texas A&M are committed to excellence. So if one does excellent work in African American studies, that's one thing. If you are an administrator doing hiring and arguing that you have been laboring under an unfair hiring quota, that's another thing, I think that's what Texas is presently concerned about.

"What I think Frederick Douglass would have objected to is the effort to write out of our history or censor the idea that there were some bad things like racism, like slavery, that happened in this country. The argument that's made in Florida and Texas that bringing this kind of thing up is inflammatory and leads to polarization and, thus, shouldn't be promoted, is, in fact, a form of censorship.

"What I liked about my experiences at Texas A&M is that people there really are open to learning about history. Certainly the people in the humanities there are not lined up against DEI, but rather they're against discounting our nation's complex and sometimes troubled history.

"Given my areas of academic interest, I've always felt very welcomed at Texas A&M, and not just by the people in the English and History Departments.

"At the University of Maryland, where I have been employed for forty years now, and where I still spend most of my time, there is an interest in DEI, and I think it's mostly for the good, but it's not a hundred percent for the good. Maryland is a blue state, and there's a lot of diversity here, and people think that that's important.

"Yes, there are issues in Texas, Florida, and other red states that trouble me as someone who closely studied the lives and beliefs of people like Frederick Douglass. But I work on a number of writers, not just Douglass, and I'd like to think the majority of my work is on a human scale. And on a human level, my interactions at Texas A&M have been very good. They were good when I was there as a Fellow, they were good when I gave my lecture there in 2022. And in between, I made another visit and that was a very good time for me, as well.

"When Jerome Loving retired from the Department of English back in 2017, he invited me to give a lecture in his honor. He, too, was a Guggenheim Fellow who specialized in nineteenth-century American authors, people like Walt Whitman, Theodore Dreiser, Mark Twain, and Emily Dickinson.

"During my lecture to honor Jerry, I probably talked about some aspect of Frederick Douglass's career, and, you know, I got a very warm reception for that. I hope to return to Aggieland to celebrate the publication of my next book: *After Uncle Tom's Cabin: Harriet Beecher Stowe, Black America, and the Quest for Interracial Democracy.*"

14

In Search of Truth

Wolfgang Schleich, 2013–2014

Wolfgang Schleich's Hagler Institute story begins at a place you wouldn't expect and a time that goes a number of years into his past.

"I was a postdoc at the other Texas university in Austin in 1985," Schleich says as he begins to tell his Hagler Institute tale during his visit to Texas A&M in March 2024.

"I studied with a gentleman by the name of John Archibald Wheeler. He's a very famous theoretical physicist. He was involved in the Manhattan Project, creating the plutonium for the atomic bomb. And he was one of the developers of the hydrogen bomb. By the time I was a postdoc under him at Texas, he was seventy-three, and he was very much interested in quantum mechanics and quantum information. In some sense, you can call him the father of quantum computing and today's quantum information revolution.

"One day he came to me, and he said, 'Would you mind coming with me to Texas A&M? One of my former students retires there today and I'm giving the retirement speech.'"

The young Schleich eagerly said yes, dropped what he was doing—or more accurately, put things carefully away—and offered to drive Wheeler's car to College Station.

"The thing I remember most about that trip," Schleich says, "was when Dr. Wheeler took me into the physics laboratory at Texas A&M. He said to me, 'Wolfgang, this is a very important place in physics.' When I asked why, he answered, 'This is where Ed Fry did the experiment that demonstrated that quantum mechanics violates the Bell Inequalities.'"

A BBC News story published in 2014 labels John Bell as "the Belfast scientist who proved Einstein wrong."

The story offers these details: "Bell's theorem—more formally known as 'On the Einstein-Podolsky-Rosen paradox'—demonstrated that Einstein's views on quantum mechanics—the behavior of very small things like atoms and subatomic particles—were incorrect. [Bell's] theory continues to have a lasting impact on modern physics, and is said to have laid the foundation for quantum information technology."

Ed Fry built on Bell's groundbreaking work. He also, says Schleich, had a strong and lasting impact on the Texas A&M Physics Department.

"When I came here in 1985," Schleich says, "the physics department had maybe twenty-five people, much smaller than the same group at the University of Texas. Under Ed Fry's leadership, the department grew substantially. Fry convinced George Mitchell to build the physics and astronomy building. With Mitchell's support, Fry attracted top-tier people, and the physics department grew and grew and grew.

"Today I would say there is a physics department in Austin, but the one at Texas A&M is much stronger. Things have changed. And a lot of that, I think, has to do with the Hagler Institute because the Hagler Institute brings in people from the outside. It's always important to be open and bring in people from the outside, because no matter how bright your people are, you need stimulus from the outside."

As a Hagler Fellow himself, Schleich provides that much-needed stimulus in a warm and inviting fashion. He's a tall man with an engaging and lighthearted personality, at least outside the working environment of a research laboratory.

Schleich's full-time academic appointment is at the University of Ulm in Ulm, Germany, which happens to be the city where Albert Einstein was born. Following in the footsteps of Einstein, Bell, Wheeler, and Ed Fry, Schleich is a professor of theoretical physics and director of the Quantum Physics Department at the University of Ulm. He joined the faculty there in 1991.

He became a Fellow of the Texas A&M Institute for Advanced Study in 2013, nominated by Dr. Marlan Scully, director of the Texas A&M Institute for Quantum Science and Engineering. Of Schleich's qualifications to become a Fellow, Scully wrote in his nomination letter, "His presence will enrich the intellectual and academic experiences at Texas A&M. [He will] help to promote interdisciplinary and collaborative research within quantum science and engineering, to further develop and enhance the intellectual community among faculty, graduate students, and advanced undergraduate students; stimulate interaction by means of lectures and colloquia, all of which will create a further dissemination of knowledge generated as a result of his interactions with TAMU as a whole."

Tall order for a tall man, but Schleich delivered, and he continues to maintain an association with the Hagler Institute, which brings him back to College Station frequently.

"You know, there's a very old saying that John Wheeler always used," Schleich says. "'Nobody is anybody without somebody around.' I feel like that describes what the Hagler Institute is doing, both for its Fellows as well as the Texas A&M academic community as a whole. You need people to provide 'constructive interference,' which is also the essence of quantum mechanics. So one plus one is not necessarily two, but four, because two people interact with each other. They get new ideas, and suddenly they interfere constructively.

"Of course, sometimes people cannot stand each other, and then nothing will come out. I think it's good that people within the university nominate candidates for the Hagler Institute, people they feel certain they can work with in a positive way."

A common theme that permeates the stories of many Hagler Fellows is the relationships and friendships they have formed with other Hagler Fellows. Those bonds, as Schleich explains, cross scientific fields of endeavor.

"The Hagler Institute is probably unique in the sense that it encourages and facilitates interdisciplinary research. The Gala which is held every year is a unique opportunity to meet everybody, the new Fellows as well as the previous ones. I've attended almost every Gala since I became a Fellow just to meet new people, sit with them, and talk and learn."

Schleich has particularly fond memories of two deceased Hagler compatriots: Harold Adams and Christodoulos Floudas, both of whose stories you will encounter later in this book.

"I have talked with, not all, but many of my fellow Fellows," Schleich says, bemused at his play on words. "A lot of them are not in my field of quantum optics, but some of them are, thanks to Marlan Scully. He has been very active in bringing in famous people, which reminds me of another quote, this one from Victor Weisskopf, an Austrian-born

American theoretical physicist at MIT. He said, 'First-rate people hire first-rate people. Second-rate people hire third-rate people.' So many of the people involved here with the Hagler Institute, from John Junkins to Marlan Scully and many of the faculty and Fellows I've met, are definitely first-rate people."

In spending time with Schleich, it doesn't take long to realize that he fits right in with the Aggie community. While his English is spoken with a pronounced German accent, in many other respects he embodies the core values of Texas A&M: respect, excellence, leadership, loyalty, integrity, and selfless service. His first taste of the Lone Star State may have been in Austin, but it turns out he became an "adopted son" thanks to the purchase of a second home in Denton.

"While I was doing my postdoctoral study in Austin, I met my wife, Kathy," Schleich says. "She's originally from North Carolina. Her father worked for the US State Department and traveled with his family all over the world and eventually settled in Texas and Kathy chose to attend the University of Texas to get a teaching degree.

"After her father retired, he started building houses. At that time, if you would drive from Dallas to Denton, there would be Dallas, and north of Dallas would be fields, and lots and lots and lots of cows, then a sign saying 'Denton.' So Kathy's father bought a big piece of land south of Denton. And today, there are no more cows, just houses.

"Back in Germany, Kathy and I live in an apartment. For years, Kathy would say, 'We will never own a house here in Germany, because land is extremely expensive.' So eventually I said to my father-in-law, 'Why don't you build a house for your daughter?' He was eager to do that and I told him I would pay for it, but that he would have to build it the way she wanted. 'No way,' he told me. 'I'm not going to build a house that my daughter wants. I'm going to build a house that anyone would want.' That puzzled me. 'But you call yourself a "custom" home builder.' I said 'Yes, that's true,' he replied, 'but I tell the customer what he likes.'

"So that's how we got the house in Denton. We spend a lot of time in Denton: Christmas, Easter, summers, and sometimes on weekends when I'm at Texas A&M for my work with the Hagler Institute."

Which may explain why Schleich fits in so well in Aggieland . . . and why he's taken to Texas customs.

But the way things are done in Texas isn't always the way they're done in Germany.

"After I informed Ulm University I had been offered a visiting fellowship to the Texas A&M Institute for Advanced Study," Schleich says, "they told me I could not accept any money. 'Because you're a German citizen and a German civil servant,' they told me, 'you cannot accept that money.'

"After a bit of back and forth with university officials, I finally asked if I could accept the money as a grant. They agreed to let me do that, but that meant I had to ask John [Junkins] and Clifford [Fry] if they could accommodate that arrangement. They told me it wouldn't be a problem—although I think it might have been. In any case, the Institute sent the University of Ulm my grant money and from that I can book my flights, book my hotel rooms, pay for meals, and even rent my tuxedo when I come back for the Gala! In that way, I never receive any personal money, which makes the German government happy."

As it turns out, Wolfgang Schleich's relationship with Marlan Scully goes back beyond his time at Texas A&M as a Hagler Fellow, even beyond his time with John Archibald Wheeler at the University of Texas.

Schleich first met Scully in 1980 at the Ludwig Maximilian University of Munich. Scully taught there while employed with the Max-Planck-Institut für Quantenoptik, located in nearby Garching, Germany. Scully and Schleich worked together for parts of four years, from 1980 to 1984, as Schleich moved toward his *doctor rerum naturalium*, which translates to "doctor of the things of nature," encompassing studies in physics, chemistry, biology, geology, and other natural sciences. As part of those studies, Schleich spent time with Scully at Scully's Institute of Modern Optics at the University of New Mexico in Albuquerque.

Marlan Scully is a restless academic soul. Even today, in his mid-eighties, Scully has joint appointments at Texas A&M and Princeton University.

Today, the Max Planck Institute for Quantum Optics is one of nearly ninety research facilities in Germany operated under the umbrella of the Max Planck Society. Planck was a German theoretical physicist who won the Nobel Prize in Physics in 1918. Twice, Planck served as president of the Kaiser Wilhelm Society, founded in 1911. On the occasion of Planck's death in 1947, that organization was renamed in his honor.

"In 1900, Planck proposed the first part of quantum mechanics," Schleich says. "He called it 'the quantization of energy.' It was really quantization of angle, when he discovered the quantization of radiation.

"During World War II, Planck lost his son. Erwin Planck was killed by Hitler because he was associated with the assassination attempt on Hitler in 1944. Planck had written Hitler a letter asking him to spare his son's life, but he was executed nonetheless. In part, this was the reason the German government chose to rename the Kaiser Wilhelm Society in Planck's honor after his death."

"Through the years, the Max Planck Society has always picked outstanding people to head the various arms of the Institutes," Schleich continues. "These were not necessarily excellent teachers, but top researchers, many of whom eventually received the Nobel Prize. And so, why am I telling you this?"

That phrase frequently comes up in conversation with Schleich. He's a good storyteller.

"Well, it was in the 1980s when lasers became very important, and the Max Planck Society decided to start a new Max Planck Institute for Quantum Optics. They hired a man by the name of Herbert Walter, who started to build up this institute. Walter then hired Marlan Scully to head up the theory group. Walter was a professor at the University of Munich. And I met him, took a class from him, and he kind of drafted me to do my thesis in his institute.

"Through that experience, I met Marlan, and Marlan immediately insisted that I work directly with him. And so I did."

Forty years later, thanks to Schleich's involvement with the Texas A&M Hagler Institute for Advanced Study, he's still collaborating with Marlan Scully.

"We did many things during my Hagler Fellowship," Schleich says, "but, not everything is published."

While interviewing Schleich for this book, we met in the conference room of the Hagler Institute on the eighth floor of Rudder Tower, next door to the Memorial Student Center and a block from the towering structure of the university's football stadium, Kyle Field. As Schleich answered questions, he noticed a poster across the room promoting the Institute. The poster features a photograph of a lone man holding a book in front of a chalkboard. The photograph is the embodiment of "deep thoughts," perfectly suited for its purpose to build the Hagler Institute brand. The poster reminded Schleich of yet another story.

"That's me standing there at the blackboard in that photo," he said. "See the book in my hand? What happened there was that I had copied out a formula from the book and wrote it on the blackboard because Marlan and I couldn't agree on something. What you don't see is that there were students sitting in that room, and Marlan was sitting there, too!

"When the two of us couldn't agree, I finally said, 'That's it! I'm going to get the book!' The book contained research work done by another Hagler Fellow, Roy Glauber."

Glauber was a Harvard professor who won the Nobel Prize in Physics in 2005 for his contribution to the quantum theory of optical coherence.

"When I found what I was looking for, it turned out funny because neither Marlan nor I had remembered Glauber's formula correctly.

"I can't even tell you what the problem was."

Schleich laughed at the end of his story.

One of the many, many interesting ideas that has bubbled to the surface of Marlan Scully's thought processes through the years had to do with something he called a "quaser." Not a "quasar," which is a star, but rather the "quantum amplification by superradiant emission of radiation.

"His idea," Schleich says, "was to take two modes with slightly different frequencies. You drive this difference with a microwave, but each mode has a frequency that could be in the X-ray regime. As a result of these two modes being driven, they would both go up in energy. Now that's very surprising, because somehow you would think that you're violating energy conservation.

"Researchers had seen this effect in numerical calculations, and it was a very strongly debated phenomenon. When Marlan told me about it, I started thinking that it reminded me of something I had read in an old paper by Roy Glauber. Previously, Marlan and I had put together the collected works of Roy Glauber and I remembered the formula we needed was in that book.

"So even though Roy wasn't there when that picture was taken, the book I'm holding contains many of Roy's most famous papers, which makes that photograph a perfect representation of the Hagler Institute story!"

Schleich points out that when you go through many listings of his own recent research work, you'll see his Hagler affiliation mentioned. "Every paper that I publish has the Hagler Institute on it. Why? Because a lot of the work started while I was here with different people, different groups.

"In 2015, Roy Glauber, Marlan, and I put together a special issue in the international science journal *Physica Scripta* on 'The Year of the Light.' Much of that work was done while Roy and I were Hagler Fellows."

And, as Schleich points out, there is more work that has yet to be published.

"I'm a very slow writer. It sometimes takes years before a paper is finished. Unfortunately, I learned this from Roy, who sometimes is also very slow.

"Somebody gave a talk recently, and I said, 'Gee, that's exactly what we did ten years ago.' I gave the presenter the draft of our paper, and he says, 'Why did you guys never publish this?' Well, it's because Glauber and I always wanted to fix this part and fix that part. Marlan always yelled at us and said, 'Look, you guys should finish it and send it in!' I always told him, 'Hey, Glauber has tenure. I have tenure.' It's like Wheeler always said, 'You have to write a paper so that you're not embarrassed about it twenty years later.'

"That's the approach I take."

When Schleich turned sixty in 2017, Scully and one of his Princeton peers, Vince Poor, hosted a quantum physics workshop in Schleich's honor on the Princeton campus. In 2018, Poor became a Hagler Fellow, at the urging of both Scully and Schleich. You'll read more about Poor in the pages to come.

In 2014, Schleich was inspired to start his own institute at the University of Ulm.

"It's called the Center of Integrated Quantum Science and Technology," Schleich says "Both Vincent Poor and John Junkins were on my advisory board. It's a slightly different idea than here."

Schleich goes on to explain . . . in his sometimes meandering, but always interesting and informative way: "Ulm is a town of about two hundred thousand people. It is about halfway between Munich and Stuttgart. If you take one of the fast trains from Munich, it takes you an hour and ten minutes to get from Munich to Ulm. It then takes ten minutes by a street car from the train station to get to the university.

"The University of Stuttgart is forty-five minutes by train. But two years from now, it will take just twenty minutes because they are building a new fast train line that goes underneath mountains and through tunnels. Just goes straight.

"About twelve years ago, I had the idea that we should bring together the university in Stuttgart and the university in Ulm, particularly in quantum physics, because of their close proximity and the fact there's a lot of quantum physics in Stuttgart. My idea was to bring both together in a kind of virtual institute, a little bit like here. It is not one big institute with permanent people. It's people coming in and staying and then leaving again.

"So after my Hagler experience, I wanted to have something similar to here. I got the money for it. We have memberships, but unlike John Junkins, I didn't want to be the director of it, because I wanted to continue my research work and continue writing papers.

"John told me how much time heading up the Hagler Institute took from his academic opportunities. So I asked John to be an adviser, and he agreed. I asked Vince Poor to be another adviser, and he agreed. Both of them gave me really good advice for many years. I am very grateful for their contributions."

As the interview with Schleich neared an end, I asked him what his Hagler experience has meant. He thought for a moment, then shared a story that, at first, seemed tangential to the Hagler Institute, but in which he came full circle to make his point.

"I was once detained briefly by a customs officer in Canada. I was on my way to give a lecture at a school in Ottawa and must have given him some wrong answers, because the guy became very skeptical about me. At the end, when he was finally willing to let me into his country, he said, 'We don't mind scientists like you hanging around here, but we sure don't like them working here.' He must have been very afraid that I would eventually be staying in Canada.

"I do remember well the last question he asked me. 'What will be left of you when you leave this country?'

"I told him, 'I hope knowledge that I brought into the country will be left with the young people here in this country.'

"I think that's the most important thing that we can give to the next generation and to society as a whole. You know, I feel like there is so little that I have actually learned. I still have many more questions than I have answers. But I feel like it's my responsibility to give the next generation the knowledge that we have learned, and to make sure they understand it may not be the absolute truth.

"It is their responsibility to go out and find the truth."

15

New Energy

Christodoulos Floudas, 2013–2014

The corner office near the reception desk on the third floor of Texas A&M University's Frederick E. Giesecke Engineering Research Building is vacant, and has been for nearly eight years. Occasionally, visiting VIPs will be located there, but most of the time it sits empty, a shrine of sorts to a departed visionary, leader, and friend.

"Chris Floudas had a beautiful mind," says Stratos Pistikopoulos, director of the Texas A&M Energy Institute since 2016. "He was a great thinker, a great researcher, both inspiring and motivating. He was also my closest friend, best man at my wedding, and traveling companion to academic conferences all over the world."

Both a Hagler Fellow and Permanent Fellow from Princeton University, Christodoulos Floudas was director of the Texas A&M Energy Institute when it moved into the Giesecke Building in the spring of 2015. Fifteen months later, Floudas died of a sudden heart attack while on vacation with his family in his home country of Greece.

"After his passing, I became even more committed to continuing the work that he and I came to Texas A&M to accomplish," Pistikopoulos says while sitting in his longtime friend's old office. "Too often, people forget history and forget the founders, the people who made things happen. Chris was a person who made things happen.

"On a personal level, I still miss him greatly, but when I'm here in this room, I can still feel his presence and he still inspires me."

In addition to a desk, computer, and sitting area, Floudas's office also contains a display case presenting eighteen of his more than four hundred published research papers. Against another wall sits an open shelving unit. There, a visitor sees a number of books with titles like *Mathematical Programs with Equilibrium Constraints* and

Systematic Methods of Chemical Process Design, most of which were written by Floudas's colleagues and friends about subjects in which he was keenly interested. More noticeably, the bookcase also features some three dozen hardbound doctoral theses from students who are part of the Christodoulos Floudas "academic tree."

In his *in memoriam* tribute after Floudas's death, Pistikopoulos wrote,

> Chris was indisputably one of the very best minds of his generation—the finest ambassador of the "systems thinking" philosophy, vision and practice. With his extraordinary talent and enthusiasm, and his endless number of seminal and unique contributions, he substantially influenced and greatly shaped the field of process & systems engineering and the research minds of contemporary and younger generations around the world.
>
> His academic tree is most impressive, spanning over four generations with 43 PhD students of his own and a total number of 186, more than 20 of whom became academic leaders themselves. With his students and research associates, he made key contributions in: (i) theory and algorithms in deterministic global optimization and derivative-free optimization, (ii) robust optimization under uncertainty, (iii) planning and scheduling of complex systems, (iv) process synthesis and global optimization for multi-scale energy systems and their supply chains, and (v) protein structure prediction, refinement and de novo protein design via deterministic global optimization.
>
> As a colleague, collaborator and friend—it has been truly a blessing, a privilege and a great honor to be closely associated with such a great man, a great mind, a great scientist, a gentleman, a man of principle, a truly inspirational person—a visionary and a leader who radiated optimism and determination to those around him, resonating excellence—impacting the lives of so many people, both intellectually and as a role model.

Both Pistikopoulos and Floudas attended the Aristotle University of Thessaloniki in Greece. Both pursued chemical engineering degrees, but because Floudas was two years older, the pair were destined to become acquainted not in a classroom but on an intramural soccer pitch.

"The game was Year-One students, like me, against Year-Three students," Pistikopoulos recalls. "Chris was the star of the Year-Three team, and I guess you could say I was the best player on the Year-One side.

"Back then, as well as in later years, Chris was an extremely passionate type of person. In that game where we met, he became rather frustrated with my high level of play. I remember he kept yelling out to his teammates, 'Kill the tall guy!' He didn't mean that literally, but he wasn't very happy when the 'tall guy'—me—scored the game-tying goal. He was even more displeased when the final score wound up 1–1. Afterward we talked, got acquainted, and that launched the beginning of our lifetime friendship."

Those two former college soccer rivals put the Texas A&M Energy Institute on the map both academically and within the industry's corporate sector. Another University of Thessaloniki product—in fact, a classmate of Floudas—has also contributed much to Texas A&M and its College of Engineering.

In his more than thirty years in Aggieland, Dimitris Lagoudas has held a long list of important positions at Texas A&M. A distinguished professor in aerospace engineering, Lagoudas has been head of the aerospace department, chair of the materials science and engineering graduate program, deputy director of the Texas A&M Engineering Experiment Station (TEES), and associate vice-chancellor for engineering research for the Texas A&M University System.

"Chris and I both entered university in 1977," Lagoudas says today. "Although he pursued a degree in chemical engineering and I was a mechanical engineer, we took a number of the same classes during our first couple of years and that's how we became well acquainted. After we graduated, we stayed in touch, but didn't have the chance to become colleagues because we were in different fields."

After leaving Thessaloniki, Lagoudas obtained his PhD in applied mathematics at Lehigh University in Bethlehem, Pennsylvania. He did postdoctoral studies at both Cornell University in Ithaca, New York, and the Max Planck Institute in Stuttgart, Germany. His first full-time faculty position was at Rensselaer Polytechnic Institute in Troy, New York, where he taught in the Civil and Environmental Engineering Department from 1988 until 1992.

He then moved to Texas A&M.

"At the time, the US economy was shifting from the Northeast to the South," Lagoudas says. "Texas A&M was a growing university in a state that had a growing economy. I visited here and loved the place.

"My host for the visit, David Allen, took me to Chicken Oil Company for lunch, and I immediately knew I belonged here. While we were there, David told me I had to have a 'Texas name.' He said his was 'Harvey.' Since Dimitris is sometimes translated as 'James,' I decided my Texas name would be 'Jim-Bob.' When I told our waiter that day, he misunderstood my heavy Greek accent and called me 'Zimba.' So that became my 'real' Texas name, 'Zimba.'

"I just love those kinds of stories that are unique to this place," Lagoudas muses.

Christodoulos Achilleus Floudas had already adopted his "Texas name" well before arriving in the Lone Star State. Upon receiving his undergraduate degree, Chris Floudas did his doctoral studies at Carnegie Mellon University in Pittsburgh. In 1986, he joined the faculty at Princeton University, where he remained until accepting the position as director of the Texas A&M Energy Institute in 2015. Prior to that, Floudas had been named a Fellow of the Texas A&M Institute for Advanced Study's Class of 2013–2014. At the same time and not coincidentally, Stratos Pistikopoulos arrived at Texas A&M on sabbatical from his position at the Imperial College of Science, Technology and Medicine in London.

"Chris had previously taken a sabbatical in London, and I had taken one at Princeton," Pistikopoulos says. "We were both due sabbaticals in 2014 and initially were talking about going to Zurich, Switzerland, together. Then Chris called me.

"'Listen,' he said, 'they approached me from this place, Texas A&M, where they have this institute for advanced study. They've made me an offer to become a fellow. What if you also take your sabbatical there at the same time? You know, we are quite well known in chemical engineering, so I think we could do this.'

"As it turned out, the chemical engineering department jumped to invite me for my sabbatical here," Pistikopoulos says, "and Chris and I were able to come together in January of 2014. Not long after that, Chris was approached about the Chancellor's Research Initiative, which Texas A&M had just started."

Laughing, Pistikopoulos adds, "And that's how Chris and I wound up here at a place that, originally, I had never heard of."

After Floudas's death on August 14, 2016, tributes to the man poured in from all corners of the academic world. The Princeton University Office of Engineering Communications cited his "exceptional contributions" to chemical engineering:

> As a researcher, Floudas employed the discipline of optimization, in which mathematicians find the best solutions among an often-complex array of alternatives to grapple with a wide range of problems. He wrote key papers on the structure of proteins and the behavior of protein fragments called peptides.
>
> In 2012, his research played a key role in a white paper from the American Institute of Chemical Engineers that proposed a method to form a national system to create synthetic fuels.

Elected to the National Academy of Engineering in 2011, Floudas received high praise in the NAE announcement of his death, calling him "a highly respected world leader who set very high standards, goals, and challenges in research."

"Chris Floudas' death," wrote Ignacio Grossmann, an NAE member and professor of Chemical Engineering at Carnegie Mellon, "is a huge loss to the mathematical optimization and process systems engineering communities. He was not only a major intellectual leader but also a very good friend. His absence will be felt for many years. The only consolation is to know that his legacy will continue through his students and that his work will inspire new generations of researchers in optimization and process systems."

Of Floudas's too-brief time in College Station, Drew Thompson of the Texas A&M College of Engineering communications staff wrote this tribute:

> Floudas was instrumental in the growth of both the Energy Institute and the Artie McFerrin Department of Chemical Engineering at Texas A&M. Floudas took the helm as the director of the Energy Institute on Feb. 1, 2015. During his 18 months as director, the Energy Institute took a prominent position in two

major U.S. Department of Energy (DOE) initiatives. In May, it was announced that a nine-university coalition, which included the Energy Institute, would participate in a six-year, $20 million fossil energy research project. In June, the Energy Institute was selected to lead the Gulf Coast Regional Manufacturing Center, one of five regional hubs established as part of the $140 million Smart Manufacturing Innovation Institute through the DOE. The number of faculty affiliates of the Energy Institute also grew significantly during Floudas' time as director, growing from around 70 associated faculty members, to more than 245.

The Texas A&M Energy Institute was founded in 2009, initially called the Energy Engineering Institute, part of the Texas Engineering Experiment Station. ("Texas A&M" was added to the name in 2012 under a system-wide rebranding effort initiated by Chancellor John Sharp.) Today, the Institute—with the word "engineering" dropped to differentiate it from the state agencies known as the Texas A&M Engineering Extension Service and the Texas A&M Engineering Experiment Station—is a flagship-wide endeavor encompassing representatives from nine colleges and schools and more than twenty Texas A&M University departments, two Texas A&M University branch campuses, and two Texas A&M University System member institutions. Those partnerships include several outside the College of Engineering at the A&M Flagship, such as Agricultural Economics and Plant Pathology and Microbiology from the College of Agriculture and Life Sciences, the Mays Business School, the Bush School of Government and Public Service, the Texas A&M University School of Law, and the College of Liberal Arts. Other partners include Texas A&M University at Galveston Marine Sciences, the Texas A&M University–Corpus Christi College of Science and Engineering, Texas A&M University–Kingsville, and Texas A&M at Qatar.

Dimitris Lagoudas was head of the College of Engineering's aerospace department when the Engineering Energy Institute was first established in 2009. He says a group of A&M Engineering leaders met with Kem Bennett, then dean of the College of Engineering, to pursue the establishment of an institute devoted to what matters most in Texas.

"The idea of creating energy institutes at Texas colleges became a popular one by the late 2000s," Lagoudas says. "There was one at Rice University. They brought in the Under Secretary of Energy to run the program. The University of Texas established one. They brought in some high-level person as well. So when we met with Dean Bennett, we told him we thought an energy institute at Texas A&M would be a good idea and we wanted to put it in the Department of Petroleum Engineering. So he gave us some resources, told us to put in additional resources from our departments, and we established the Institute in 2009."

Theresa Maldonado served as the first director of the Energy Engineering Institute (EEI), appointed by Bennett. Maldonado was educated as an electrical engineer at Georgia Tech and came to Texas A&M in 2003 as associate dean for research in the College of Engineering and associate director of the Texas Engineering Experiment Station. She left her EEI job at the end of 2010 to become a division director of the National Science Foundation in Washington, DC. Today, Maldonado is vice president for research and innovation at the University of California Office of the President while also leading the American Association for the Advancement of Science as that organization's president.

Steve Holditch, then head of the Petroleum Engineering Department, succeeded Maldonado as EEI director in 2011. Holditch earned his undergraduate, master's, and doctoral degrees from Texas A&M and served on the school's petroleum engineering faculty from 1976 until his retirement in 2013. Highly regarded as both an academic and a researcher—he pioneered much of the technology that refined fracking in the oil and gas fields—Holditch was elected to the National Academy of Engineering in 1995 and became president of the seventy-thousand-member Society of Petroleum Engineers in 2002.

Upon Holditch's retirement, John Pappas was named interim director of the newly christened Texas A&M Energy Institute. At the time, Pappas was director of the Texas A&M Wind Energy Center, the first center formed within the Energy Institute. While a member of the A&M faculty, Pappas was better known for his years spent in the private sector as a registered professional engineer. In looking for a permanent director, the search focused, according to Lagoudas, on finding a "renowned scholar" who also had a legacy of transitioning academic research into engineering practice.

"The institute-building experiment at Rice and the University of Texas didn't work very well," Lagoudas says. "They brought in high-level people from outside education. The academic environment has different requirements. An inspirational leader that could bridge the gap between academic research and practical applications was needed."

As an aerospace engineering peer to John Junkins, Lagoudas had closely followed the formation and birth of the Texas A&M Institute for Advanced Study and strongly supported the concept. In 2004, as associate vice president of research at Texas A&M, Lagoudas shared Junkins's initial concept for an advanced study institute with his boss at the time, Vice Pres. of Research Dick Ewing.

"Ewing thought it was a great idea," Lagoudas says, "and I think he would have encouraged the university to pursue the concept, but he died before we could get the ball rolling."

Like Floudas, Ewing died unexpectedly, suffering a fatal heart attack in 2007.

Lagoudas gives credit to the Hagler Institute—then known as the Texas A&M Institute for Advanced Study—for finding the right person at the right time to assume the helm of the Texas A&M Energy Institute. That person turned out to be a TIAS Fellow . . . and, for Lagoudas, a personal friend.

"Chris's name came up because he was already on campus as a part of the advanced study institute," Lagoudas says. "I didn't know he was here, but when I found out, I told the rest of the search committee that I had known Chris for a long time. I was then given the privilege of making the phone call to ask if he would be interested in the Energy Institute position.

"'Chris,' I said when he answered my call, 'It's Dimitris Lagoudas. It's been a long time! How are you? There's a position open on campus here, and I thought you might be interested in it.'

"As it turned out, Chris already knew about the Texas A&M Energy Institute.

"'Would you be interested in the job?' I asked him, and he immediately said yes, which surprised me. He was doing so well at Princeton back then. Why would he want to leave there to come to Texas A&M? Well, as he told me during that phone call, he had always wanted to establish a big 'umbrella' institute, and the Energy Institute was the perfect opportunity for him."

"The Energy Institute was attractive to both Chris and me," Pistikopoulos says today. "We saw it as our opportunity to build this 'powerhouse' that we had been discussing long before we got to Texas A&M.

"After we arrived here in 2014, we began crafting an Energy Master's program, and we quickly were able to put that into place. We also had considered a lot of the other pieces that could shape a new energy institute even before Chris was offered the job. That offer made it easy for him to leave Princeton and move here."

The package that brought Floudas and Pistikopoulos to Texas A&M full-time was made possible through the Chancellor's Research Initiative (CRI) program. Alan Needleman—a materials science engineer and, like Floudas, a TIAS Fellow—was the first academic to be lured to Texas A&M through the system-based initiative. Floudas was the second. And as Pistikopoulos describes, the CRI package was generous. Very generous.

"In order to get our attention," Pistikopoulos says, "they offered a lot more than just chair positions. Listen, if the offer had only been chair positions in the department, they knew we wouldn't come. We already had endowed chairs where we were at Princeton and Imperial College. We were very well established at very good schools. The catalyst was the Energy Institute and the opportunity to shape it to our vision. The Chancellor's Research Initiative made a strong, strong offer which enabled me to join Chris at the Energy Institute full-time.

"A lot of the credit for all this," Pistikopoulos adds, "goes to John Junkins. He was pivotal, not only in Chris initially becoming a Fellow, but in paving the way for us to become a part of the Energy Institute. John could persuade a dog to become a cat! If it wasn't for him, Chris and I wouldn't have been a part of the Hagler story here."

Says Junkins today, "As it turned out Chris and Alan Needleman were such prominent figures in their fields that they attracted other National Academy engineers and scientists. Chris and Alan were important catalysts that were crucial in the decisions made by over one hundred future academic notables to come to Texas A&M under the Institute for Advanced Study program."

Before Floudas took the helm of the Texas A&M Energy Institute in early 2015, the agency's principal focus had been on fossil fuels, not surprising given the role oil and gas play in the Texas economy and Steve Holditch's prominence in the petroleum engineering field. Floudas broadened the scope dramatically. Eight years after his death, the Institute's website still reflects Floudas's initial vision: "to transform the energy research environment through the study of both fossil and non-fossil-based technologies for energy and the promotion and development of multi-scale energy systems engineering by defining and enhancing the roles of energy economics, law, and public policy; and maintaining a commitment to the societal impact of energy initiatives on a global basis."

In the Spring 2016 issue of *Spirit*, the Texas A&M Foundation magazine, Chris Floudas wrote about "A New Era of Energy."

> Growing up in Greece, I learned to appreciate the value of pursuing excellence through diligence and precise work. In Greek, the word for "work" is ἐνέργεια, which happens to be the root word for the English word "energy." Although I didn't realize it at the time, this Greek-English connection foreshadowed my life's work.
>
> My passion for energy research stems from the contributions that my students, colleagues and I make in our field and the opportunity to make a difference in a world that has an urgent need for more effective and sustainable energy solutions. At the Energy Institute, we focus on scientific and technological energy challenges that impact society—ranging from storing power in electric cars, to recovering oil or gas from new sources, to removing contaminants from water, gas or other liquids used in energy generation and processing. With more than 225 faculty affiliates from across the College Station, Galveston, Fort Worth and Qatar campuses, the institute is truly a large-scale interdisciplinary collaboration poised to address the world's most pressing energy issues.
>
> A key component of our collaboration is our ability to pay special attention to how energy interacts with engineering, science, the environment, economics, law, health and public policy. We also

recognize the importance of all types of natural resources, whether it is fossil-based energy sources, renewable energy innovations, non-fossil-based energy options or water resources management. No matter the type, as we enter a new era, it is increasingly important that our society uses natural resources wisely, integrates them efficiently and preserves them steadfastly.

Visitors to the Texas A&M Energy Institute can't help but notice Chris Floudas's picture hanging on the wall near the reception desk. The formal portrait is framed alongside yet another tribute to the man, this one a poem written by Hagler Institute Founding Director John Junkins:

We can shed tears that he is gone,
Or we can smile because he lived.
We can close our eyes and pray this is all a bad dream,
Or we can open our eyes and appreciate all that he left.

We can feel empty because we can't see his smile,
Or we can be filled with resolve like he always was.
We can turn our back on tomorrow and relive yesterday,
Or we can be inspired by him and live our lives to fulfillment.

We can remember him and dwell only on his death,
Or we can cherish his memory and help make it live on.
We can close our mind, feel empty and fret about our loss.
Or we can do what he would want:
Smile, and bring our passion for life's possibilities.

Junkins says he was never more shocked or deeply saddened than at the news of Floudas's death.

"His departure touched my heart profoundly," Junkins says. "When I was informed of his passing, I could not speak, literally, and I am seldom at a loss for words.

"As I reflected on the shocking tragedy and realization that we would all be living in the past tense of Chris's beautiful life, I wrote the poem to capture these intense feelings. At first, though, I could not find the words.

"To gain inspiration to fulfill the task, I read several poetic eulogies and literature on the sudden loss of life. The works of Margaret Mead, Christine Rossetti, and, in particular, David Harkis were especially meaningful in that journey.

"I can tell you the finished poem consisted of sentiments painfully extracted from my heart."

16

Home Away from Home

Jim Singleton and Stephanie Sale

Jim Singleton and Stephanie Sale have been married for nearly three decades, since 1996. They've known each other much longer, but that's another story . . . for later in this chapter.

The couple is a fixture in the local community. One is tempted to label them "Aggieland royalty." For nearly one hundred years, Stephanie's family owned the local Coca-Cola Bottling Company in an era when the soft drink of the same name was an integral part of the American way of life. She is a lifelong artist and continues to champion women artists as a board of trustees member of the National Museum of Women in the Arts in Washington, DC. Jim, meanwhile, is a Texas A&M Distinguished Alumnus and a member of the College of Fellows of the American Institute of Architects. As a precursor to his business career, he also played football for Texas A&M in the mid-1960s.

In addition, their list of civic accomplishments together is impressive, as written about in the Bryan *Eagle* newspaper on the occasion of Jim and Stephanie receiving the City of Bryan's 2021 Downtown Impact Award.

> Sale and Singleton have been involved with several projects across the Brazos Valley over the years, including helping make Bryan a Texas Cultural District, revitalizing the Queen Theatre, relocating Twin City Mission, expanding Downtown Bryan's Lights On! Event, completing the Brazos Valley African American Museum, refurbishing the landscaping of St. Andrews Episcopal Church with the help of the Gloria Stephan Sale Trusts, constructing Sale

Park and the Depot and creating a new location for the Boys and Girls Club of the Brazos Valley in Downtown Bryan.

Stephanie and Jim are also wine people.

"We have been known to have a wine date from time to time," Stephanie says with a smile.

"At a bar," Jim interjects.

At home, the couple entertains frequently, and good wines are a staple of those occasions, as anyone fortunate enough to receive an invitation to their home can attest.

"It's always an opportunity for great conversation," Stephanie says of sharing drinks with friends or making new friends. This is how their affiliation with and advocacy for the Texas A&M Institute for Advanced Study began.

"We were at the old Veritas Wine and Bistro," Jim says. "This would have been in 2013, as I recall. Stephanie and I were on one of our wine dates, and as the evening progressed, we entered into conversation with two gentlemen unknown to us sitting at the bar."

"Who happened to be John Junkins and Clifford Fry," Stephanie interjects. "They told us they were neighbors, and that night, as it happened, they were exploring a partnership in which Clifford would join John in his work on campus, but initially they didn't go into any details."

Instead, as often happens when men meet for the first time, the conversation turned quickly to sports.

"That took up the first thirty minutes," Stephanie laughs, "over the first glass and a half of wine."

"John's a pretty big guy, big hands," Jim reflects. "We got to talking back and forth. 'Where are you from?' 'Where did you go to school?' I said to John, 'You look like you probably played a little football or basketball somewhere along the line.' And he said, 'Yeah. I played football in high school.' And he told me, 'I was even given a couple of scholarship offers to play football in college.' And then he said, 'But I really had it in mind that I wanted to be a rocket scientist.'"

"At that point, we knew we were in for a really interesting conversation," Stephanie says. "After a great deal of football talk, John moved on to other topics, and we sat there totally enthralled with his wide range of experiences. When he eventually got around to talking about the Texas A&M Institute for Advanced Study, I remember thinking, 'This is the greatest idea I've ever heard.'"

"I was impressed hearing about it," Jim recalls, "but I didn't get it like Stephanie got it. Stephanie really grasped the magnitude of the idea. I didn't get that right away. To me, John was just this good guy; we both used to play football; and then he got into rockets and things like that. But the idea for the Institute went completely over my head.

"When we got home that night . . ."

Stephanie interrupts. "I told Jim, 'Do you realize what this is going to mean for Texas A&M?' I knew Princeton had something similar," she says, "but it was not connected to the university.

"'To have this connected to Texas A&M,' I told Jim, 'how can that not be a win-win-win proposition for the university?' It was just fabulous to try to wrap my head around the concept. I mean, it was just a wonderfully ambitious idea."

"When we left Veritas that night," Jim remembers, "I said, 'Hey guys, let's meet here again sometime soon and talk more about this.'"

Jim and Stephanie met several more times with John and Clifford, the four of them quickly becoming friends. Each time they met, the excitement surrounding the concept for the Institute grew.

"They were the first people in the general community to really get behind our idea," Junkins says. "And they've never wavered in their support."

Stephanie and Jim attended the Texas A&M Institute for Advanced Study's first Gala in 2013. Immediately, they saw a way they could help the cause.

"We realized," Stephanie says, "these esteemed academics who are going to do so much for our university, fly in for the Gala on Friday afternoon, put on their tuxedos, and attend the Gala that night. Then on Saturday morning they board their return flights back to wherever they came from.

"We thought that people should be able to come in here and know where they are, and feel welcomed by the community—and that their time here at Texas A&M shouldn't be in an academic vacuum."

So Jim and Stephanie suggested they could unleash one of their most formidable social weapons on future groups of new Fellows. What they had in mind was to host a "Jim-and-Stephanie brunch" at their home, known locally as "Turkey Creek."

The brunch, indeed, has become a fixture the morning of the Institute's annual black-tie event held at the A&M Memorial Student Center. But it's hard to describe the brunch and their Turkey Creek compound without delving into Stephanie and Jim's backstory.

Stephanie is always a little restrained in sharing the story of how she first met Jim. She suggests the details can get a little "out of hand" when Jim speaks of it. But for the purpose of this book, the two sat together in the sun-filled living room of their remarkable home willing to relive their personal history.

They met more than thirty years before they married, in the summer of 1965. Jim was set to embark on his junior year at Texas A&M. He was not only a football player—soon to be the starting center for first-year head coach Gene Stallings's Aggie team—but also studying to become an architect; Stephanie had just graduated from the Holton Arms School in Washington, DC, and was preparing to attend Skidmore College, at the time an elite all-women liberal arts school in Saratoga Springs, New York.

"I was going to summer school," Jim says. "I was in class from eight to twelve every day. I got a part-time job at the Coca-Cola Company, delivering Coca-Colas from one to six in the afternoon. That's how I met Stephanie. Her family owned the local Coca-Cola Bottling Company."

"We had common interests," Stephanie says today. "I wanted to become an artist and he was studying to be an architect. So frequently, we would take to the country roads and draw landscapes from all over the area."

"On one of our dates," says Jim, "we went over to the community of Anderson and sketched the courthouse. Now how many guys and girls have dates like that, you know? But we did.

"Then we went up to Calvert one day. There are a lot of old nineteenth-century buildings left there in what was once the community's thriving downtown area. We also once took a memorable, if itchy, trek through Booneville Cemetery."

"Did some great rough sketches there," Stephanie says.

"That area is a very nice park now," Jim adds, "but back then it was wild and overgrown. You had to deal with chiggers and thorns and poison ivy and maybe an occasional copperhead. But we persevered and we came out with some really good grave rubbings.

"That's the kind of friendship we had. I thought she was so talented and so great, and we had fun. I don't know what she thought of me."

Life moved quickly for the pair. After college, Stephanie lived for a few years in New York and later married, moving north again. She spent twenty years in the province of Québec, Canada. Jim, in turn, was a member of the Class of '66, but took an extra year to receive his degrees in both architecture and architectural construction. After his playing eligibility ended, Stallings hired Singleton as a student coach. The Aggies' 1967 season culminated with Texas A&M's Cotton Bowl upset of Alabama. What was significant about the 20–16 win was the fact Stallings had played for Crimson Tide coach Bear Bryant at Texas A&M and then coached under Bryant when the Bear left College Station for Alabama. Stallings later returned to Alabama as their head coach, leading the Tide to a national championship in 1992.

For those unfamiliar with Bear Bryant's legacy, he won six national championships as head football coach at the University of Alabama.

"We had eighty guys quit the team when Coach Stallings first came to Texas A&M," Singleton says of events leading up to the 1965 season. "He was one of the famed 'Junction Boys' when Coach Bryant took over the A&M program in 1954. Coach Stallings was tough, just like Bear Bryant. We had 'Junction' right there in G. Rollie White Coliseum on campus, but we didn't have it for just ten days in the summer. We had it for an entire season. Eighty guys quit. I don't know how many guys quit in Junction, but, I don't think it was that many."

According to ESPN.com, seventy-six players quit the Texas A&M football team during the infamous "Junction" training camp under Bryant.

After completing his undergraduate studies in the spring of 1968, Jim Singleton joined the army. Singleton had gone through school as a member of the A&M Corps of Cadets, earning the honor of "Distinguished Military Graduate." As it was the late '60s, an American war raged in Southeast Asia. Today, Jim says, "I was the right age at the wrong time." As a commissioned second lieutenant out of the Corps of Cadets, Jim trained to become a helicopter pilot. He was quickly called to active duty.

Singleton spoke of his early days in Vietnam for Bill Youngkin's column on Aggie war veterans, which ran in the Bryan *Eagle* newspaper for many years.

> My first impression of my outfit was how young all of the pilots were. Of the 26 pilots in the unit, I was about the oldest. There were three college degrees in those 26 pilots and I had two of them. Most were 19 or 20 years old and were fresh out of high school and helicopter school.
>
> Just after midnight of my second night with my unit, I was in the front seat of a Cobra gunship that had been called out to help some long range reconnaissance patrol guys that [were] in trouble. It was also the first time I was shot at. After a successful mission we flew back to camp and I went to bed. I woke the next morning and realized that I had not been dreaming. I was in fact in Vietnam, and this is what it is all about.
>
> My group was a good bunch with multiple personalities, just like a football team. But on mission, those different personalities were replaced by duty to mission, just like a football team.

After the war, Singleton returned to Texas, launching his career as an architect in Dallas. He married, started a family, and ultimately divorced. Then more than three decades after they had first met and became friends, Jim reconnected with Stephanie Sale.

Jim has designed medical buildings, banks, office buildings, restaurants, and shopping centers throughout his half-century architectural career. Through the years, he has specialized in educational facilities, with more than three hundred K–12 projects built in more than one hundred school districts throughout Texas and surrounding states.

A school project in Normangee, Texas, brought Jim from the Dallas suburb of Irving back to the Brazos Valley in the mid-1990s. On a visit to Bryan to find a civil engineer for the job, he met Stuart Kling. Leaving Kling's office, Jim noticed Kling's Aggie diploma on the wall. Jim asked what year Kling had graduated. "Class of '64," was the reply. "Are you from around here?" Jim asked. "Yes, I am," was Kling's reply.

"You wouldn't happen to know Stephanie Sale, would you?" Jim inquired.

Kling smiled. "All the boys knew the Sale sisters back then. Their house is where you went to watch color television and have a cold Coca-Cola."

Kling went on to tell Jim that Stephanie had married, moved to Canada, gotten divorced, and had relocated back to Bryan with her two children.

After leaving his meeting with Kling, Jim quickly made a phone call to the local Coca-Cola Company office. It was still owned by the Sale family, and as a result of Jim's inquiry, he was soon on the phone with his old friend.

"I've got a meeting with the Normangee school board tonight," Jim told Stephanie. "Would you be interested in having an early dinner with me?"

Jim knew his invitation was probably a long shot. But she accepted his offer.

"And we've been together ever since," Jim says, beaming at his good fortune.

"Well, we weren't exactly 'together' at the start," Stephanie corrects. "You were still in Dallas. But yes, that evening at dinner, we reencountered our friendship."

After seeing each other for two years, they married. Jim opened a new office in Bryan / College Station.

Once Jim relocated his practice, the couple's top priority as newlyweds was to build a home that would enable them to successfully blend their families. Both had two children from their previous marriages. To accomplish that, Jim sat down at the drawing board and mapped out what became their Turkey Creek residence.

That home is adjacent to "Fiddler's Green," a horse farm that serves as headquarters for the Texas A&M Corps of Cadets' Parsons Mounted Cavalry. Jim and Stephanie's seventy-three-acre spread is heavily wooded but mostly bereft of the "chiggers and thorns and poison ivy" they ran into on their youthful artistic escapades.

Turkey Creek is a wonderland for outdoor enthusiasts.

"We moved into this house twenty-two years ago," Jim says, looking through the glass wall that brings nature into their living room. "We thought it would be best for each of the children, my two sons and her son and daughter, to have their own space."

To accomplish that, part of Jim's design for the new family's new home included four stand-alone two-story cottages on the south side of the main house.

"The children's first question when they realized what we were doing," Stephanie says, "was to ask if they could decorate the inside of those cottages. The answer was no. We told them, 'This will be where you spend nights and can get away by yourself if you choose, but they're not *your* cottages.'"

The couple laughs at the recollection.

Over the last decade, after Jim and Stephanie's children moved out and started lives of their own, those cottages frequently have been occupied by visiting Hagler Fellows.

"We had an ongoing discussion with John and Clifford," Stephanie says, "about how difficult it was for some of the Fellows to find a suitable place to stay when they came into town."

While the average fellowship is an academic year in duration, many academics choose to come in for two weeks or a month at a time, thus spreading their commitment to the Institute over multiple years. That can make securing accommodations difficult outside the mundane existence that comes with a hotel room.

"Jim and I looked at each other and had the same thought at the same time," Stephanie says, continuing her story. "'We might be able to help you with that,' we said."

Stephanie and Jim estimate more than a dozen Hagler Fellows have stayed with them at Turkey Creek. Their first guest was Yale mathematician Roger Howe. "He loved it out here," Stephanie says. "He was a walker, and we have lots of trails for people who enjoy a walk in the woods."

Howe eventually left Yale and in 2015 became a Permanent Fellow of the Texas A&M Institute for Advanced Study. One can imagine that his time at Turkey Creek may have set an inviting precedent that eventually led him to relocate to the College Station area full-time.

"We're a part of the recruiting team," Jim says knowingly. "Stephanie won't admit that, but we really are."

"Without question," John Junkins confirms.

The morning-of-the-Gala brunch Stephanie and Jim host is a special occasion. Invitees include not only the incoming class of new Fellows but also their spouses or guests and members of the institute's External Advisory Board. This brunch is an opportunity for Fellows in the current class to meet and converse.

There is no real "front" door at Turkey Creek. The main entrance is on the woodland side of the home facing a plaza with a reflecting pool, the cottages, and a pond beyond. The gravel drive from the front gate meanders for a quarter mile past a hay meadow, fenced paddocks, and dense woods that lead to the house. "What's fun about the brunch," Jim says, "is when Fellows come and they see either familiar faces or are meeting each other for the first time. 'I've been wanting to talk to you,' is a refrain you frequently hear. We begin at 11:00, and no one ever leaves early. By the middle of the afternoon, around 3:30, we have to start shooing them off with the reminder, 'Hey guys, you better get back to the hotel and put your tuxes on.'"

"It makes us smile that once they get here, they really don't want to leave," Stephanie adds with a twinkle in her eye.

As was the case with John Junkins and Clifford Fry, a chance encounter with Jim Singleton and Stephanie Sale oftentimes turns into a lifelong friendship. Such is the case with several of the Hagler Fellows who have entered the sphere of Jim and Stephanie's home.

"We become good friends with our scholarly guests," Jim says. "They're just really good people that happen to be extraordinarily intelligent."

Stephanie gives Jim a quizzical look with that remark, but the point is well taken and reciprocated.

Turkey Creek is a quiet retreat and welcoming respite, nestled within nature and away from the hustle and bustle of busy schedules. Frequently, Stephanie and Jim act as conduits on behalf of their guests, linking them to community events, eateries, and little-known haunts where a glass of excellent wine can be had.

"We enjoy sharing this community with other people," Stephanie says. "It's a shining example of a hospitable Texas culture with a lot of different aspects to it. It's two towns with two fire departments, two mayors, two city councils, two school systems, and two completely different histories."

"I call it a peanut," Jim adds. Stephanie nods, having heard this from Jim before. "One shell with two nuts.

"The shell is the community. And the two nuts are the cities of Bryan and College Station."

That peanut analogy could also be used to describe Jim Singleton's experience so many years ago as a student athlete at Texas A&M. And just as "Twelfth Man," "Junction Boys," and "Gig 'Em" are terms that resonate across generations, Jim has a story from his playing days that goes far in capturing "the Aggie Spirit."

"Jim, I don't think you *cannot* share this very interesting story," Stephanie observes. "I don't know how you would encapsulate it in this particular book, or how you would draw a parallel with the Hagler Institute, but I think you should tell this story about what it was like to play football for Gene Stallings, a hall of fame coach, and study architecture under Ed Romieniec, the renowned dean of the College of Architecture.

"Two giants in their fields."

"My first spring practice under Gene Stallings was during the spring semester of my sophomore year," Jim says, beginning his story. "I was determined to get the maroon jersey awarded to starters.

"Football practice started at five o'clock in the afternoon. I had an architecture lab, which ran from one to five. So, to make sure I made it onto the field for the first day of practice, I snuck out of the lab at four o'clock, ran all the way across campus, got taped, dressed, and out to practice a little before five.

"Again, keep in mind," Jim adds, "that maroon jersey was waiting for me.

"Well, it turned out Dean Romieniec happened to be wandering around, going from lab to lab that day. He was my professor, and my regular class was a small one with only about twenty students. He knew I was the football player who was studying architecture. When Dean Romieniec noticed I was absent from my lab, he asked about me and was told that I was at football practice.

"The next day, I was informed that Dean Romieniec wanted to see me . . . immediately! Never in my life have I been cussed out like he cussed me out that day. Dean Romieniec stood maybe 5′6″, and weighed maybe 160 pounds. He told me that if I ever slipped out of class again—even if it was just one minute early—I could go study something else because I would be kicked out of architecture.

"Man, that scared me to death.

"So the next day, right at five o'clock, I bolted out of lab, ran all the way across campus, got taped, got my uniform on, and ran out to football practice about twenty minutes late. Coach Stallings was up in his little crow's nest on the side of the practice field, so he had a bird's-eye view of me coming out of the locker room late.

"From that tower, I heard him growl, 'Singleton! Where you been, boy?'

"I looked up and answered, 'Coach, the prof won't let me out of lab early to make it here on time.'

"'What???' Stallings responded angrily.

"I repeated, 'The dean told me I gotta stay for lab the whole time.'

"'Well, you get your butt out to practice! I'll talk to that guy.'

"The next day, I was at the football team's mandatory lunch, and Stallings walked in and immediately headed in my direction. He walked up to me and said, in the gravelly Bear Bryant impersonation he used his first year at A&M, 'Jimmy, I talked to that Romieniec guy.'

"I replied, 'Great Coach! Thanks! I really appreciate it!'

"Coach Stallings paused for a moment, lowered his voice, and said, 'Well, you're going to have to go to that lab.'

"We looked at each other for more than a moment, and as we did, Coach Stallings's face turned red. He stood there for a time, speechless it seemed, before finally leaving.

"As he made his exit from the dining hall, walking past all of my teammates, I thought, 'Dean Romieniec slayed the giant!' It was unexpected, but I was impressed. My 5'6", 160-pound dean kicked my 6'3", 220-pound coach's you-know-what. Much later, I looked at it a little differently: They had 'agreed to disagree.'

"Coach Stallings let me come late to practice when I had labs, and I got my maroon jersey as a starter. I kept that jersey for two years. And then, for the rest of my life, I got to be an architect."

Giving back to enterprises like the Hagler Institute for Advanced Study is how Jim Singleton and Stephanie Sale pay forward the good fortune life has bestowed upon them.

17

"The Tribe He Touched"

Harold Adams, 2014–2015

Bill Peel is not a Hagler Fellow. Peel is the executive director of learning environments for the Mays Business School at Texas A&M University. If you thought that title suggests he might have a background in architecture, you would be right. Peel received his undergraduate degree in architecture from Texas A&M in 1974, and after a multifaceted career in various architecture-related pursuits, he returned to Aggieland in 2017 and has held a series of administrative roles within the Mays Business School since that time.

While not a Fellow himself, Peel was close friends with the Hagler Fellow—later Permanent Fellow—Harold Adams, who, like Peel, was a product of the A&M College of Architecture.

"Harold was such a gracious person," Peel says. "It was always delightful to be around him. He was just a consummate gentleman. He always dressed meticulously and was perfectly poised in terms of the comments he would make, Don't get me wrong, Harold was a master politician. If he wanted to get something done, he knew how to get it done. He learned a lot of that from his early years working in and around the Washington, DC, area on projects for the federal government. He definitely had an aura of authority.

"You remember the old advertisement, 'When E. F. Hutton talks, people listen'? Well, when Harold Adams spoke, people listened intently to what he had to say. I admired him so. He was such an inspiration as to what it meant to be a leader."

Adams died in the spring of 2022 at the age of eighty-two. In the summer of 2021, he was honored with another in a countless list of accolades received during his distinguished career: the National Academy of Construction (NAC) Ted C. Kennedy Award. And no, that's not the Ted Kennedy you may be thinking of, but Adams did have a close relationship with the Kennedy family known so well.

More on that in a moment.

On the occasion of Adams receiving the Ted C. Kennedy Award, the Austin-based NAC's press release summarized his remarkable life:

> Adams, chairman emeritus of RTKL Associates, has influenced the world of architecture for more than six decades on a global scale. He has been honored by the AIA [American Institute of Architects] with two of its highest recognitions: as a member of its College of Fellows and as a recipient of the Edward C. Kemper Award. He served as chancellor of the [AIA] College in 1998. He is a member of the Royal Institute of British Architects and was one of the first Americans to become a member of the Japan Institute of Architects. A strong advocate for collaboration, he was the first architect elected chair of the Design-Build Institute of America, exemplifying his efforts in bridging the design and construction disciplines. He was a licensed architect in 44 states.
>
> From his humble beginnings in a small Texas town, his career took him from his alma mater, Texas A&M University, to some of the most noteworthy architectural projects ever built: the John F. Kennedy Eternal Flame, the redevelopment of Lafayette Square, the U.S. Naval Academy master plan, the rebuilding of the Pentagon after 9/11, the U.S. Capitol Visitor Center, the Shanghai Science and Technology Museum, and many more.
>
> Adams' selection as the Ted C. Kennedy Award recipient bolsters the prominence of the honor, according to Wayne Crew, NAC General Secretary. "Harold Adams stands apart in the world of architecture and in the design and construction industry," Crew said. "His career, his leadership, and his contributions have made the world a better place for us all."
>
> The award honors late industry leader Ted C. Kennedy, founder of BE&K who built the company into a worldwide leader.
>
> Today, Adams is a professor of practice at Texas A&M, where he is a Distinguished Alumnus and holds an Honorary Doctorate of Letters. He also is a Fellow in the Hagler Institute of Advanced Studies at the university. He has endowed four professorships and one scholarship at A&M as well. His creation of a week-long

collaborative class for students from each department in the architectural college presents them with real-world issues to resolve.

Bill Peel says he first met Harold Adams in the late 1980s at one of a number of alumni events held by the Texas A&M College of Architecture. Peel then was senior vice president of business development for the architecture firm Caudill Rowlett Scott in Houston. Adams, meanwhile, was CEO of RTKL, named for its early principals Archibald Rogers, Francis Taliferro, George Kostritsky, and Charles Lamb. He joined the Baltimore firm in 1967 and assumed the role of president four years later at the age of thirty-one. He became CEO of the company in 1987.

"By the time I met Harold, he had built RTKL into a formidable competitor," Peel says. "They had gone from one office in Baltimore to doing business in more than fifty countries worldwide. Later, when I became president of Pierce Goodwin Alexander & Linville, also known as PGAL," another Houston-based architectural firm, "we won a huge commission to redo Washington National Airport in the early 1990s. We realized the job was too big for one firm to undertake, so we interviewed a number of other companies to partner with us on the project.

"I met with Harold in Maryland, and I remember him talking about how they would approach the project if it was awarded to them. Of course, we already had a rapport because we were both Aggies. He remembered that we had met before and knew that I would have a large say in the selection of his firm. We ultimately did bring them into the project and it was a very successful relationship."

Not long after the Washington National Airport project was completed and the airport renamed to honor former Pres. Ronald Reagan, Peel and Adams began serving on the advisory council for the dean of the Texas A&M College of Architecture. Their involvement continued for many years.

In that time, Adams became a Hagler Fellow in 2014. Two years later, he, Peel, and Jim Thompson—another graduate of the A&M College of Architecture—led a committee to develop a system of leadership principles. Given their zeal in doing so, Adams, Peel, and Thompson became known as the "Three Amigos," and their end product was called the "Aggie C8," with each of the eight leadership principles beginning with the letter "C."

- Character
- Communications
- Creativity
- Competition
- Competence
- Collaboration
- Courage
- Community

"We were seeking to create and codify a leadership model for individuals who were in the built environment, whether it was design or construction or planning," Peel says. "We came up with the concept that we needed to develop a leadership minor in the College of Business, which ultimately incubated and remains in the Construction Science Department."

For two years, from December 2016 to December 2018, the College of Architecture Development Advisory Committee produced a series of nine videos spotlighting the Aggie C8 model. Included in that series was an hour-long interview with Harold Adams, who would join the College of Architecture faculty as a professor of practice and Permanent Hagler Fellow in 2018.

What follows are excerpts from Adams's C8 talk, hosted by Robert Riggs, a Peabody Award–Winning journalist and Texas A&M grad.

• • •

Robert Riggs: Our inaugural guest is Harold Adams, Class of 1961. Harold, you are a Distinguished Alum, which is the highest award that is bestowed upon a graduate here. You're also an Outstanding Alum from the College of Architecture. How did you go from the son of a barber and a seamstress in the little town of Palmer, Texas, to leading RTKL, a global leader in design, architecture, and planning?

Harold Adams: Well, it happened one step at a time. I've been very fortunate throughout my career in that I had some wonderful mentors. Here at Texas A&M. I had a professor, Ed Romieniec, who encouraged me, about my third year, to get out of Texas and to get some experience elsewhere. He assisted me in finding a summer job in New York City. So I worked the summer before graduating in New York City.

Riggs: Let me stop you right there. One of our C8s is "Courage." For a kid from a small town in Texas in that age, did it take some courage for you to go to New York City from here?

Adams: I'm amazed at the courage my parents had to allow me to go. A classmate of mine wanted to do the same thing, and the two of us drove to New York City over spring break and pounded the pavements interviewing at firms. And each of us found a job in New York.

Riggs: So what I'm hearing is that, innately, you were a risk taker.

Adams: Yes. That's served me well, taking chances, stepping outside the box.

Riggs: Tell us what you did in New York and how some of these C8s—perhaps "Competence"—came into play.

Adams: Well, we were very, very well prepared here at A&M. We were in an office where we were doing primarily hotels, and they had a group of summer interns from various universities around the country, places like Cornell, Harvard, MIT, and the University of Pennsylvania. It ended up that I was one of the few that could really draw. So the strengths that I had and the ability that I had developed here at A&M ended up helping to put together presentations for major hotel projects around the world. That summer I worked on a proposal for five hotels in Pakistan. It was to be presented to the president of Pakistan while he visited the United States. That was a very important project that I worked on, doing the drawings for it.

One of the things that I often tell students in my lectures here at Texas A&M is the importance of maintaining relationships. And especially if you leave a job, to always leave on a friendly basis and to go and speak to the owner when you're leaving and explain why you're leaving. So it was developing those relationships that have served me throughout my career from the very beginning.

Riggs: One of our C8s is "Collaboration." The New York City experience eventually takes you to Washington, DC, where you eventually find yourself working with Pres. John Kennedy and his wife, Jacqueline.

Adams: Well, that was almost a fairy-tale kind of story. When I graduated, I wanted to go back to New York and applied for jobs and, in fact, got three job offers in New York. The one that I really wanted was with I. M. Pei. I received an offer from Pei, but with a contingency. They didn't know when they would have enough work that they could justify my coming on board, so I had to keep contacting them to find out if they had work. It was a slow time in the economy, and they kept putting me off. So I decided to go back to where I worked the summer before. I had a standing offer to rejoin them, so I took that job and waited for the Pei job to open up. That resulted in me getting a call from Mr. Romieniec.

At that time he was the assistant dean at Columbia University in New York. He said, "Harold, would you be interested in working in Washington, DC?" I said, "Well, I've been there once." That was when we drove from Texas to New York for our summer jobs. We stopped in DC, and I found it very interesting, a beautiful city. I'd always been interested in politics and had just voted for President Kennedy in the 1960 presidential election, my first time to vote. So I was excited about the opportunity to land a job in Washington, DC, even though I didn't know what the project was about.

It turns out that one of the individuals that had interviewed me at I. M. Pei's office was changing jobs and was going to work in Washington, DC, and had been challenged to put together an office quickly to do a major project. He couldn't tell me the name of the project or the location of the project because it was still in negotiation. So I talked to him. Mr. Romieniec said I should ask for $10,000 a year in salary, which was a lot of money then. In those days that was what a senior architect with maybe ten years' experience would be paid.

So I asked for $10,000, and the person that I was dealing with was Ty Sutton, who had been the senior architect with I. M. Pei and had worked on a number of major projects for Pei. He had been a student of Mr. Romieniec when Mr. Romieniec was in Oklahoma teaching at Oklahoma A&M, which is now Oklahoma State. He sort of cleared his throat and said, "Well, I don't know that we can go that high." But he came back with an offer of $8,500, which was still unbelievable. It was about double what most of my classmates were paid. And so I agreed to join, loaded up all of my stuff, and took the train down to Washington and started the new job.

The location of the office was right on Lafayette Square. It was in a government office space on the north side of the White House.

You could hit the White House with a rock from there. Looking out the window, you could almost touch the White House.

The project was to redo Lafayette Square.

When President Kennedy came into office, he inherited a plan that was very controversial. The plan was to tear down the historic buildings around Lafayette Square. This was the "president's neighborhood," which included houses where many of the former presidents' cabinet members had lived and worked, famous residents in all of the houses. And they were going to be torn down under this plan. [In their place] on the west side of Lafayette Square [was to be] a large white marble, somewhat fascist-looking building. That was going to house the Office of the President, the Bureau of the Budget, and many of the aids to the president. And on the east side was going to be a similar white marble building. And that was going to house two federal courts.

The community in Washington was very upset over destroying all these beautiful and historic houses. They had gotten to Jackie Kennedy and talked to her. She took up the cause and had been lobbying her husband to do something to cancel that project and to do something to save the historic buildings.

Riggs: And this is a big part of her legacy today.

Adams: It still is a part of her legacy. The project was very exciting. The architect that I went to work for was John Carl Warnecke. He was a well-known San Francisco architect who had designed buildings on the campus of Stanford University that were modern buildings, but blended in with the history of the campus. He had grown a reputation for doing this kind of sensitive project, and had explained to the president that he would do the same thing to save the buildings at Lafayette Square. The Kennedys loved it.

While Jackie was away in India on a state visit, the president hired Warnecke as a surprise for her when she returned.

By the middle of the summer, Mr. Warnecke came back to Washington and went to the White House. He met with the president and the First Lady, and was told that they really wanted to see this project move much faster. He came back to the office and discussed it and said, "I really don't have enough people here in Washington to do this project, so I'm going to move it to San Francisco." There, he had a two-hundred-person office, a very large office with all the capability.

I celebrated my seventieth birthday with Jack Warnecke, who was ninety at the time. At that birthday party he talked about why he picked me to stay in Washington to act as the coordinator. He told me that I was being paid so much, he assumed that I was an experienced architect.

Riggs: And you didn't even have your license yet.

Adams: No. Just out of school. Two months, three months out of school.

Riggs: So what among the C8s did he see in you?

Adams: I think he saw my ability to "Communicate" and my ability to "Coordinate." And that I was very "presentable." I was always well dressed and an individual that came across as being much older than I was at the time.

Riggs: So you're the son of a seamstress. Tell us what you mean by "well dressed."

Adams: I went through A&M without owning a pair of blue jeans. Blue jeans were too expensive. My mother was the manager of a men's clothing factory that made trousers and jackets and suits. At the end of each season of making the suits for the next season, they would have a big box of samples and leftover fabric. All the women that worked there could go through the box and pick out what they wanted. On weekends, the women would each work on projects for their families. So I had wonderful slacks and suits and sport coats that were all made by hand in this factory. And they cost a dollar, two dollars at the most. So I was always well dressed. At A&M people thought I was a wealthy person because I was so well dressed. In Washington, I was always in a fresh suit and shirt and looked more experienced and older than I really was at the time.

Riggs: So eventually on the Lafayette Square project, you find yourself working directly with the First Lady. What comes into play then? There's certainly the communication skills, you're navigating politics, and you're having to find collaboration between all these competing agencies. Walk us through that.

Adams: Well, it was a very complicated project. On the west side of the square we had the executive office building. We met with the occupants there and that was fairly easy. On the east side of the square, we had two federal courts, and they were both the highest courts in the land for very technical cases. One of them dealt with patents and customs issues. The other one dealt with claims from the Indian tribes. Neither of these courts have ever been overruled by the Supreme Court, so their decisions are viewed as the final statement on the law in these complicated areas. The judges in both courts were very technical judges, and both, as it happened, were Texans. In fact, one was an A&M graduate. His name was Eugene Worley.

I remember one time I was telling Judge Worley that he couldn't do something, and the judge leaned back in his chair and said, "Harold, where'd you grow up?" I said, "Texas, sir." He said, "Where'd you go to school?" I said, "Texas A&M, sir." He said, "Didn't they teach you that you never say 'You can't do it'?" I've always remembered that. It's been a wonderful lesson.

This is a project that the president and the First Lady had blessed, It required numerous approvals through other organizations in Washington, DC. You would think it would be a rubber-stamp, but much to everybody's surprise, these two judges balked.

They said, "The plan doesn't work. It doesn't work for us. The building is just not functional." So I told my boss—and here's an area of 'Courage'—"I know the requirements of the court, and I think I can solve this problem." So I worked all weekend redrawing the building within the envelope that had been developed in the overall design scheme, but making the function of the building more what they needed. On a Monday or Tuesday morning I met with Mr. Warnecke and he said, "Harold, you know the program, you make the presentation." So at the meeting with the judges, he did an introduction and turned it over to me.

Robert Riggs: And is that something unheard of?

Harold Adams: Unheard of for a young architect four months out of school. I made the presentation. The judges excused us out into the hall to await their decision. After they invited us back in, they said, "We've unanimously approved the plan." They were very complimentary that I had remembered all the details of their requirements. And from that point on, I was John Carl Warnecke's man in Washington. I was managing his office. I was his personal and

> administrative assistant firm-wide, helping him to write letters to the president and the First Lady. In fact, there's a letter that was used in the press release on Lafayette Square that the president signed, and I personally wrote much of that letter. We sent it over to the White House as a draft. The president added one sentence, saying that this was an example of saving history that he hoped could be applied elsewhere in the country. He added those few words, but otherwise, it's the letter that I wrote.

• • •

Adams's work on Lafayette Square brought him into the inner circle of President Kennedy and his wife, leading to multiple projects for the Kennedy family. Among the designs he executed were an addition to Robert and Ethel Kennedy's house in McLean, Virginia, and renovations to several homes at the Kennedy compound in Hyannis Port, Massachusetts. Adams also attended early meetings to discuss possible sites for the John F. Kennedy Presidential Library and Museum. He also served as the project manager for the John F. Kennedy Memorial at Arlington National Cemetery, known as the "Eternal Flame," following Kennedy's assassination in Dallas in 1963.

Adams is the only Texas A&M alumnus to date to be named a Hagler Fellow. After retiring from RTKL in 2004, Adams joined the School of Architecture, Planning, and Preservation at the University of Maryland in College Park. There he established the Harold L. Adams Practice Leadership Fund, dedicated to thwarting the issue of environmental degradation by educating "creative and thoughtful students who believe that physical design and social environment are interrelated." He also taught as a distinguished professor at Maryland.

Upon the occasion of Adams joining the Texas A&M College of Architecture faculty as a professor of the practice in 2018, Executive Associate Dean Dawn Jourdan said of his time as a Hagler Fellow, "[Harold] has been instrumental in invigorating the interdisciplinary activities that bring all the college's students and faculty together. We welcome the infusion of his ideas and energy into the college in this formal capacity."

"I think Harold was a natural teacher," Bill Peel says today of his late friend. "I mean, he was teaching anytime you were with him, whether you were in a board meeting or having cocktails together. Harold was always teaching. He had an innate ability to not just communicate but also connect with people. And he was flush with ideas. He was a true visionary and a deep thinker. And he loved students.

"You know, it's one thing to 'teach,'" Peel adds. "In the academic world there are some who tolerate teaching in order to do research. There are others who like to teach. But there are also those, perhaps the rare ones, who really crave the classroom and crave their interaction with students.

"Harold had a deep love for Texas A&M. He had a passion for sharing his ideas. You might think that a guy of his stature would be unapproachable, but Harold was the exact opposite of that when it came to his students. He literally ran to his student meetings or any time he was called to be in a student setting. He made himself transparent and vulnerable and available. And he listened to what his students had to say.

"I remember him telling me, 'You know, Bill, I learn more from my students than they are learning from me.' He was probably in his late seventies by that time, and he was still learning. The tribe that he touched—not just here at Texas A&M, but throughout his entire life—was quite large."

With those words, Peel stops and reflects, and then adds, simply, "I miss him every day."

18

The Bond of Fellowship

Robert Skelton, 2014–2015

Bonds developed within the Hagler Institute for Advanced Study at Texas A&M University frequently become strong. There is a synergy and kinship among those who have accomplished much in their scholarly pursuits. While the journey toward noteworthy achievements and lofty accolades is often wide and varied—as you're learning from the stories contained in this book—the satisfaction and responsibility that comes with creating a noteworthy career often breeds familiarity among associates and, ultimately, friendship.

The relationship forged between Hagler Institute Founding Director John Junkins and Hagler Fellow—later Permanent Fellow—Robert Skelton became of significance to both men as they grew to know one another after Skelton became a member of the Hagler Class of 2014–2015. At the time, Skelton was a Distinguished Professor Emeritus of Mechanical and Aerospace Engineering at the University of California, San Diego. Academically, Skelton was best known for research that focused on integrating system science with material science. In later years, he did pioneering structural design and dynamics in the field of tensegrity, a term originally conceived by the architect and futurist Buckminster Fuller to describe a design aesthetic developed by him and one of his students, the artist Kenneth Snelson. "Tensional integrity" is a structural system that combines cables and rigidity of elements to create an integrated whole. The roof of the Georgia Dome football stadium in Atlanta—opened in 1992 and demolished in 2017—was an example of the application of a tensegrity system.

Skelton moved to Texas A&M full-time in 2016, becoming a Texas A&M Engineering Experiment Station Eminent Professor. He died in early 2023, and his loss was hard on those within the Aggie community who knew him, including John Junkins.

"Our lives had an amazingly similar trajectory," Junkins says. "I once told him, 'You stole my resume,' to which he replied, 'No, you stole mine. Remember, I'm older than you.'"

While Junkins grew up in Dalton, Georgia, near the Tennessee border in the far north central part of the state, Skelton was raised 180 miles to the east in the town of Elberton, not far from South Carolinian soil. Both men were products of blue-collar families, and both ultimately ventured outside the state of Georgia to attend colleges that were namesakes for the communities in which they were located: Junkins to Auburn, Alabama, while Skelton attended Clemson University in Clemson, South Carolina. There, Skelton majored in electrical engineering.

"It took Bob seven years to get his undergraduate degree," Junkins says, "because he kept running out of money for school. After his first term at Clemson, he couldn't afford his tuition, so he joined the Army Reserve for a number of months. He would save his money and half his KP rations during intermittent stints of active duty so he could continue his education. He did this all throughout his time at Clemson."

By the early 1960s, both men wound up in Huntsville, Alabama, although their paths never crossed during their mutual time there. After graduating from Clemson, Skelton went to work at the Marshall Space Flight Center a year after it was established in 1960. Junkins arrived in Huntsville on an Auburn University work-study program in the spring of 1963. Both men took classes through the University of Alabama Research Institute, Skelton toward his master's degree in electrical engineering and Junkins toward his aerospace engineering undergraduate degree.

"It took Bob eight to ten years to get his master's as he continued working full-time," Junkins says. "At Marshall, he worked for Lockheed, a NASA contractor involved in some really important problems. Among other things, Bob was a part of the team doing design studies for what eventually became the Hubble Space Telescope.

"One of the big challenges in designing a space vehicle that would double as a telescope was that you had to be able to point it to anywhere in the sky. And as you went from point to point you have to torque this truck size spacecraft. Thus, it had to be both lightweight and rigid. In the 1960s, these requirements were not compatible from a design perspective. In theory, every time you maneuvered this thing, it had to settle down and stop vibrating so it could gaze clearly at the stars. And so, Bob was working on trying to model this large flexible structure, but he was an electrical engineer and this was a physics of structures and materials problem. He did some really good work, but he was flying by the seat of his pants without the kind of knowledge of physics and mechanics required to do the job properly. At that point, he knew he needed to be properly educated.

"Curiously," Junkins adds, "I was on the structural mechanics side of controls in my aerospace studies, while Bob was on the electrical engineering side doing controls, but he wanted to do structures more rigorously and I wanted to be stronger in controls. It turned out we sort of passed each other in the night on the crossing of our discipline boundaries."

Skelton ultimately pursued his PhD with the same group of professors at UCLA that had guided Junkins toward his own doctoral degree. Junkins completed his graduate studies in December of 1969, while Skelton came to UCLA ahead of the first quarter of the spring semester of 1970. Skelton studied under Peter Likins, the same professor who had inspired Junkins toward his future academic career.

In hindsight, Skelton's work on the Hubble Telescope may have been at the root of his later interest in the field of tensegrity. Buckminster Fuller first coined the tensegrity term in 1955. He composed an essay entitled "Tensegrity" in December 1958, detailing a system of structural solutions based on Snelson's rod-and-cable sculptures. Fuller's patent application was submitted a year later.

Skelton ramped up his own tensegrity pursuits during his time at Texas A&M, as written about by Jan McHarg in a 2019 *Texas A&M Today* story entitled "Building a Habitat for Sustainable Life in Space":

> As NASA and private companies race to send humans to Mars and beyond, the need for a program providing sustainable life for long periods in space becomes apparent. Dr. Robert Skelton has proposed a growable habitat with 1 G of gravitational force to solve this need.
>
> Skelton, a TEES Eminent Professor in the Department of Aerospace Engineering at Texas A&M University, is wrapping

up Phase II of his space habitat project funded through NASA's Innovative Advanced Concepts Program.

Skelton aims to answer four yet-unsolved needs for maintaining long-term life in space: gravity, radiation protection, agricultural space for food production, and a growable technology for expanding the habitat as economic conditions allow.

Skelton argues that the only way to sustain life in space long term is to provide a habitat for humans that provides a gravity environment similar to what they experience on Earth. The habitat he envisions is a settlement in space with research capabilities to determine what level of gravity is required for human health before launching long-term space travel to Mars and farther.

He proposes to start the habitat at the size of 20 meters radius, enough to sustain about 20 people, with the final structure being built over time out to 225 meters, housing 8,000 people with 300 square meters of agricultural space per person.

The initial habitat would be small and built from materials launched from Earth. Because material resources are costly in space, Skelton suggests using tensegrity systems for the design of the growth adaptable space structure. These minimal mass structures make the habitat easy to change in shape, and it's very strong and lightweight. Subsequent growth stages of the habitat would rely on tensegrity robots mining materials from the moon and asteroids.

Team members that contributed to the project include Drs. Manoranjan Majji, Raktim Bhattacharya and Bonnie J. Dunbar from the Department of Aerospace Engineering at Texas A&M, as well as Anthony Longman from Skyframe Research and Joel Sercel from Transastra, Inc.

Of all the members of the Texas A&M University community who came to know, respect, and befriend Robert Skelton, no one may have been closer to him than aerospace engineering professor Manoranjan Majji.

"I first met Bob as I was wrapping up my PhD work in aerospace engineering at Texas A&M in 2009," Majji says. "He had just retired from his full-time position at the University of California, San Diego, and had begun work there as a distinguished professor emeritus. He had just finished his book called *Tensegrity Systems*, and he came to A&M to give a seminar based on the book. When I had the opportunity to talk to him then, we discussed my own dissertation work, which I had recently completed. He told me that he had done research before on a similar problem but had not solved it completely. So we found that we shared some things in common."

Majji was born and raised in India. He calls his parents "working class" and takes pride in his public-education background there. "Public school education in India is very interesting," Majji says, "because sometimes you don't even have a bench to sit on. You're sitting on the floor."

After earning a scholarship to gain admission to a private high school, Majji attended Birla Institute of Technology and Science in Pilani, a small town near the city of Jhunjhunu in central India. There he majored in mechanical engineering. When it came time to consider his options for graduate school, he turned his attention to the United States.

"Mechanical engineering is quantitative, but it's frequently not very precise," Majji says of his decision to alter his course of studies in graduate school. "I always wanted the ability to mathematically model complex things like rockets and jet engines. That really fascinated me because you need a good bit of model accuracy and high precision to do that kind of work. I wanted to apply to graduate school here because the United States is a leader in aerospace engineering. So I contacted all the leading professors across the country, and one of them responded back to me in just five minutes! That was John Junkins. His email to me said, simply, 'If you're interested in graduate studies, you should apply here at Texas A&M University.'

"I didn't hear back from anyone else for a long time. Even today, I'm sometimes guilty of not responding to emails as fast as I should. I don't know how Professor Junkins does it, but he's very good at responding to emails in a timely fashion. And he's the reason I came to Texas A&M in 2003."

Majji says Junkins was also the reason he stayed at A&M to work on his PhD.

Skelton's visit to the Texas A&M campus for the tensegrity seminar came five years before he became a Hagler Fellow. Prior to his ten-year stay at UCSD, Skelton spent twenty-two years at Purdue University as a professor of aeronautics and astronautics. Purdue has produced twenty-four American astronauts, more than any other public institution in the country. Fifteen of those hold degrees from Purdue's aerospace department, including several who studied under Skelton.

In a UC San Diego Jacobs School of Engineering tribute written after his death in 2023, Skelton was described as "an academic mentor and advisor and teacher to many, Skelton's gentle, approachable and humble manner was driven by his irrepressible and infectious scientific curiosity. Though UC San Diego and the larger scientific community mourn his loss today, at the same time they celebrate his diverse, substantial, and lasting impacts, as a brilliant scientist, as an innovative colleague, and as a true friend. There are many who have been shaped by his compassionate guidance, and will endeavor to carry forward his remarkable legacy."

Few were shaped more by Skelton's "compassionate guidance" than Manoranjan Majji.

After receiving his PhD from Texas A&M, Majji took a teaching position at the University at Buffalo, State University of New York. There, he remained in contact with Skelton. While it may not be all that unusual for older academics to mentor younger colleagues—Majji also remained in close contact with Junkins after leaving Texas A&M—Majji was particularly influenced by Skelton's approach to being a good teacher.

"Bob had a very unique talent," Majji says. "When he talked about academic or technical matters, he said a lot of things, but he had this uncommon ability to spot whether or not somebody actually understood what he was saying, and at what level they understood him."

Think about that comment for a moment. How many times do people try to communicate something only to make the mistake of assuming that their message is actually being communicated? Majji says Skelton made sure that his words made sense to his audience before going on to elaborate on the topic at hand.

In describing Skelton as a "good mentor," Majji goes on to say that Skelton taught him the importance of taking both students and younger colleagues "under your wing."

"When I first met him, he was basically retired at the time," Majji says, "so we mostly just shared and discussed interesting papers we came across. Then when he came here as a Hagler Fellow, he saw the opportunity to transform aspects of the aerospace department and essentially bring tensegrity back to life. His vision was to create an academic program that would sustain tensegrity and disperse it throughout the academic and technical worlds."

Majji returned to Texas A&M in 2016 to present a paper at a Society of Engineering Sciences conference. The paper, Majji says, was about the topic that Skelton had told Majji he had been unable to resolve when the two men first met in College Station in 2009.

"My paper was on a time-varying version of the algorithm that he had not been able to make work quite right," Majji says. "After my talk, he came up to me and was excited that I had more or less resolved the matter."

By this time, Skelton had joined the A&M faculty full-time as a Hagler Permanent Fellow. After Majji's talk, Skelton felt empowered to drop a bit of a bombshell on his young friend.

"'We have a lot in common,'" Majji says Skelton reminded him. "'Would you like to come and work with me here as a professor in the aerospace engineering department?' And I said, 'Do you remember that I'm a product of the department?'"

That question mattered, as you'll see shortly, but there was an even more pressing issue to be considered, at least in Majji's mind.

Did Skelton have the authority to hire someone like him?

"He did," Majji says. "He was one of the early hires under the Chancellor's Research Initiative program, so he had the wherewithal and the budget to bring in associates like me."

Thus was Majji reunited with Texas A&M and given the opportunity to collaborate closely with one of his field's most preeminent scholars.

As for what attracted Skelton to the Central Texas campus of the state's oldest public university, Majji offers, "I think he liked the environment in the aerospace engineering department here. We like to try different things. We embrace things that are very dynamic, even in the way we approach teaching. We try different models and see what works. We are not rigid and fossilized as many programs are.

"We are a different kind of university. I think he saw that."

As for leaving Buffalo, Majji smiles and says, "Yes, that was something very sad. I knew I would miss the winters . . . sarcastically speaking."

After Skelton formalized his offer, Majji became concerned that he might not be welcomed if he returned to Texas A&M.

"I was not sure if the provost and others would approve hiring a product of the department to work here. I had proven myself as an academic at Buffalo, but many places frown upon hiring their own because when you come back, you work with people who taught you, and the thinking is they may never see you as a true equal.

"Fortunately, that wasn't the case here."

Majji says that it wasn't until he began working together with Skelton full-time that he "truly got to know the man" . . . and, as a result, become a member of the Skelton family.

"After I settled in here, Bob and his wife, Judy, basically adopted me as a new son," he says. "One of the first things Bob told me was, 'You are here on your own merit. I have always hired people who are better than me. So you don't have to think you're here because you were Junkins's former student or anything like that. We are not doing a favor for you. You are actually doing us a favor by being here.'"

As soon as Majji became a member of the Texas A&M aerospace department, Skelton began including him as a collaborator as he prepared research proposals and submitted grant applications. Skelton also offered Majji continuing advice about the art of mentoring students.

"He always thought that students are really an extension of yourself as an educator," Majji recalls. "The only way you can permeate and procreate important ideas you want to impart to society as a whole is through students, which I have found in my own experiences to be very true. What I believed, before Bob steered me in the right direction, was that it was more important to get research funding first, and that the students' well-being could wait. Bob changed my perspective on that. It's very important to do both. The students have more immediate needs, and attending to them is extremely important.

"The other thing I learned from Bob was that we really need to bring different disciplines together in very fundamental ways. I got a dose of that as a graduate and PhD student here because Professor Junkins knows a lot of things about a lot of things, Both John and Bob helped me realize that it is not productive to say that something challenging 'is not in my area.' Now if something important arises as a barrier and isn't in my area of expertise or experience, I go and study and make it my area.

"Bob once told me, 'We are supposed to be engineers, and if engineers need to know something, we identify the area we need to understand and go learn it.'"

Soon after joining the Texas A&M aerospace department in January 2017, Majji began spending more and more time at the Skelton home. He was frequently their dinner guest, so much so that Majji's own parents took to calling him from India while he was at the Skeltons' house for an evening meal. At 9 p.m. in Texas, it's 7 a.m. the next day in India. Thus did Majji's two sets of "parents" get to know each other.

In late 2018, Skelton underwent a minor medical procedure in Austin. As was his way, according to Majji, he had scheduled the procedure the day after a seminar he would be holding at Rice University in Houston. Since Judy would be along with him for both events, under normal circumstances, she would have been able to drive the couple home from Austin. But Judy had recently been in a car accident and couldn't drive. So Bob reached out to Majji.

"I moved my meetings around and agreed to meet Bob and Judy at their house when they got back from the seminar on Thursday," Majji says. "Then I would drive them to Austin where Bob would have his procedure done first thing on Friday.

"But things didn't go quite as planned."

Majji says complications arose and Bob was hospitalized over the weekend and into the early days of the following week. While he eventually recovered, health matters became a constant concern in the Skelton household.

"After Bob recovered from his stay in the hospital in Austin, Judy's health went into decline," Majji says. "She developed cognitive impairment that was eventually diagnosed as slow-onset Alzheimer's. As you can imagine, this really weighed on Bob. Judy's sister had passed away from Alzheimer's complications, and all of us were concerned about both Judy and Bob."

Then in 2019, Bob suffered a stroke, which paralyzed half of his body.

"Judy had always managed the household and finances, and when her health declined, she was no longer able to do that," Majji says.

"Bob tried to step in, but he was overwhelmed. From what I saw, he sort of stopped taking care of himself, and this is what led, I believe, to his stroke.

"Bob and Judy were always very good at hiding things, particularly from loved ones, and none of us really knew the situation that Bob was getting into."

After the stroke, Bob and Judy moved into an assisted-living facility located north of the Texas A&M campus in the city of Bryan. There, his Aggie family, which included not only Majji but also John and Elouise Junkins as well as others with ties either to Texas A&M or the Hagler Institute, began rallying to offer moral and spiritual support to the Skeltons. These friends and associates did their best to make sure that Bob and Judy never ate dinner alone, and Majji began bringing work from various research projects to Bob so he could see the progress being made on those endeavors.

When the COVID-19 pandemic struck in 2020, Bob's health fell into further decline. Because of the greater risk of contracting the disease living in the communal setting of an assisted-living environment, the Skelton family decided to move Bob and Judy to Fairhope, Alabama, where Bob's daughter, Leigh, lived.

Despite the greater distance, Majji continued to be Bob Skelton's faithful friend and colleague.

"I would visit him every two months," Majji says. "Usually, I brought one of his graduate students with me. Those trips seemed to lift his spirits. Initially, he was still able to contribute a lot, and I was always inspired by how dedicated he was.

"Most of the time on those trips, I would bring with me something new that we had done. Bob was particularly enthused when I would bring a new student dissertation that had just been finished. He always expressed disappointment that he couldn't be there with the student in person, but he was proud of the work his students were producing."

Skelton's stroke triggered hemispatial neglect issues, which meant his vision was obstructed in such a way that he could not see all the documents and slides that Majji would bring to Alabama. "We frequently had to move our laptop around to help him visualize all of what he wanted to see," Majji says.

Despite his limitations, Skelton would frequently join Majji for Zoom meetings online. At the time, Skelton had four PhD and three master's students, and despite having a like number of students of his own, Majji agreed to support those under Skelton's watch as well.

Majji's dedication to his friend and mentor went on like this—with trips to Alabama every two months—for nearly two years. At one point, Skelton showed signs of improvement, but in mid-2022, his kidneys began to fail. Soon after, he was in a fight for his life.

In early February of 2023, while attending an aerospace engineering conference in Breckinridge, Colorado, Majji received a phone call from Leigh.

"Listen, we don't have much time," Majji recalls Leigh telling him on that call. "If you want to see Bob and talk to him, you better get here soon."

"She said Bob was on hospice again," Majji says. "I had sort of gotten used to this, because in the last several months of his life, Bob was frequently in failing health. That's why I visited them so often, because we never knew when the end was going to come. That call while I was in Colorado was maybe the third or fourth time he was placed into hospice care. But each time before, he would rally and start making jokes again."

Majji left the conference immediately, caught a flight to Pensacola, Florida, and then made the one-hour drive to Fairhope.

"He was still talking when I got there," Majji says, "but I couldn't make out much of what he said. I told him, 'Bob, I'm going to try my best to preserve your legacy. We have done a lot of work in tensegrity since you left, and I'm going to try to keep it up.' By that time, the aerospace department had declared it would name its best PhD dissertation award in his honor. I told him about that, and I remember he squeezed my hand.

"After about an hour with him that day, he sort of disappeared, but he was still alive when I left."

A week later, on February 15, 2023, Bob Skelton died.

"I went to Fairhope to celebrate his wedding anniversary about two months before he died," John Junkins says. "To be honest, it was like a wake. So many of his friends were there, people from UCLA, from UC San Diego, and others from across the country. Majji and I and two or three other people from Texas A&M went to see him to show him how much we cared about him.

"Toward the end of the party, with Bob and Judy sitting at a table in the front of the room, we all came up to them one at a time to give them our regards. As I walked up to greet him, I saw that Bob was crying. I think he knew this was near the end of the line. It was a good thing that we were all there with him, but it was a painful thing, too. That was the last time I saw Bob.

"I went to his funeral in Fairhope, and a few weeks later, back in College Station, we had a postfuneral event, a wake, if you will, with so many of his friends.

"That was an emotional few months for all of us who loved Bob," Junkins adds. "It was a long goodbye, but it was a good goodbye, too."

PART 3

The Namesake

19

"A Fortunate Life"

Jon Hagler

After graduating from Texas A&M University in 1958 with a degree in agricultural economics and then spending three years in the US Army, Jon Hagler set his sights on Cambridge, Massachusetts, and the Harvard Business School there. Despite the way things worked out in his life, his intent in attending Harvard, he says, wasn't to get rich.

"When I was a kid growing up, I had restrained ambitions about money," Hagler says today from his home in Dover, Massachusetts, just twenty miles south of the Harvard University campus. "I knew I didn't want to be poor because I had experienced that as a youth. My career ambition, really, was to make something of myself, take care of my family, and leave the world a little better place."

Hagler admits attending graduate school at an elite Ivy League institution might have seemed like a stretch for someone with no real affiliations with the upper crust. Keep in mind, in the late 1950s Texas A&M's academic bona fides were not quite what they are today.

"Yeah," Hagler laughs, "I think I was a beneficiary of Harvard's 'affirmative-action' program. They looked at me as a kid from Texas and probably thought, 'Admitting him will help acculturate our class of scholars.'"

To pay for graduate school, Hagler took out loans. He remembers making a trip to a local "S&L" in Cambridge to borrow money, bringing to mind the actor Jimmy Stewart as George Bailey in the classic 1946 Christmas movie *It's a Wonderful Life*. That film represented simpler and more gentle times before savings and loan institutions—the aforementioned "S&L"—suffered a calamitous demise during the eponymous financial crisis of the 1980s. It was while Hagler was in the process of going deeper in debt to fund his graduate education and support his family in the early 1960s that he chose to reflect on—among other things—what it meant to him to be an Aggie.

"My time at Texas A&M was important to me," Hagler says. "My father, John, was a vocational agriculture teacher, and I grew up an FFA kid."

In fact, Hagler was state president of the statewide Future Farmers of America chapter his senior year of high school in LaGrange, Texas. As for what led him to College Station to attend school, "It would be hard to underestimate proximity," he says. La Grange is eighty miles southwest of the Texas A&M campus. "I was one of four kids in our family and I knew my parents didn't have the wherewithal to fund each of our college educations. Thanks to my affiliation with FFA, I was able to get several scholarships, and I layered those to help me cover my tuition at A&M. Once there, I took a job in the mess hall, which pretty much took care of my room and board."

As was the case with many Aggies who came of age within the Corps of Cadets, Jon Hagler credits that experience and lessons learned there with shaping his future self.

"I was a pretty self-confident kid at that time," says Hagler, who eventually became a Ross Volunteer, member of the student senate, outstanding junior cadet, and Corps commander his senior year. "I liked the environment of the Corps, the discipline, and the leadership-training experience.

"It was at Texas A&M I met an individual who became one of my most important mentors. J. Wayne Stark ran the Memorial Student Center program. He saw it as his mission in life to put his hand on the shoulder of a few people who he thought might benefit from a graduate education. He encouraged me to go to graduate school, and I thought about both law school and business school. That's the principal reason I ended up going to the Harvard Business School. J. Wayne was on me for a long time about what I was going to do with my life."

A year after Hagler graduated from A&M, Stark became a sponsor for the Experiment in International Living program, which enabled hundreds of Aggie students to travel overseas to live and work.

A Memorial Student Center art gallery today bears his name.

It was at the end of his first year of graduate school in the Boston area, in June of 1962, that Jon Hagler made the decision to give back to his alma mater. After conferring with his wife, Jo Ann, he pulled out his checkbook and authored his first donation to the Texas A&M Alumni Association . . .

. . . in the amount of $2.50.

"That would have been like making a twenty-dollar contribution today," Hagler points out, laughing. "Despite my financial straits at the time, I felt it was important to indicate to the university that I intended to be a supporter of the school. I thought it was important to make a small contribution as opposed to succumbing to my circumstances and make no contribution at all."

Nearly half a century later, Hagler ran across that old canceled check. Amused, he decided to send it to Eddie Joe Davis, friend and president of the Texas A&M Foundation. In 1996, Jon and Jo Ann Hagler made a $5 million commitment to Davis as the lead donors for a new building to house the foundation's offices. Today, that building, located on the south side of the A&M campus, is known as the Hagler Center.

"From what Eddie Joe tells me," Hagler relates, "he saw some significance in that first check of mine and used it frequently to show people how small things can grow into big commitments."

Hagler goes on to say the current president of the foundation, Tyson Voelkel, also carries a laminated copy of that check to pull out at opportune times.

What enabled Hagler to begin making larger financial commitments to Texas A&M had to do with, in his words, "being in the right place at the right time."

The Dr. Tyrus R. Timm Registry of the Texas A&M Department of Agricultural Economics honors distinguished former students. Included is Jon L. Hagler, whose registry entry reads in part,

> Mr. Hagler received his B.S. degree in agricultural economics from Texas A&M in 1958 where he also served as Cadet Corps Commander his senior year. While a student, he also was a member of the Ross Volunteers, earned many academic honors, and was active in many organizations. After graduation and marriage to his wife, Jo Ann, he served three years in the U.S. Army as an artillery officer. Upon completion of his military service, he entered

> Harvard Business School and earned an MBA degree. While advising fellow students, he found that he possessed a special gift for managing money.
>
> His first position after Harvard was that of research analyst with United Funds, Inc., a mutual funds organization. Since then, his entrepreneurial skills have supported a meteoric career in investment management. Four years after receiving the MBA, he was managing an investment portfolio of $1.2 billion and founded two investment management firms—in New York and Boston. By the age of 40, he was chief investment officer at the Ford Foundation. Mr. Hagler currently serves as a member of the Board of Directors and as a partner of Grantham, Mayo, van Otterloo & Co., LLC in Boston. This institutionally oriented investment management firm handles over $80 billion and has offices in Boston, San Francisco, London, and Sydney. He is also a trustee of the Hagler Foundation.

"As I've already stated, I never really had aspirations for great wealth," Hagler says. "During my second year at Harvard Business School, they asked us to put down our goals and objectives. I don't recall much of what I wrote down, but I do remember that back in 1963 I was hopeful that by the time I was forty years old, my compensation would be equal, in thousands of dollars, to my age.

"For a guy coming out of Harvard Business School, those were pretty modest ambitions. But I joined an industry—investment management—that was in its infancy. I got in at the right time and was able to ride that wave."

The $5 million gift for the construction of the new Texas A&M Foundation building was Hagler's first major donation to the university, but not his last.

"That happened fairly early in my engagement with the foundation," he says. "It seemed like such a natural thing. I thought there was an extraordinary level of possibilities to impact the school by leveraging a contribution through the foundation. I happened to attend a meeting where they announced the proposed project and afterward I told Eddie Joe that I was very interested in supporting the effort."

A year later, Hagler was co-chairing the *Vision 2020* campaign alongside his former Corps roommate, A&M Pres. Ray Bowen.

"After I got substantially involved in the A&M Foundation," Hagler says, "there was a group of people there who were advising President Bowen on the leadership structure of *Vision 2020*. Somehow, my name got into that mix. Then in a meeting in Dallas that both Ray and I attended, he pulled me to the side and asked if I knew about the project and if I would be interested in helping him lead it. I said to him, that depended on what he wanted to accomplish. If it was a serious effort to really make a difference in Texas A&M's future, I'd be tickled to death to help. I knew Ray wouldn't embark on something of this magnitude if it wouldn't potentially be a game-changer. He told me he was serious, and I told him, 'I'm your man.'"

Bowen, Davis, Hagler, and Robert Harvey, an A&M graduate and management consultant, formed what Davis affectionately labeled "the kitchen cabinet," leading the 250-member *Vision 2020* Advisory Committee, which met more than a dozen times in College Station between 1997 and 1999. In those meetings, questions were asked and suggestions made that ultimately formulated the document to help guide the university's aspirations into the twenty-first century. And even though the project was the catalyst for John Junkins to begin defining what a Texas A&M Institute for Advanced Study might look like—and that institute today bears the name of Jon Hagler—the two men did not become acquainted during that time.

"I don't know that we transformed the place," Hagler says, "but I think that *Vision 2020* was a really important recalibration for the mission of the university, its attitude about itself, and its place in the world."

As for when the Hagler-Junkins connection was finally made, that took place in 2002 as the university sought to find a new president to succeed Bowen. John Junkins was named chair of the faculty committee to lead that search.

"I have it on pretty good authority that he was a pivotal player for good in that presidential search," Hagler says today. "In my opinion, selecting Bob Gates to lead the university was a tipping point for Texas A&M."

Of that pursuit, Junkins says, "As we began our efforts to find a new leader, none of us on the committee realized that Gov. Rick Perry had already promised his support for the job to a former A&M economics professor who had left the academic world for a successful career in politics."

That person was US Sen. Phil Gramm.

Junkins had vowed he would not advocate for or against any candidate in the search, but many on the committee believed Gramm would be a polarizing figure as Texas A&M president. Junkins defended the committee's view and fought hard against extreme political pressure to ensure the efforts of his search committee remained fair and balanced. When word of his opposition to Perry's handpicked choice became public, Jon Hagler took notice.

"I don't remember communicating directly with him," Hagler offers about his first interaction with Junkins, "but I was very anxious and concerned about that search. Looking back, it's not surprising to me that I would have sent a letter or email about that to John. He was emphatic and decisive in his role as committee chair, insisting that the school advance its quest for academic excellence above all other things.

"Robert Gates was a terrific president."

Several years passed before John Junkins was able to advance his Institute for Advanced Study to the full attention of Texas A&M's top administrators. After securing both funding from Chancellor John Sharp and a commitment from Norm Augustine to be chair of the external advisory board, the Institute began to take shape in 2012. Because of his support for the concept while president of Texas A&M, Ray Bowen was an obvious candidate to become a member of the external advisory board, and he quickly agreed to help. Then just as he had done to launch *Vision 2020*, Bowen reached out to his old college roommate to gauge his interest in helping get the new Institute for Advanced Study endeavor off the ground.

"My first question to Ray about becoming a member of the TIAS board," Hagler says, "was whether or not this was a project of 'substance,' basically the same thing I asked him at the beginning of *Vision 2020*. He assured me he believed it was, and when he told me that John Junkins had created the concept and would serve as its first director, I readily agreed to lend a hand."

John Junkins likes to say that he does not ask people for money, but the fact remains that he was a one-man fundraising juggernaut in the Institute's early years. That ability is based in part on his uncommonly cordial demeanor as well as a long history of writing grant proposals as an academic.

"Once I got onto the board," Hagler says, "John and I got better acquainted, and at some point he came to me with the idea that maybe I'd be the guy who could be the lead donor for the organization. We talked about that for a year or two. Initially, I was not sure I knew exactly where TIAS was headed, and I remember having the feeling that it was going to take a lot more money than I was prepared to give.

"I remember specifically one conversation with John where I said, 'You know, even something on the scale of $20 million is not going to get this job done.' He agreed, saying, 'It's going to take more than that over a long period of time. But what we need to do now is get the ball rolling. What we need is an initial donation that will ensure that it is perpetually secured.'

"After we talked several times, I realized TIAS was shaping up to represent a lot of the reasons why I had become a donor to the university. John is an exceptional individual. He's a remarkable person, a visionary, a doer, and although he may not think of himself in this way, a terrific entrepreneur. He's just a good human being. But if I were to try to give some order to my constellation of reasons for supporting the Institute for Advanced Study, the most important one is that I've had a very, very fortunate life. I did a few things right, I think, but I've also had a lot of good luck. Texas A&M has been really important in my development and in my success."

Hagler adds, "I also have this feeling about education, this strong affinity for the many benefits of a good education. The whole idea of America and the idea of American democracy has always been something that has resonated with me, probably because I was born in the middle of the Great Depression. And right after the Great Depression, while I was still in my formative years, we had the 'calamity' that became World War II, where millions and millions of people lost their lives. I look back and can still remember feeling that the world was burning down right in front of my young eyes every day.

"So in addition to wanting to give back and pay forward, I see education as a critical component of democracy. I thought that the Institute for Advanced Study could combine my love of the university and its educational mission with my desire to see it become a little less insular. That's one of the things I worry about, and my contribution to the Institute gave me a chance, at least in my mind, to become a meaningful player in saving American civilization."

On February 20, 2017, the Texas A&M Foundation received a $20 million gift to rename the Institute for Advanced Study in honor of Jon L. Hagler '58.

In its press release announcing the donation, the Texas A&M Foundation reported,

> Hagler's gift will provide endowment funding for the institute, the income of which will be used to attract and support academic visionaries from around the world. These scholars and leaders will continue to elevate the national reputation of Texas A&M. Additional funding allows the institute to continue providing a gateway for esteemed academics to encourage groundbreaking research and support students in their quest for knowledge. Hagler is contributing the gift in two parts, with the second part distributed through his estate as a planned gift. He has already funded a substantial portion of his gift.
>
> "The Hagler Institute for Advanced Study will continue to elevate the quality, impact, and reputation of Texas A&M on national and international playing fields," said Texas A&M President Michael K. Young. "Jon's gift will ensure the permanency, stability, and future growth of this prestigious endeavor."

PART 4

Profiles

20

Of Stephen and George

Rob Kennicutt, 2016–2017

Robert Kennicutt's nomination form for admission into the Hagler Institute for Advanced Study was submitted on January 28, 2016, and authored by Nicholas Suntzeff, University Distinguished Professor and head (since its establishment in 2006) of the astronomy group in the Department of Physics and Astronomy at Texas A&M.

Of his department's goals in bringing Kennicutt aboard as a Faculty Fellow, Suntzeff wrote,

> The Astronomy program at Texas A&M is less than a decade old. Our goal is to achieve a top-10 program by the first light of the Giant Magellan Telescope in 2021. Prof. Kennicutt is a titan in astrophysics. The fellowship will allow our faculty, postdocs, and students to interact with him, which will increase the quality of research through his experience, acumen, and resources. The fellowship offers another possibility to have Prof. Kennicutt become emotionally invested in Texas A&M. There is a chance we can recruit him to join our faculty, which would propel the reputation in research to new levels.

This is high praise when you consider Suntzeff's own credentials. In 1998, he cofounded the High-Z Supernova Research Team, whose work in the discovery of the accelerating expansion of the universe through observations of distant supernovae won the Nobel Prize for Physics in 2011.

Kennicutt is perhaps best known among astronomers and astrophysicists for his work concerning the empirical relation between surface gas density and star-formation rate. The pioneer in that field of study was Dutch-born American astronomer Maarten Schmidt, whose 1959 paper, "The Rate of Star Formation," led to the creation of the Schmidt law. Kennicutt's paper "The Global Schmidt Law in Star-Forming Galaxies," published in 1998, refined the work done by Schmidt, thus establishing what has come to be known as the Kennicutt–Schmidt law.

"For years, when I thought of Texas A&M, I thought of it as a football school, not an academic institution at all," Kennicutt says today with a smile on his face. "I knew the name R. C. Slocum [former Aggie football coach] but not much else about the place."

Kennicutt received both his master's and PhD degrees from the University of Washington, and during his graduate studies days in Seattle, the Huskies displayed their own football-playing prowess, beating Michigan 27–20 to win the 1978 Rose Bowl.

Astronomy, not football, is why Kennicutt enrolled at the University of Washington after receiving his undergraduate degree from Rensselaer Polytechnic Institute in 1973. Texas A&M wouldn't turn its collective eyes to the heavens for another thirty years but eventually did so in a big, big way thanks to a big, big donation from an Aggie alum with a lifelong passion for the stars.

And as it turns out, the late Texas oilman and billionaire George Mitchell, Class of 1940, also played a key role in getting Rob Kennicutt's academic attention focused squarely on Texas A&M.

Upon George Mitchell's death in 2013 at the Galveston hotel he owned and called home in the years after the passing of his wife, Cynthia, Texas A&M University System Chancellor John Sharp reflected on the man's enduring legacy:

> George Mitchell was a great example of what makes this university so special. . . . He truly loved the school and believed in paying forward to benefit the next generation of great Aggies with his time, talent, and treasure. The George P. and Cynthia Woods Mitchell Institute for Fundamental Physics and Astronomy will be a visible legacy, but his impact and influence runs deep throughout Aggieland—as Texas A&M's largest benefactor ever—and the world. The fracking technology he made available to the world will pay huge dividends, not just to Texas, but also the world. We will never forget him and his family.

Mitchell grew up in Galveston, Texas, and his life and business career may have been blessed by the fact that he did not carry his father's original surname. Savvas Paraskevopoulos emigrated to America from Greece and became "Mike Mitchell" when a railroad paymaster—potentially the inspiration for the American moniker as some stories suggest—threatened to fire him if he didn't simplify his name. Mike's youngest son, George, began his career in the oil business—which garnered him his immense fortune—at the age of seventeen, working on a Louisiana rig owned by his brother, Johnny. George then followed Johnny to Texas A&M, graduating in 1940 as valedictorian of his class with a degree in petroleum engineering. It was on a train ride from College Station to Houston after a football game a year later that George met Cynthia, a native New Yorker who came to Texas to study art.

By December 1946, George Mitchell became a partner in a small drilling company, which eventually became Mitchell Energy and Development Corporation. Six years later, Mitchell's purchase of a ten-thousand-acre oil lease in North Texas, tagged the "frustration fields" for its spotty production record, ignited the growth of the company. Boonesville Bend spawned thirteen consecutive producing wells after Mitchell gained control and became a steady source of revenue for Mitchell's company for decades.

In the 1980s and 1990s, Mitchell Energy perfected the use of hydraulic fracturing, a well-stimulation technique to elicit hard-to-extract oil and natural gas deposits. Refinement of that technology, which dates back to the 1860s, cost Mitchell Energy an estimated quarter-billion dollars. That commitment, coupled with the fortune the protocol later generated for his company, earned George Mitchell the title "Father of Fracking."

Given fracking's sometimes negative impact on the environment, Mitchell's legacy in the oil industry is not celebrated by everyone. But his largesse is beyond reproach.

In the book *George Mitchell: Fracking, Sustainability, and an Unorthodox Quest to Save the Planet,* author Loren Steffy, himself a Texas A&M graduate, wrote, "When the world's largest telescope project needed funding, he paid for it. When his alma mater needed a new physics building, he underwrote it. When his hometown languished in economic decay, he led the effort to revitalize it. When researchers wanted to build the world's largest particle accelerator, he helped them launch it. When he felt promising biotech research needed more funding to unlock breakthroughs, he set up a venture capital company to fund it. He believed that business leaders had a responsibility beyond their own balance sheets."

George and Cynthia Mitchell gave away more than $175 million during their lifetimes. Of that sum, nearly $100 million went to Texas A&M, making them, as John Sharp pointed out, the school's largest benefactors. Word of George Mitchell's investment in Texas A&M first caught the attention of Rob Kennicutt during a winter retreat in 2003 at the Aspen Center for Physics.

"It was at a before-dinner reception," Kennicutt recalls. "I was talking with a friend of mine, Wendy Freedman, a very well-known astronomer with whom I had collaborated. She had spent some time that fall in College Station chairing a committee to help design a future role for Texas A&M in deep-space studies. She told me that a wealthy and very prominent alumnus sat in on the proceedings. And that person was George Mitchell."

Little did Kennicutt know at the time that the same George Mitchell had been a frequent visitor himself to Aspen Center for Physics events. After purchasing much of the land at the base of Aspen Mountain, Mitchell launched his Aspen Alps development there in 1963. Mitchell and his family loved Aspen and frequently spent time in the Rocky Mountain enclave. As a man of wide-ranging interests, Mitchell would have been keenly curious regarding some of the activities that went on at the Aspen Center for Physics. And given his prominence in the local community, he would have always been a welcomed guest.

By the time of their meeting in 2003, George Mitchell and Wendy Freedman were already well acquainted. Earlier that year Freedman, a Canadian-born-and-educated astronomer, had been named director of the Carnegie Observatories in Pasadena, California. And not long after taking that position, Freedman launched a quest to develop the Giant Magellan Telescope (GMT) as a means to advance the scientific search for extraterrestrial life. That project caught the attention of Mitchell, and he and his wife, both astronomy buffs, soon became the lead donors to the GMT through a $1.25 million gift given to Texas A&M and specifically earmarked for Freedman's venture. That came on the heels of the founding of the George P. and Cynthia W. Mitchell Institute for Fundamental Physics at Texas A&M in 2002. In fact, the Freedman-led conference to further determine Texas A&M's role in the GMT was held in a conference room of the newly designated Mitchell Institute.

Kennicutt laughs at the memory of when George Mitchell first appeared on his radar screen: "Now that I know how Texas A&M makes use of its money to make more money, I understand exactly what Mitchell was doing with the school twenty years ago . . . and he did it incredibly well! George was intent on leveraging his investment in the GMT as a way to build an astronomy program at Texas A&M. And given his deep pockets, George soon got what he wanted."

As a boy, George Mitchell promised his mother, Katina, he would become a doctor, but after building his own telescope while a student at Galveston Ball High School, he wondered if he should pursue the field of astrophysics instead. Although he didn't—citing an even stronger desire to make money—his interest in the heavens led to his commitment to the GMT and his positioning of his alma mater as a founding partner in the endeavor along with the likes of Harvard University, the University of Chicago, the University of Arizona . . . and the University of Texas.

In fact, Mitchell's initial GMT donation was contingent on UT's involvement. In 2011, Mitchell enabled Texas A&M to make an additional $25 million donation to the project.

"This gift not only brings the dream of the Giant Magellan Telescope much closer to becoming reality, but also helps propel Texas A&M and the entire state of Texas to the forefront in the important fields of physics and astronomy," said then–Texas A&M Pres. R. Bowen Loftin in announcing Mitchell's gift. "We are extremely grateful to Mr. Mitchell for his profound vision and his significant investment in our shared future."

In addition to their initial endowment to launch the Mitchell Institute in 2002, George and Cynthia made several more substantial contributions to bolster the school's reputation in the fields of scientific study. In 2005, the Mitchell Foundation pledged $25 million for the construction of the George P. Mitchell '40 Physics Building. George and Cynthia threw in another $10 million to accommodate project-cost overruns incurred as a result of the use of Cynthia Mitchell's favorite architect, Michael Graves, whom the Mitchells had handpicked for the project.

Kennicutt's path would eventually cross that of George Mitchell in 2006 as a result of their mutual relationships with one of the great scientific minds of modern times.

After earning his PhD from the University of Washington, Kennicutt became a postdoctoral fellow at the Carnegie Institute's Hale Observatories. Then following a nine-year stay at the University of Minnesota as an assistant and associate professor, Kennicutt moved to the University of Arizona, where he became an associate professor/astronomer at the school's Steward Observatory in Tucson. In 1992, he became a full professor at Arizona and remained at the school until 2005.

As a result of Kennicutt's established and respected reputation in the field of astronomical study, in 2005 he was appointed as the Plumian chair of Astronomy and Experimental Philosophy at the University of Cambridge in England. This highly coveted position was founded by Christs College astronomer Thomas Plume in 1704 and has been held by a number of prominent astronomers, including Roger Cotes, a student of Isaac Newton; George Biddell Airy, for nearly fifty years the Astronomer Royal to the king of England; and George Darwin, son of Charles Darwin, the famed English naturalist, biologist, and geologist.

Another of Cambridge's best and brightest at the time of Kennicutt's arrival in London was the Lucasian professor of mathematics from 1979 to 2009 and director of research at the Centre for Theoretical Cosmology, Stephen Hawking. He is best known for his study of the physics of black holes and for the progressively disabling effects of his amyotrophic lateral sclerosis, an incurable degenerative neuromuscular disease that ravaged his body but not his singularly gifted mind.

Soon after his arrival in Cambridge in 2005, Kennicutt learned that Stephen Hawking had a close relationship with the same Texas oilman who had helped fund Freedman's telescope. How Hawking and George Mitchell came to be close friends is detailed by *Houston Chronicle* reporter Eric Berger in a story he wrote for the paper shortly after Mitchell's death.

George Mitchell had an epiphany as he watched PBS a decade ago.

There was Stephen Hawking, the world's most famous living scientist, being asked about his greatest disappointment. His answer? The U.S. government's failure to complete construction of the Superconducting Super Collider, near Waco, after spending $2 billion.

Mitchell, whose wealth played a key role in persuading the United States to build the collider, could relate.

So as Mitchell watched a physicist whom he idolized, awakening within him was a renewed desire to plumb mysteries at the edge of modern physics. His deep pockets and enthusiasm helped Texas A&M University launch an astronomy program and gave the state a stake in the construction of the world's largest telescope, and his eventual friendship with Hawking elevated the profile of Texas among the exclusive club of cosmologists, some of the world's most brilliant thinkers who use mathematics to try to understand the earliest moments of the universe.

When A&M physicist Peter McIntyre called him out of the blue in 1983, Mitchell listened as the timid caller on the other end of the line explained about quarks, and superconducting magnets, and 100-mile tunnels.

"I was expecting that at any moment I would hear a click on the other end of the line," McIntyre recalled.

But Mitchell didn't hang up, and after a face-to-face meeting he gave McIntyre $1.5 million to move forward with a proposal to build a large collider in Texas to probe the most fundamental particles in nature. The federal government later agreed to build the massive collider before canceling the project in 1993 due to rising costs.

Mitchell was bitter, which is why Hawking's comments in 2002 resonated with him. After that TV program, he phoned McIntyre to talk about it, at which point McIntyre suggested they go and meet the great physicist, then visiting the California Institute of Technology, in person.

Hawking, who was skeptical of oilmen and had taken some delight in the fall of the energy company Enron, at first was uncertain about meeting Mitchell. But after the initial face-to-face meeting in 2002, the relationship quickly warmed.

Hawking had visited Texas once, in 1995, at the invitation of Ross Perot. During that trip he had stopped to visit Chris Pope, a physicist at A&M who had trained under Hawking.

Under an agreement between A&M and Cambridge University, where Hawking taught, the physicist would visit Texas more than half a dozen times after the 2002 meeting in California, giving several sold-out lectures [at Texas A&M]. Hawking came to love the 5,650-acre ranch owned by Mitchell in Montgomery County, Cook's Branch Conservancy, where he and other elite physicists could gather in private and natural beauty to ponder the universe.

Mitchell's last gift to A&M, in fact, was $20 million to ensure such meetings would continue.

"I first met George Mitchell in 2007, about two years after I moved to Cambridge University," Kennicutt says. "In the years to follow, I had the opportunity to visit Cook's Branch, initially for reasons I didn't fully understand. I'm an observational astronomer. I don't do theory equations, which is what Stephen and his kind mostly discussed at the Mitchell-family getaway."

Kennicutt says he first met Hawking—in his time, and despite his illness, a scientific rock star on par with the Rolling Stones' Mick Jagger—soon after he assumed the Plumian chair at Cambridge.

"He came to my inaugural lecture," Kennicutt recalls. "I paid homage to him in my remarks, grateful to have him there. Communication with Stephen, because of his condition, was a one-way street. He was physically unable to talk, but as far as incoming messages were concerned, he was entirely alert."

Hawking's "voice" from the 1980s until his death in 2018 was the product of computer-generated text-to-speech algorithms developed by the MIT scientist Dennis Klatt. In fact, it was Klatt's voice that people heard as Hawking's. Despite later advances in technology, Hawking remained loyal to Klatt's voice generator, which is closely tied in the public consciousness with his continuing scientific achievements.

Being in the room with Hawking, Kennicutt says, meant accepting the time lag required for him to fully communicate.

"The first time I saw this was at a big dinner held at Trinity College in London," Kennicutt says. "Sitting at the table with him there, he was still capable of facial gestures that gave you an idea of how he felt about the ebb and flow of the conversation around him. When he heard a clever remark that caught his attention, he would start pecking at the computer keyboard that was his ever-present companion. The temptation was to stop and wait, but those who knew Stephen encouraged everyone to continue the conversation. Then maybe ten or twenty minutes later, from completely out of the blue, he would engage his voice generator and share his very clever response to something that had been said much earlier.

"People got used to that. It was an accommodation you made."

Imagine a world in which Stephen Hawking existed—and thrived—alongside the recent advances made in generative AI technology. Who knows where that would have taken his scientific thinking . . . and global stardom?

Interplanetary dimensions, perhaps.

"Seeing Stephen and George together," Kennicutt recalls, "reminded me of a doting father and son. George's daughter, Sheridan, liked to remind people that in his later years, her father was half deaf, while Stephen couldn't talk. Yet that didn't stop them from fostering their remarkable and important connection."

Three years into his time at Cambridge, Kennicutt became director of the school's Institute of Astronomy. Three years after that, in 2012, he was appointed head of the school of physical sciences, a post he held for three years. He retired in 2017 on the precipice of Cambridge University's enforced mandatory retirement at age sixty-seven, a policy that exists, according to Kennicutt, to ensure the faculty does not become "fossilized."

"I retired a year before that, when I was sixty-six," Kennicutt says, "for reasons having to do with taxes over there and Medicare over here.

"I was pretty keen to move back to Arizona, where I had raised a family. The University of Arizona gave me a desk but no job at the time. When Nick Suntzeff contacted me to tell me about the Hagler Institute, I told him I would be interested."

Thus culminating a lengthy courtship.

As far back as 2010, during a trip to Cook's Branch, Texas A&M had made overtures to Kennicutt about becoming director of the Mitchell Institute for Fundamental Physics and Astronomy.

"At the time, I was just five years into my job at Cambridge," Kennicutt says. "I had just taken the directorship of the department and the observatory, so I was content with my work there."

Typically, Hagler Faculty Fellows spend an academic year on campus, the equivalent of nine months. As Kennicutt was wrapping up his duties at Cambridge, he knew he couldn't make that kind of time commitment to Texas A&M. So he asked if he could do his fellowship in two-week intervals. Given Kennicutt's reputation in the astronomy world, that was a deal Hagler Institute Director John Junkins was willing to make.

Anything to finally bring Rob Kennicutt to College Station.

"It was not obvious to me at the time, but I should have figured it out," Kennicutt says, laughing. "Hagler Fellowships are 'honey traps.' The whole idea is to get you on campus and hope you fall in love with the environment and the incredible resources you're offered. They'll let you percolate in all this 'sweetness' for about six months or a year, then they hit you with the Chancellor's Research Initiative, and offer you a package that can be well into the millions of dollars of pay, facilities, and personnel."

That full-court press to bring Kennicutt to Texas A&M as a Permanent Hagler Fellow started almost immediately after he began his initial fellowship.

"Back in London, I joked with my colleagues that becoming a Hagler Fellow made me feel a little like David Beckham," Kennicutt says. Near the end of his soccer-playing career in 2009, the London-born Beckham shocked the global sports community when he left the prestige of playing with European powerhouse Real Madrid to join the ranks of the Major League Soccer's Los Angeles Galaxy . . . for a boatload of money.

"I felt like a footballer who's no longer good enough to play in the Premier League," Kennicutt explains, "then discovers there's still a lot of money to be had from the mere value of my name."

Kennicutt accepted the CRI offer with one stipulation. He would not work full-time at Texas A&M. For his five years as Mitchell Institute director, he's split his duties between College Station and Tucson. In fact, as he admits, he's contractually obligated to spend half his time at the University of Arizona.

That's nothing out of the ordinary for academics the caliber of a Rob Kennicutt.

When asked about his "biggest contributions" as a member of the Texas A&M faculty, Kennicutt offers, "I really think that it is up to others to judge, but a few things do come to mind. I have a very high national and international profile in my subject. I am a member of the National Academy of Sciences and also a fellow of the Royal Society in the UK, and served as co-chair of the NAS 2020 Decadal Survey of Astronomy and Astrophysics, where national priorities were set for the coming decade in astronomy as well as federal agency spending priorities [amounting to tens of billions of dollars collectively]. The visibility and reputation of TAMU astronomy and physics have benefited considerably from my presence here.

"I was given an embarrassingly large start-up package of funding from the chancellor, and I have used most of it to join projects and hire postdoctoral researchers who work with the astronomy group as a whole—as opposed to all of them working with me. This has had a visible impact on the research and career trajectories of the early to mid-career members of our group, far wiser than squandering all of it on a seventysomething like me. One result of that, not all due to my efforts, is that we have hired four new faculty members to the group—five if you count me—out of a total faculty of ten full-time employees.

"The program of postdoc hires has proven to be a resounding success. To date we have hired about a dozen individuals, who have published more than 120 papers since we began about four years ago. They have energized the research of our junior faculty. Most importantly, they have enlivened the academic atmosphere in the building—now considerably outnumbering the faculty—and bolstered our scientific output.

"For years the Mitchell Institute operated without a permanent director of national standing, and there was considerable friction between the donors and the university in that regard. My appointment as executive director solved that problem, and since I took the job, we established an advisory board and formalized reporting lines within the university without disrupting the day-to-day management of the Institute, which was functioning fine before I arrived. We continue to be the most productive unit within the department, with a steady stream of high-profile discoveries and a strong funding profile.

"Virtually every combined Physics and Astronomy Department I know suffers friction and competition for resources between the physicists and astronomers, and we are no exception. I have made a point from the beginning to establish strong communication with faculty across the department and with both departments' heads under whom I have served. This has paid off with smoother relations and stronger support from physics faculty and department leadership for our growth and needs.

"More widely within the university I have become increasingly involved in campus-wide activities, partly the result of my Hagler ties and more so since my appointment as a University Distinguished Professor. I serve on the Faculty Advisory Board for the Hagler Institute. One of my proudest achievements was to float the idea of cross-campus gatherings of academics to help break down the 'stovepipes' separating departments and colleges. I used the collegiate structure at Cambridge and Oxford as examples. This was taken up by John Junkins, Jorge Steiner, and others to initiate two activities: biweekly lunches of the permanent Hagler Fellows and visiting Fellows and once-per-semester 'Community of Scholars' evening events attracting high-ranking academics, administrators, and their early career 'protégés' from across campus to meet each other and network.

"I doubt whether anyone else remembers where those ideas were hatched, but who cares—both are alive and flourishing."

21

"Marketing Guru"

V Kumar, 2016–2017

Growing up with only one name and an initial is fairly common in many parts of India, particularly in Tamil Nadu, where Hagler Institute Fellow V Kumar was born and raised. The practice was originally adopted to dispel telltale signs of caste associations. The caste system has existed in some form in India for at least three thousand years. It is a social hierarchy passed down through families that can dictate the professions a person can work in as well as impact aspects of their social lives, including whom they can marry.

Such, though, is not the reason Kumar does not have both a first and last name. It's more a matter of tradition and the belief in his family that all are equal.

"I was born into a Hindu Brahmin family," Kumar says. "Although my family never focused on that. I do have a very humble background."

Kumar's use of the first initial "V" is taken from his father's name, Viswanathan.

"My parents were both office workers," Kumar says. "My dad in railways and my mom in telephones. Growing up, my family did not have much money. We fell into a lower socioeconomic status, and therefore high-quality education was a luxury. Yet my parents encouraged me to pursue education to advance myself, as they believed that knowledge is both power and wealth."

Kumar started his formal schooling at just two-and-a-half years of age.

"I was given opportunities. People would say, 'I will give you a chance. Do you want to take it?' And I took every chance I was given. Hard work took care of the rest."

By the time he reached college, Kumar had set his sights on obtaining a degree in engineering. He attended one of the five campuses of the state-managed Indian Institutes of Technology (IIT), first majoring in metallurgical engineering, which specializes in the extraction of metals, formation of alloys, and casting of products. He then obtained his master's degree in industrial management, also from one of the IITs.

During his time in college, something profound occurred that changed the future course of Kumar's life.

"I had an opportunity to listen to a lecture by Dr. Philip Kotler, who is known around the world as the 'father of modern marketing,' when he visited India," Kumar says. "This was maybe forty years ago. Kotler told us that the marketing profession needed more science because at the time marketing was primarily considered an art. He told us firms were recruiting engineers to work in their marketing departments to build statistical models to answer questions such as 'If one spends X dollars on advertising, how much will that impact sales?' or 'If you increase distribution by 20 percent, how much will sales go up?' He said companies desperately wanted answers to questions like that.

"I was good in math and statistics. So I immediately decided marketing [would] be my area in which to pursue my doctoral education."

Kotler taught marketing at the Northwestern University Kellogg School of Management in the Chicago area for most of his career. Thus, Kumar set his sights on obtaining his PhD in marketing somewhere in the United States. That place turned out to be the University of Texas at Austin.

"I received a full scholarship there," Kumar says. "The program was ranked among the top ten in the country."

Among Kumar's many, many professional honors, awards, and distinctions, two are of particular note. He was conferred the title of "Marketing Guru" by the Indian Institute of Technology Alumni Association in 2006. And in 2012, he was recognized as a "Marketing Legend" by Dr. Jagdish Sheth for publishing his work in ten different volumes with commentaries from scholars worldwide.

Apparently, no grudges were held at IIT when Kumar left the engineering field.

A marketing guru the man truly is, as evidenced by many of his other professional accolades:

- Received "Bharat Gaurav Award," and "Hind Rattan Award" from the Union Ministers of the Government of India in 2009
- Conferred with "Pravasi Achievers Gold Medal" at the House of Lords, UK, in 2010; and "Glory of India Award," UK, 2010
- Conferred with "Best Professor in Marketing" Trophy and certificate by the Asia's Best Business School Awards Program organized by the CMOAsia.org in Singapore 2010 (Repeat Winner—Kuala Lumpur 2013)
- Winner of the 2012 Paul D. Converse Award (For Advancing the Field of Marketing)
- Editor-in-Chief of the Journal of Marketing, April 1, 2014–June 30, 2018
- Serving on Editorial Review Boards of twelve scholarly journals
- Held several Distinguished Faculty positions for many years in the United States: at the University of Houston, the University of Connecticut, Georgia State University, and St. John's University.
- Recognized as a Distinguished Fellow, MICA, Ahmedabad, India
- Recognized as a Fellow, Hagler Institute for Advanced Study, Texas A&M University, College Station, Texas
- Recognized as a Fellow of the American Marketing Association (AMA)
- Recognized with the highest honor from the Ministry of Education, China, as the Chang Jiang Scholar, for the period 2013–present
- 2019 AMA Higher Education SIG's Lifetime Achievement Award
- Recipient of the 2022 AMA Foundation's Charles Coolidge Parlin Award (oldest and most distinguished in marketing)
- Selected as a "Marketing Legend" and featured in the "Legends in Marketing" series, Jagdish Sheth (editor) as one of the Eight Legends along with Phil Kotler, Paul Green, Jerry Zaltman, Jagdish Sheth, and Jerry Wind

- Created over $2 billion in incremental revenue through implementation of marketing science models and methods in the private business sector

In addition to having published over three hundred articles in journal papers, conference proceedings, research reports, book chapters, and case studies, Kumar has also taught more than two dozen courses worldwide across bachelor's, master's, and doctoral programs. He lays claim to having lived in sixty-two different countries.

"And by 'living,'" he laughs, "I mean that I've spent at least a couple of weeks there."

At the time of his nomination to the Texas A&M Institute for Advanced Study in 2016, Kumar was a member of the faculty at Georgia State University in Atlanta. Recently, Kumar became a professor of marketing in the Goodman School of Business at Brock University, St. Catherines, Ontario, Canada, helping to launch doctoral and executive doctoral programs in business there while also setting up various academic-industry collaborations.

After receiving his PhD in marketing from the University of Texas in the mid-1980s, Kumar began his academic journey by looking for a teaching job at one of America's top business schools.

"I interviewed at several of the top schools, including Northwestern, Stanford, and the Wharton School at the University of Pennsylvania," Kumar says. "But I chose the school where I could get a lot of data and time to solve business problems. That school was the University of Iowa, and here's why I went there: Back in those days, there were two companies that collected information about what consumers bought, how much they bought, and how much they paid. We call that 'scanner-panel data.' One of those companies was AC Nielsen, and the other was called IRI, Information Resources Inc. IRI cofounder Jerry Eskin was a marketing professor at the University of Iowa. He told me, 'VK, if you join us here at Iowa, I'll give you all the data of all the brands over a several-year period. Would that be of interest to you?'

"That was a gold mine of information, so I joined Eskin at Iowa and eagerly looked forward to my work there. Unfortunately, I could not breathe the air in Iowa City because of pollen from all the corn grown in that part of the state. In fact, my sinus condition got so bad, at one point I had to be hospitalized.

"Jerry was very understanding, and ultimately told me, 'You graduated from Texas, and I think you should go back to Texas where you felt a little more comfortable.' When I reached out to my PhD adviser, Bob Leone, he said that the University of Houston was still interested in me joining them. And so I returned to Texas."

In due time, Kumar was given tenure at Houston. Then, in one of the defining moments of his career, he began a collaboration with two of the most respected scholars in their fields to publish a seminal textbook that is still going strong into its fourteenth edition thirty years later.

"David Aaker was a branding pioneer from the University of California at Berkeley," Kumar says, "and George Day was a strategy guru from the Wharton School at Penn. They approached me about being a coauthor on their book, *Marketing Research*. 'Given your expertise on the methods and applications side,' they told me, 'you would bring in a new perspective to our work.' So I joined them for their fourth edition, which came out in 1995. And we've been updating the book ever since."

Kumar points out the recent addition of cocontributor Robert Leone from Texas Christian University to the publication.

"It's one of the standard marketing research books that is used at all the major schools."

Kumar may very well be the most prolific Hagler Fellow in the program's dozen-plus years of existence, at least when measured by research papers published while involved with the program: He lists fifty-one. Ten were in collaboration with Amalesh Sharma, Kumar's former doctoral student at Georgia State, who in 2017—during Kumar's Hagler residency—was hired as an associate professor of marketing at the Texas A&M University Mays Business School.

One of their papers, "Investigating the Influence of Characteristics of New Product Introduction Process on Firm Value: The Case of the Pharmaceutical Industry," was featured in a *Harvard Business Review* story published in July 2018:

> Early this fall, if tradition holds, Apple will introduce one or more new iPhones—an unveiling that's among the year's biggest events in consumer electronics. The smartphone helped make Apple the world's most valuable company, even though Samsung and other

rivals introduce new products much more frequently. That paradox led V. Kumar, a marketing professor at Georgia State University, and his colleagues Amalesh Sharma and Alok Saboo, to wonder: If a company wants to maximize shareholder value, what's the optimal number of new products to launch in a given time frame? Does it matter whether the launches are spread out or bunched together, and whether a new product is similar to the rest of the company's current product portfolio?

Managers don't need an academic study to recognize that launches take a toll on many parts of a company, from design and development to manufacturing and marketing. On the basis of previous research into how companies can quickly incorporate learning from product launches, Kumar's team believed its questions involved more than just costs and resource constraints. "Firms introducing products at a rapid pace have little time to evaluate their products, learn from them, assimilate their experiences, and deploy them to commercial ends," they write. In theory, optimal pacing allows firms to use the lessons from one launch to improve subsequent ones. And if that's true, the researchers believed, they could prove it empirically.

To do so, they looked to the pharmaceutical industry, where new products are especially important to growth in revenue and market value. Using various databases and studying 73 publicly traded U.S. firms from 1991 to 2015, they identified when each of 1,904 new drugs was introduced. They then calculated the pace of the launches (the average number of products introduced over a period of time) along with the irregularity in pacing (the variance in timing between launches). They also looked at whether each new drug fit into a therapeutic class and treated a specific ailment already represented in the company's product portfolio or whether it was outside the firm's existing scope. They gathered data on company stock prices and compared the returns to industry benchmarks.

The results largely confirmed the researchers' hunches. Firms that launched many new products saw their increase in value diminish over time, as did those introducing products just loosely related to their current offerings. Companies whose launches came at irregular intervals did worse than the industry average: They

> saw their market value fall, and the drop was greater in the case of complex products and for firms with large R&D budgets relative to their marketing budgets (a high ratio of R&D to marketing may signal that a firm is more focused on innovation than on sales). "Our results indicate that there is an optimal level of pace and scope of product introductions that managers must consider," the researchers write. "Managers need to spend time learning from the products they have already introduced and incorporate these insights into their subsequent products."
>
> Kumar recognizes that managers face numerous pressures—from investors, customers, the media, and competitors—to introduce products faster. But he says the perceived need for speed is often misguided. "Our study highlights the importance of looking at the entire portfolio instead of focusing only on the next product," he says. It's also a mistake to focus too much on when competitors will launch products. Kumar likens companies that launch a product quickly in the hope of beating competitors to investors who try to time the market—which, research has shown, usually backfires. "There's something to be said for spacing out launches," he says. "You need to make sure you've learned enough from the last one and that you're not constrained by lack of resources."

Looking back in wonderment at his good fortune in the marketing world and all that he has accomplished, Kumar points out the similarities between his own life and that of the principle protagonist depicted in the Academy Award–winning film *Slumdog Millionaire*. In the movie, the title character, an eighteen-year-old who survives the mean streets of Mumbai, India, to gain a spot on the Indian version of the TV series *Who Wants to Be a Millionaire*, ultimately overcomes death threats and accusations of cheating to win the show's grand prize.

Be sure to look up the video clip of the movie's end-credits dance scene, "*Jai Ho!*"

That level of "grand prize" in life came for V Kumar when he learned the "father of modern marketing" had been watching his career.

"About twenty years ago, Philip Kotler listed a handful of people whom he thought were candidates to assume his mantle," Kumar says. "Mine was one of the names he mentioned. Then just recently, he offered that I'm the only one of those people who is still active in the field."

And it was with that in mind, no doubt, that Kotler agreed to collaborate on Kumar's most recent book—his thirty-fifth—published in July 2024 and entitled *Transformative Marketing: Combining New Age Technologies and Human Insights.*

The publisher, Springer Nature, says,

> This book gives an indispensable guide to navigating the shift in customer behavior and discovers how to rally their resources, cultivate capabilities, and forge strategies that harness cutting-edge technologies. In today's tech-centric world, customers crave lightning-fast digital experiences and demand instant solutions. In response, firms are changing the way they do business by accelerating the application of new age technologies, revamping processes, building new organizational structures, and innovating new business models.
>
> The authors unveil the secrets of integrating diverse data sources, principles of Marketing 5.0 and employing advanced techniques to unearth profound insights about the customers.
>
> This work is the ticket to the latest in AI, machine learning, drones, and other game-changing technologies. Stay ahead of the curve by learning not just what tech to use, but how, when, and why to deploy it in this digital age.
>
> For the trailblazers with the influence and resources to reshape marketing strategies, this book is the essential read. Executives climbing the corporate ladder will find it a compass, unraveling how new age technologies dance with both traditional and emerging marketing practices. And for MBA students hungry for insights on navigating the digital era's competitive landscape, this book is the treasure trove of tools and real-world cases. Dive in and chart the course in the tech-driven marketing landscape!

"We took an approach where every company in the world will have to transform themselves adopting the marketing practices we describe," Kumar says of his collaboration with Kotler. "We addressed how marketing would look using new-age technology tools like artificial intelligence, blockchains, robotics, and drones. Our belief is that if you don't adopt these new advances, your business will become irrelevant."

Concerning the corporate world, Kumar knows about what he writes. Through the years, he has provided thought leadership to numerous major companies. One experience, though, stands out for him above the rest.

"In 2006, Procter and Gamble—known more commonly as 'P&G'—issued a call for proposals," Kumar says, "to build a model using datasets on two of their detergent products in Asia to enable them to determine the right pricing and distribution strategy for the market. I was surprised because P&G normally worked with companies like McKinsey and Boston Consulting Group, but this time, they were putting the project out for bids to a broader audience. I entered the competition with one of my doctoral students at the University of Connecticut, and we were shortlisted and asked to come to corporate headquarters in Cincinnati to make our presentation."

And they got the job.

"The market turned out to be India," Kumar goes on to say. "They gave us all the data not only on their two brands but for competing brands as well. Previous consultants had told them to lower their price, but based on the model we built, we found the market to be 'inelastic,' which meant that our research suggested they needed to raise the price. We told them that the price of each package size should be raised by different amounts, ranging from 10 to 20 percent.

"One year later after making our suggested changes, they made $43 million in the detergent category with their two brands, Tide and Ariel, and the study was published in *Marketing Science* magazine."

Perhaps the best way to summarize the remarkable career of V Kumar can be found on his own website, www.drvkumar.com. The words may be his, but as a marketer presenting the facts, he does an outstanding job highlighting his "points of differentiation" in both the academic and "real" worlds.

> Possibly the simplest yardstick of Dr. Kumar's productivity within the field of Marketing and Business would be his contributions in terms of academic publications within the domain, stemming from his ability to generate a continual stream of research ideas, newer methodologies and techniques, strategies, and solutions.

Dr. Kumar has published over 300 articles in leading academic journals including the *Harvard Business Review*, *Journal of Marketing*, *Management Science*, *Marketing Science*, *Journal of Marketing Research*, *Operations Research*, *Sloan Management Review*, *California Management Review*, *Journal of the Academy of Marketing Science*, *Journal of Advertising Research*, *Journal of Business Research*, *Journal of Retailing*, *International Journal of Forecasting*, *International Journal of Research in Marketing*, *Journal of Direct Marketing*, *Marketing Letters*, and the *Journal of International Marketing*, and in many other international journals such the *European Journal of Marketing*, and the *Journal of International Business Studies*, as well as in the proceedings of many national conferences in the US and abroad, books and book chapters, research reports, and case studies.

Over the last two decades, Dr. Kumar's research on newer, more innovative and efficient methodologies has resulted in a rich accumulation of marketing techniques and research frameworks. In this regard, Dr. Kumar's ever-growing list of contributions to Marketing has been extremely influential in fostering a number of much-needed paradigm shifts. Further, through his research findings and the theories that often break the conventional logic and their subsequent implementation in different industries, Dr. Kumar has enabled multi-million-dollar gains for many firms, as well as advancing the thought leadership in this area for the academic world.

And the "Marketing Guru" continues his quest to change the world for the better.

22

"My Story Goes Something like This"

James Hubbard, 2017–2018

James Hubbard pulls no punches. He calls it like he sees it. And he relishes losing sleep over challenges in the laboratory.

To spend time with Hubbard is to encounter a man of honesty, humor, and—as the first Black person to receive a PhD in mechanical engineering from the Massachusetts Institute of Technology, a cofounder of four companies, a holder of twenty-four US and global patents, and someone recognized as a pioneer in the field of adaptive structures—an enduring spirit.

He's also a heckuva storyteller. For that reason, one is compelled to let James Hubbard tell his own story.

• • •

Everywhere I've been, most of the people didn't want me there. Most of the people thought I was an affirmative-action hire. When I did good in engineering, they would say, "Well, you just got lucky."

But Texas A&M is different.

I will tell you this: I've been everywhere in the world, and I've done everything. I even made it past the first cut to become an astronaut, in NASA Astronaut Candidate Group 10. You can look it up at SpaceFacts.de.

But of all the places I've been, I've never seen a place like Texas A&M.

It's just wonderful.

After I first got here, people would come up to me on the street and just start talking to me. When they found out what I did, they would say, "Thank you so much for coming to College Station." At first, my response would be, "Hey, don't come walkin' up on me like that because where I come from, that means you're probably trying to rob me!"

But now, six years later, I'm just as friendly as they are.

When I walk across campus now, the squirrels come up to me saying, "Oh my God, Jim! How you doing? It's so nice to see you!" And I'm like, "I'm all right, squirrel. How you doing? What's shaking?" "Oh, Jim, ain't nothing shaking but the leaves and the trees, man. You know?"

Everybody in this town is so happy that I'm here. What the hell is that about?

Well, now I know why. I'm the only one like me out of thousands of faculty at Texas A&M. There's no other Black National Academy members here, but that's OK.

I'm going to do my best to change that.

The only thing I ever wanted in my life was freedom. And that's always been nonnegotiable. I'll come to your place, but I gotta be free.

Before they hired me at Draper Laboratory back in the mid-'80s, early in my career, my boss there, Don Fraser, the executive vice president, said, "Jim, I got twenty-two hundred engineers here. I can't have two of 'you' here. It'll be too much, too much trouble. But I do need one because you are a guy that likes to play in the future, and we need somebody here like that."

By "you," he wasn't alluding to the color of my skin but the tenacity of my mind.

As for what brought me to Texas A&M University to become a Fellow of the Hagler Institute for Advanced Study . . . I'm glad you asked.

My story goes something like this.

I was giving a keynote address at an SPIE conference back in 2016. SPIE is the International Society for Optics and Photonics.

First, let me be clear about something. I don't like speaking in front of people. I actually prefer to be on a desert island alone, but given what I do, that's not going to happen. I have this rule that, first thing I do when they invite me to speak at a conference, I say no. And then when they press me or bring in somebody like John Junkins who says, "Come on, Jim!" I tell them, "I'll do it, but I'm going to present a topic so controversial and scary the attendees are going to sleep with the lights on at night."

I gave that kind of talk at the SPIE conference. What I spoke about was the fact that there are a lot of people practicing in the field of smart structures today, but none of them know the history. Only like two or three people know that. So I decided to give a talk on the history of smart structures. Was it truly scary? Probably not, but maybe it was to some.

After I finished my talk, I was eager to get home. My wife had recently suffered a stroke, and I was very concerned about her. As I'm getting ready to leave the conference, my phone rings. It was Kathy Banks, who I had never heard of before, from Texas A&M, which I had barely heard of.

At the time Kathy Banks was vice-chancellor and dean of the College of Engineering at Texas A&M. She later became president of the university. But when Kathy first called me, she started telling me that she wanted to offer me a distinguished position as a Hagler Fellow at the Hagler Institute. I'd never heard of that, so I didn't know what it was. She said, "Dr. Hubbard, you'll be given a generous financial stipend, and over a three-year period you can come here anytime you want and do anything you want."

I cut her short.

"Ma'am, this sounds like a part-time position," I told her, "and right now, I'm not looking for a part-time position."

The phone line got very quiet. I wasn't looking for a job or position or anything and what she was suggesting made me uncomfortable. I was at the University of Maryland at the time in the aerospace engineering department and also doing work for the National Institute of Aerospace as a Langley Distinguished Professor. All I wanted when Kathy Banks called me was to get home and take care of my wife.

But Kathy wasn't going to take no for an answer. She said, "Why don't I have somebody email you and explain the position?" And so, N. K. Anand, the dean of faculty, wrote me a long email. I'd never heard of him, either. It turned out he was Kathy's right-hand man.

Didn't read his email, and they didn't hear back from me.

Then John Junkins called me.

He said, "Jim . . . blah, blah, blah." But then I realized John Junkins, the founding director of the Hagler Institute, was somebody I had heard of and did know.

I first met John back in the eighties. Every time I would see him, I would go over and greet him. Back in those days we had honor and respect, and we didn't hesitate to show it. So I listened more closely to what he had to tell me when he called.

"Jim, why don't you just come out here and look around?"

"John, I appreciate the offer, but I've got other problems. Just not interested right now." And then he said something that would change my life.

"Jim, Herb wants to have dinner with you."

I have had three mentors in my life, and Herb Richardson was the first. I was a Black boy who grew up on the mean streets of Baltimore, and Herb Richardson took a chance on me. I'd tried to reach Herb so many times through the years, but after he left MIT, I lost track of him. And now, John Junkins was telling me that Herb wanted to have dinner.

"What did you say?" I asked to make sure I was hearing John right.

"Herb Richardson is here now," Junkins said. "He rose through the ranks and became chancellor of the Texas A&M University System. When he found out you weren't interested in talking with us about being a Hagler Fellow, he told me to tell you he'd like to have dinner with you."

"John," I replied, "I'll fly down the night before. I'll have dinner the next day and then I'll fly right back home. That's the best I can do right now."

So I flew to Texas, got to the restaurant, and encountered a group of about ten people who seemed really glad to see me. None of them was Herb.

I started sweating. These people were all trying to talk to me at once, telling me all the great things about Texas A&M. And I didn't hear a word. I was just wondering where Herb was. I kept asking, and they kept telling me, "Oh, he'll be here." And then finally, in the middle of it all, Herb Richardson walked in. I stood up and walked over to him. Before he could even say hello, I grabbed and hugged him, and I started crying like a baby. I couldn't stop crying in front of all those people.

When I finally regained my composure, I said, "Herb, I did good! I turned out OK!" Of course, by then, he already knew that, given that I was being invited to become a Hagler Fellow.

Over dinner, Herb told me more about the Hagler Institute. When he finished, I replied, "I don't think I'm really interested in doing anything like this right now." He said, "Jim, I want you here. I think you'll be good for Texas A&M, and Texas A&M will be good for you."

That was it. I told him right then and there that I would come.

Later, Dimitris Lagoudas, who as senior associate dean for research at A&M became a good friend, jokingly asked me, "Do you do everything Herb Richardson tells you to do?"

I said to Dimitris, "Absolutely!" and I wasn't joking. "I do everything Herb Richardson tells me to do,"

Now, let me explain why.

I grew up in the Deep South, a place called Danville, Virginia. Danville was the last Confederate capital during the Civil War years. What I experienced in my youth there, almost a century after the Civil War, was "American apartheid." My mom, Lillie, was a civil rights worker. In that capacity, she was beaten, she was jailed, and eventually my family was torn apart. As a result of her "civil disobedience," Mom was locked up for a year. During that time, my sister went to live with relatives, and I went to live with my grandmother.

I didn't see my mother for a year. When I finally did, while she was still in jail, she had lost more than thirty pounds. After she got out, she was intent on moving us north of the Mason–Dixon Line.

Mom had a brother who lived in Baltimore, so we moved in with him there. I was ten years old, and I had never seen a big city.

From the beginning, I didn't like the way our new "big city" smelled. Not only did it have a stench about it, but back then, Baltimore was a violent place.

There was one kid in our neighborhood who was a bully. I still remember the day Mom called me over to a window in my uncle's house. "Scrappy," she said. That was my nickname when I was a kid. My grandmother gave me that name because I came out of the womb fighting. I would fight anybody, anytime. She would say, "He sure is a scrappy little boy." Anyway, Mom called me over, "Scrappy, you see that boy? He's a bully." She turned to look at me and I looked back at her. "I want you to take care of him."

Mom and I went outside and sat on the porch. Pretty soon she called the bully over. She said a few things, spoke her mind, and called him out for what she thought he was. And then she told me to get him. So I jumped on his butt.

Oh man, it was great! We must have fought for like twenty-five minutes. He ended up swinging at me, and he hit the wall with his fist. That was the end of our fight, and the end of his bullying, at least around Mom and me. In fact, he and I became good friends after that.

One of the things that got me into a lot of trouble in Baltimore was my Southern accent. Another thing that caused problems was my mother's intent for me to make something of myself. Her brother told her about Baltimore Polytechnic Institute, a famous engineering high school. In order to qualify to get in there, I had to go to a special preparatory school called Robert Poole. The namesake for that school, which was housed in the mansion where he lived in the historic Hampden District of town, was a nineteenth-century Irish-born engineer, inventor, and entrepreneur. It turns out Mr. Poole and I have a lot in common, but his school was located in a part of town that abided by white supremacist rules.

To this day, I'm not sure how Mom got me into that school. My cousin and I were the only Black kids there. We rode the bus to school, and in the first week I was there, a crowd of white people pelted me with eggs when I got off the bus. My intent was to behave in school because I didn't want to let my mom down. But after that day they threw eggs at me, I'd already had enough.

So I just started beating up on everybody almost all the time: between classes, in the cafeteria, and on the playground, being the aggressor until they left me alone.

During my whole life in Danville, I had never really seen a white person up close. Then when I started going to a school with nothing but white teachers and white students, I got scared. They looked like ghosts to me.

It was an adjustment, to say the least.

My mom taught me everything, but my dad was an important part of my life, too. After Mom went to jail and my sister and I went to live with relatives, Dad—whose name is James, just like mine—moved to Philadelphia, where his sister lived. His plan was to make a better way for his family. He took a job as a janitor in a famous boxing gym.

When he finally got settled, after Mom moved her kids to Baltimore, Dad sent for us. Mom told him she wasn't coming because she was with family and didn't want to leave. And that's how it was forever: Mom was in Baltimore and Dad was in Philadelphia. They stayed married and truly loved each other, but they lived apart. Dad just couldn't take the racism we were still experiencing in Baltimore.

In the part of Baltimore where we lived, people didn't fight much with their hands. They shot each other with guns instead. Because I was an angry kid, Mom suggested I join the Boy Scouts to keep me out of trouble. Eventually, she signed me up for the Sea Scouts. My uncle told Mom the Sea Scouts met on Fridays at Fort McHenry, made famous in Frances Scott Key's poem that became "The Star Spangled Banner." It turned out it wasn't the Sea Scouts meeting at Fort McHenry but rather the US Naval Sea Cadets, which is how I ended up wearing a Navy uniform from age eleven until I got out of high school.

During the summers, the Navy sent me to Great Lakes Naval Training Center in Michigan and the US Naval Training Center Bainbridge on the bluffs of the Susquehanna River in the far northeast part of Maryland. Also during those times, we went out on the USS *Darby*, a destroyer escort of a Naval Reserve unit.

And before I knew it, I was thoroughly brainwashed.

When I got out of high school, I couldn't wait to go to Vietnam. My high school counselor told me I was not college material. I was a C-minus student. In my mind, though, the problem wasn't me. I didn't like the people who attended my high school. I didn't like the culture. And I didn't like the teachers. I couldn't fight because I didn't want to disappoint my mom.

But if I went to Vietnam, I could fight . . . and serve my country at the same time.

As I was nearing my high school graduation date, I found out there was a shortage of engineers to run ships. The Marine Engineers' Beneficial Association (MEBA) had set up a school in Baltimore, so after I left Baltimore Polytechnic Institute, I went there. By the time I was nineteen, I had passed the US Coast Guard test and was certified as a licensed marine engineer—an officer's rank—with no restrictions on engine size, steam or diesel. So I jumped on a Merchant Marine ship to Vietnam as fast as I could and fell in love with the boats and had a wonderful time at sea.

While I loved the ocean, the crew didn't much like being there. They wanted to be ashore, so when they weren't ashore, they got into trouble a lot.

Nobody on my boat had seen a Black engineer. They didn't know they even existed in the Merchant Marine. So the crew decided to adopt me and gave me a new nickname.

They called me "Youngblood."

Now the officers on the ship were all racists and wouldn't speak to me and wanted as little to do with me as possible. Technically, I was forbidden from fraternizing with the crew because officers don't mingle with enlisted men. Still, I found a way to sneak below deck and hang out with the crew every day. And the life they lived proved to be a little too "fast and furious" for a young man like me. When we finally got to Southeast Asia and made it ashore on leave, we didn't visit the most prestigious museums or study in the library.

If you catch my drift.

I came back from the war with a drinking problem—a very bad drinking problem.

I grew up on the waterfront in Baltimore, which at the time was a dangerous place. While I was in high school, I joined a group of about thirty or forty other guys down there on the waterfront working on the restoration of the *USF Constellation*, the first warship designed and built by the US Navy. We called ourselves "the Crew." When I got back from the service, my boys couldn't wait to see me. So I rushed down to the pier to see them.

Mostly, we just sat around and drank all day.

Morgan State, a Historically Black University, was about a half mile from my house in Baltimore, so that's where I enrolled to start college. And believe it or not, I got straight A's without trying much. I did a co-op with the Navy while I was there, and things were going pretty good for me. Finally one day, one of the counselors at the school, Lena McKenna, said, "Boy, you need to be in some rich white school somewhere! You don't need to be down here at Morgan State!" I told her I wasn't interested in leaving.

I was there for three years. I got straight A's. I was a licensed marine engineer. Ms. McKenna filled out all the applications for me to transfer to a better school. As a result, I got a four-year scholarship to Drexel in Philadelphia, which is where my dad lived. He was so happy!

I also got admitted to MIT, but without any financial assistance.

I'd never heard of MIT, so I didn't show much interest in the school, until Ms. McKenna called me into her office.

"I want you to meet somebody from MIT," she said. I told her I wasn't interested.

I had a Baltimore mentality.

She insisted. Finally, I agreed.

When the day came for me to meet this guy, I went into Ms. McKenna's office dressed in jeans and a sweat shirt. I wasn't intent on impressing "Mr. MIT" at all. I was just sitting there, sort of fuming, when he finally walked in. His name was Dr. Clarence Williams. He worked for the president at MIT. He was Black. I was twenty-two years old by this time.

I'd never seen a Black man like him.

Dr. Williams was meticulously coiffed and impeccably dressed. His speech and his mannerisms were polished and dignified. He spoke to me for forty-five minutes, and I didn't hear a thing. I couldn't get over the way he looked and how he presented himself. I thought to myself, "Man, he's from another planet!"

After he left, I got to thinking, "I wouldn't mind looking like that."

It turned out one of the Crew had an aunt who lived in Boston, and he said I could stay with them while I checked out MIT. It turned out his aunt was a dean at Harvard and her husband was a judge. So there I was, staying with this accomplished Black couple, but then they got mad at me because I kept putting off my visit to MIT. Finally, they made their son take me to the MIT campus.

I was dropped off at the admissions office, and there I met this old white guy. He was trying to talk to me, but I just wasn't feeling it. Finally, he got a little frustrated and left the room. Soon, a Black man named John walked in. He read me the riot act.

"You don't seem too interested in what we're doing here," he said. "You haven't filled out an application and I'm told you have a bad attitude. So here's the deal. I'm going to bring a student in here. He's gonna walk you around campus and you're gonna talk to professors and try to get acquainted with what we do here and how we can change your life. I can tell you now you'll get credit for your time at Morgan, so why don't you make the best of your time here?

"Any questions?"

I had taken differential equations, physics, and a lot of other science courses at Morgan State, but after I transferred to MIT, it turned out all they gave me credit for were my humanities classes. So I started there as a sophomore again. That turned out to be a blessing.

Two weeks after starting classes in Boston—my father was disappointed but hopeful at the possibilities—I was drinking from a fire hose. I was flunking everything, embarrassed and humiliated by how lacking I was in what MIT expected me to know. Desperate and despondent, I called up the Crew. They didn't like the way I sounded, and told me, "We're coming up there to snap you out of this!" So they jumped in their cars after work on a Friday and drove to Boston. After they arrived, we all made the rounds of nearby drinking establishments. As was our way, we got awfully drunk and more than a little rowdy.

MIT has a church on campus with a little moat that surrounds it. After the bars closed, we ended up swimming in that moat at about 2:00 in the morning. Not surprisingly, somebody called the police, and we were still having our fun when the cops arrived. They rounded everyone up and told us, "We want you out of town now! If you're not, you all are going to jail."

Much to their surprise, one of us was an MIT student.

So the next day they sent me to the department head's office for dismissal.

At least, that's what I thought would happen.

And that's where I met Dr. Herbert H. Richardson.

I was sitting outside his office, dreading the outcome of my meeting with another clueless old white man. I knew he was going to kick me out of school and that would break my mom's heart. As I sat there, I rationalized my struggles like this: MIT felt like another planet with aliens everywhere. It was just not my kind of place. Finally, I decided, "OK, Jim, you're gonna be dismissed and it's gonna be a problem. So you don't have nothing to lose. You're gonna go in there, you're gonna meet with this white man, and if he says anything outta line, you're gonna smack the hell out of him."

When Herb called me into his office, I was itching for a fight—a real one. He offered me a chair, introduced himself as the head of the Department of Mechanical Engineering, and asked me to explain to him why I was there. I told him, "Man, I'm a licensed marine engineer, unlimited horsepower, steam or diesel. I've been all over the world sailing ships. I come here, I can't even get a D in class, and I don't understand why that is. I shouldn't be doing bad here. Granted, I don't like the people, and I don't like the place—mostly, I think, because I don't even recognize this planet."

Herb and I went back and forth for a while. Finally, I said, "And it is not just me. All the Black students are like this." He disagreed. "No, they're not," he said. "Yes they are," I responded. "They are not OK, either. I know Black students and I don't think you do, at least not very well."

"No, you're wrong," Herb replied, somewhat defensively, I thought. "Why don't you check the grades?" I asked him. "Tell me what the average grade is for a Black student in the Mechanical Engineering Department."

Surprisingly, he agreed to do that.

After thumbing through his files for a while, I got impatient: "So what's the average grade?" He was quiet for a time, calculating, and when he answered, I could tell he was surprised. "I'd say it's about a C."

At that point, I looked him dead in his soul, and I said, "You think I came here to get a C? Do you think any of your Black students came here to get C's?"

And then Herb said something that was like a horse kicking me in my face. He said, "Well Jim, why don't you and I do something to change this?"

At first, that failed to register. No white man had ever said anything like that to me. I had been treated like crap by whites down South. I had to fight my way into and through school in Baltimore. The white officers hated me on the boat.

I muttered to Herb, "What did you just say?"

"I said, why don't you and I do something to improve the climate for Black students here in the department? I'm the department head, and if you agree, I think we can do something about this obvious problem you've made me aware of."

And that's how Herb Richardson and I formed the organization called "Black ME," for Black Mechanical Engineers. Herb gave us a lounge to meet in. He provided us with letterhead to help establish our credibility on campus. He instructed faculty to drop by the lounge every evening with old problem sets like the white students had and tutor us. I went from flunking everything to getting all A's, and eventually,

I became the first African American to earn a PhD in mechanical engineering in the history of MIT.

Herb did that. But as I continued my way through school, I still had a problem disciplining myself. So Herb brought in Wes Harris to work on my discipline. It turned out Wes had shared an office with John Junkins when the two were at the University of Virginia. When I needed help with my research skills, Herb brought in Don Frazier to steer me in the right direction. These three men—Herb, Wes, and Don—became my mentors and polished me into a diamond. They all worked on me so hard for so long. And I don't ever want to let them down.

And that's why I came to Texas A&M to become a Hagler Fellow and then a Hagler Permanent Fellow.

Because I do what Herb Richardson tells me to do.

23

"A Useful Guy"

Vanderlei Bagnato, 2018–2019

In the July 2023 issue of *Revista Pesquisa FAPESP*, the monthly magazine published by the São Paulo Research Foundation, Brazilian physicist and engineer Vanderlei Bagnato told writer Marcos Pivetta how he became a Faculty Fellow of the Texas A&M Hagler Institute for Advanced Study.

How did Texas A&M University come to take an interest in your work? Were you involved in other collaborations with researchers from there?

I had never previously collaborated with them until 2018. I had once met Marlan Scully, a professor at A&M, one of the largest universities in Texas. It wasn't until 2018 that I was granted a fellowship from the university for my contributions to biophotonics. They confer this honor on senior researchers they are looking to collaborate more closely with. So I became a Hagler Fellow, which entitles me to spend anywhere from three to twelve months at the university. It doesn't have to be a continuous stay; it can be spread out over several years.

So what exactly does a Hagler Fellow do?

I was informed that I could visit the university at my convenience, to deliver lectures, conduct seminars, engage with professors and students, and establish collaborations. In total, the fellowship provides recipients with $200,000 to be used for academic activities during their time in Texas. I made two trips there during my vacation leave from USP [the University of São Paulo], once in 2018 and again in 2019. It was during these occasional visits that we started our first project, exploring ways to overcome bacterial resistance to

> antibiotics using light manipulation techniques. We had started this project at the RIDC [Research, Innovation, and Dissemination Center] in Brazil funded by the São Paulo Research Foundation and finished it there. I delivered a lecture on treating skin cancer, including melanoma, with photodynamic therapy, and we coauthored several other studies.

Vanderlei Bagnato has gained a global reputation for his work in applying the principles of optics and lasers to health-sciences research. In addition to being a member of the Brazilian Academy of Science, Bagnato is also a member of the US National Academy of Science and the Pontifical Academy of Sciences of the Vatican.

The purpose of the Pontifical Academy of Sciences—affiliated with the Catholic Church, based out of Vatican City, and whose roots date back to 1603—is to recognize excellence in science, promote public understanding of science, ensure that science works to advance both the human and moral dimensions of man, and encourage interplay between science and spiritual, cultural, philosophical, and religious views.

Bagnato was elected to the Pontifical Academy in 2012.

His profile on the Pontifical Academy website reads, in part, "He has pursued pioneering work in the field of atomic and molecular physics with worldwide recognition. He has studied Quantum Turbulence involving atomic superfluids, and time and frequency metrology with the construction and evaluation of the first atomic clock developed in Latin America, and the first pilot plant for the production of high-precision microscopes. Recently the research group under his guidance implemented for the first time in Brazil the clinical use of Photo-Dynamic-Therapy for the treatment of cancer and microbiological control."

Bagnato described that work in a Texas A&M Engineering press release announcing his addition to the Department of Biomedical Engineering staff as a Permanent Hagler Fellow in 2023.

"Photodynamic therapy uses light and ultrasound to treat skin cancer," Bagnato said. "We first treat the skin cancer with chemicals that respond to the presence of light or ultrasound waves, then we administer the light or ultrasound waves to the patient to target and eliminate the cancer. We have already treated thousands of people in Brazil and successfully eliminated 95 percent of lesions.

"We don't do science for the money," he added in the release. "We do science for the difference we make. Scientific research has to turn back to the people—otherwise, it's useless."

Vanderlei Bagnato's exceptional nature is apparent from a cursory review of his curriculum vitae. He received two bachelor of arts degrees in his home country: one in physics from the University of São Paulo and another in materials science engineering from the Federal University of São Carlos. Attaining a pair of bachelor's degrees in itself may not be too unusual, but doing so at the same time at two different universities is a unique and exceptional feat.

"I was born in 1958," Bagnato says. "I lived in San Carlos, a city in the São Paulo state of Brazil. At that time, we had a decent educational system and almost everybody was in public schools. There was a time in Brazil that being a son of a blue-collar worker, as I was, meant that you could be condemned for life to a lower social status. But when I was growing up, my parents made sure that I could go to school and then to university.

"And I wound up going to two different universities at the same time. I know it's hard to believe, but I did."

Which prompts an obvious question: Did he work day and night?

"Yeah, of course!" Bagnato replies with a gleam in his eye. "I mean, we always worked day and night."

Which, naturally, prompts a follow-up question: Where did your work ethic come from?

"You see, I am a descendant of immigrants from Italy, and hard work is one important characteristic for immigrants. When immigrants come, they know their life will be difficult for a while because they come with nothing. Right? They come with a bag, and then they want to make sure their kids can go on, because they believe their children will reach a different status than they have being new to the country.

"My grandparents came from Italy. They started working in the fields of a coffee plantation. Then like so many of the Italian immigrants in Brazil, they sold fruits and vegetables in the market. My grandparents and parents worked hard so that things could be better for the next generation, so that their children could acquire different levels of intelligence, be employed by better companies, and fit into better sectors of society. Everybody in my family were very hard workers.

"Normally, immigrants take one of two paths. One is the wrong path, the 'underground' and other bad things. The other path is the right way, improving life."

There's also a practical matter to attending two universities at the same time.

"They were like three kilometers apart," Bagnato says. "They were not in different cities. They were in the same city.

"Here in Brazil, normally what you do, you take national exams to make the choice of the public university where you're going to go. The federal system uses one type of exam, and the state system another one. I took both. And as a result of my scores, I got the first choice in both. It was hard for me to decide because I loved physics. But materials science is also something very important and it's very much related to physics. So I said, 'Well, why don't I start the two degree programs, and then I can make my decision. Right? I will be better able to decide.'

"But then, I kept going. And after two years passed, I was halfway toward my degrees from both universities, so I decided to keep going to the end.

"I concluded the materials science engineering degree in the regular time, which is five years in Brazil, and the physics in four and half. Normally you could obtain a physics degree in four years, but at the end of the physics curriculum you're taking courses that are a little heavy, like quantum science and statistical mechanics and mathematical physics. So it took me a little longer because I had to split my time between the two schools.

"I'm very happy I did that because quantum physics became my passion, but I had this ambition of application, which is something natural for many engineers."

Bagnato's next educational stop wasn't within close proximity at all. Instead, it was half a world away on the North American continent in the state of Massachusetts and the town of Cambridge.

"When I finished my bachelor's degrees, I received an invitation to stay in Brazil, but I decided to go out," Bagnato says. "I applied for MIT in Cambridge, Massachusetts, by mail. My 'interview' was a letter with a couple of questions. One of the questions was how I felt about having two different academic backgrounds. My answer was that it would be a very good decision for them to accept me as a student because of that.

"Finally, I got a phone call from one of the professors on the selection committee, George Koster, a famous theoretical physicist at MIT. He told me I had a very good score on my GRE, but I didn't do so well on the English-proficiency test. I told him, 'I'm speaking with you by telephone. Why do you believe I will not be able to follow the classes?'"

Fortunately for Bagnato, Koster was a strong proponent for the advancement of minority students at his school. His 1968 report, "Minority Recruitment in Physics at MIT," for the Council of Graduate Schools in the United States remains a seminal document advocating diversity.

Koster agreed with Bagnato's rationale during their phone conversation, thus paving the way for the Brazilian graduate candidate of Italian heritage to be accepted into the Massachusetts Institute of Technology.

"I went there, and I got my PhD there working with a wonderful group in atomic and molecular physics," Bagnato says. "So my background is in the fundamental elements of matter, atoms and molecules. My belief is that everything is made of atoms and molecules, so if you understand atoms and molecules, you are able to work in more different fields."

Which, as you already know, had become the norm for Bagnato.

"Physics is more basic science, and engineering is more applied science," Bagnato says. "The things I learned in physics always helped me to understand even better than my colleagues what was going on in the engineering part of my work, you know?"

Right.

In the *Pesquisa* article, Bagnato is asked to describe the proposal, generated by a Cancer Prevention and Research Institute of Texas grant, which brought him back to Texas A&M and made him a Permanent Hagler Fellow.

"They offered me a comprehensive research funding package and additional benefits," Bagnato told the *Pesquisa* reporter. "I would have several million dollars on hand to set up a biophotonics laboratory, plus the opportunity to hire two young professors and bring on board several postdoctoral researchers and graduate students. The yearly funding they offered to set up the lab is roughly equivalent to RIDC's entire budget throughout its life cycle. They also offered a compelling monthly stipend, and what's more, they said that if I wanted to, I could maintain my ongoing research and projects in Brazil. This particular aspect of their offer was especially compelling."

Asked in the *Pesquisa* piece why he was chosen instead of an American or European to lead these research initiatives in Texas, Bagnato responded,

> Americans are known to be pretty ambitious. They're keen on bringing in seasoned researchers to start research programs in new fields, working alongside younger talents. Texas A&M, as its name implies, has a strong tradition in agriculture and mechanical engineering. There are great biophotonics researchers throughout the United States, no doubt. But in Texas, this is a field that is still incipient. The university is looking to establish a biophotonics department that resembles our counterpart in São Carlos. They didn't ask me to change a thing. Their intention is to fully replicate the model we've adopted at the RIDC. This is my shot at creating a sort of stateside branch of our center [in Brazil].

As for the path that led Bagnato to begin his work in cancer research, he says, "Back in about 1992, I went to a meeting of the Optical Society of America"—today called Optica—"where some people were asking the question, 'If we can use light to destroy molecules at the molecular level, a cancer cell will die, right?'

"This seemed to be a good application of everything I knew: a spin of the electron, energy transfer between molecules, all crazy kinds of stuff. Everything relates to quantum process. So I started my research in Brazil and the work evolved to the point where we started to treat cancer.

"The principle is actually very simple. Suppose I put a molecule on your body, and this molecule has some affinity for cancer cells. Cancer cells are different from normal cells. They are not well formed. That's why a tumor is ugly. The cells do not touch each other in a very orderly way, like normal cells. So there are molecules that you can put into the bloodstream, and they go to the cancer. And once they reach the cancer, they cannot leave so easily because the cancer cells are so busy trying to multiply, [they don't] care what else is going on besides their own multiplication.

"This new molecule introduced into the cancer equation does nothing until it is reached with light. That is how you transfer energy to the molecule, what we call 'excitation' of the molecule. The light modifies the form and modifies the structure, and then the molecules start to react to oxygen, which produces what we call reactive oxygen species. This creates molecules that destroy everything in seconds. So the idea is to produce a very deadly molecule inside the tumor cell. And if you do that, you can destroy the cancer. And this is called photodynamic reaction.

"When you're doing this technique, you use lasers as your light sources. It's called 'dynamic' because you use light to create energy transfer with oxygen. You need to understand the molecule you are using because it needs to be friendly to the organism until it is reached by the laser light, which produces free radical species that destroy cancer cells very quickly. In fact, in real time you can see the tumor change color and shrink as it is being destroyed."

In 1997, five years after beginning his research and testing the photodynamic process with animals—using molecular compounds outsourced at the time from Canada and Russia—Bagnato implemented a program for treating head and neck tumors in human patients that did not respond to chemotherapy or radiotherapy and were not surgically accessible. Working with both oncologists and their patients, he obtained results he calls "a big success."

As a result, his group in Brazil started to grow, and his Optics and Photonics Research Center (CePOF) was established in 2000. His work zeroed in on skin cancers, and today, he estimates, 95 percent of the photodynamic procedures used in Brazil are successful, eradicating the cancer.

Bagnato is particularly grateful that the treatment has been approved as part of Brazil's socialized-medicine program, thus ensuring that it is accessible and affordable to people from all walks of life.

And now, through his work as a Permanent Hagler Fellow, Bagnato is hoping to impact the treatment of skin cancer in the United States in the same way he has in Brazil. One in five Americans, according to the Skin Cancer Foundation, will develop the disease by the age of seventy.

Within the Texas A&M Biomedical Engineering Department, Bagnato is the director of two labs, one that researches cancer treatments and another that researches the treatment of antibiotic-resistant bacteria. Through his research here, he will continue to collaborate with his colleagues at the University of São Paulo.

"I hope to create new techniques to treat melanoma," Bagnato says. "My goal will be to either eliminate the cells and cure the cancer or to make the cells attach more rigidly to the tumor and prevent the disease from spreading. In the bacteria laboratory, we will focus on treating antibiotic-resistant pneumonia to prevent the death of thousands of people each year."

The headline for the São Paulo Research Foundation's *Pesquisa* magazine article on Bagnato reads, "This Isn't a Case of Brain Drain," inferring that Bagnato could be doing harm to his nation's scientific intellectual property by taking important research work outside the borders of Brazil.

Bagnato says in the story:

> Brain drain is when someone leaves Brazil for lack of opportunities here. I've always had plenty of opportunities in Brazil. My situation is quite the opposite. I want to establish a laboratory that helps attract more people to Brazil and offers them the resources they need. We need to be in a position to retain talented individuals here. I want to build strong links with the research ecosystem in the US. I want to understand their science market, how academic research interfaces with industry, and foster a culture of science. We have to showcase Brazil's science capabilities. I intend to serve as a sort of scientific ambassador for Brazil, more than just being a visitor. I plan on always being associated with USP; I can't picture myself being away from it. I've already supervised over 130 graduate students, and I'll continue to do so.

Bagnato did admit, in his interview for this book, that his involvement with Texas A&M and the Hagler Institute for Advanced Study has been a career "shot in the arm" for him.

"I have done a lot of things in my work in Brazil," he says. "I have a lab filled with young people. Maybe it's a good time to give those young people a chance to make a name for themselves, right?

"I mean, I still have fuel to burn, so why not burn it, why not start something new, in a place that may grow even faster than what we've done in Brazil?

"As I've said before, we don't work for the money. We want to convince society that what we are doing is useful. People admire and appreciate a good football player or a good singer, right? I mean, we all like those guys. But I like when society can come to someone like me and say, 'You're a useful guy, too.'

"I remember the ceremony where I and the others were introduced as Hagler Fellows. That night was such an honor and important for me and my wife. We didn't know I had this kind of value any more.

"It is an honor to be chosen by Texas A&M. I am very glad to be a part of its team and making a difference in science for the future of humanity."

24

Evolving Interests

Vincent Poor, 2018–2019

Today's cellular phone industry—some call it "wireless"—goes back farther than you might imagine.

In 1917, Finnish inventor Eric Tigerstedt filed a patent for a "pocket-size folding telephone with a very thin carbon microphone." Decades later, Tigerstedt's vision would be called a "flip phone."

Closer to home, the first "mobile telephone service" in the United States—inspired by the radio telephony developed and deployed during World War II—was launched by Bell Labs on June 17, 1946, in St. Louis, Missouri. But not until the 1960s did researchers create the technological systems, like frequency reuse and handoff, that would enable the creation of modern cellular networks.

In 1973, the first wireless call from a handheld device was made with a phone shaped like a brick, one that weighed almost two-and-a-half pounds. Martin Cooper, an executive at Motorola, picked up the large device—one that only worked for thirty minutes of talk time after a ten-hour charge—and called Dr. Joel Engel, an engineer at Bell Labs. It took Motorola another ten years and $100 million before they were ready to roll out the phone commercially into the US marketplace.

It took that much time for the networks to which the phone could connect to be put in place.

Hagler Fellow Vincent Poor has been a professor of electrical and computer engineering at Princeton University since 1990. Like Hagler Founding Director John Junkins, Poor received his undergraduate degree from Auburn University. Poor obtained his PhD in electrical engineering and computer science from Princeton in 1977. He began teaching at the University of Illinois Urbana-Champaign that same year, and while there, he became interested in the science behind wireless networks.

"Wireless as a commercial venture was not yet a thing," Poor says. "But during my time at Illinois, there were people there who were starting to work on CDMA."

Code-division multiple access, or CDMA, is a form of multiplexing that digitally enables numerous signals to occupy a single transmission channel. This approach allows several people on different cell phones to be "multiplexed" over the same bandwidth of frequency. With CDMA's digital technology, data and voice packets are separated using codes and then transmitted via a wide frequency range.

"There was something about CDMA that was very closely related to what I'd written about in my dissertation," Poor says, "which was about trying to separate overlapping signals. Thus, I got interested in the work being done on CDMA. Eventually while I was at Illinois, I wrote a proposal on that to the Army Research Office and my research got funded."

By the time Poor moved to Princeton in 1990, CDMA was under consideration as the standard for the deployment of second generation, or "2G," wireless networks in the United States. In Europe, a rival technology, known as the Global System for Mobile Communications—called "GSM"—had already taken root, thanks in large part to the research work of the Finnish-based Nokia company. GSM was adopted as the European wireless standard in 1987, and the world's first 2G call was made on a Nokia phone by the prime minister of Finland on July 1, 1991.

This new-at-the-time 2G technology offered several advancements over what retroactively came to be known as "1G" service: improved sound quality, encrypted calls, more efficient use of the radio frequency spectrum, faster download speeds, and most significantly, the ability to transfer data from one phone to another. This introduced text messaging as a new form of human communication.

Since then, day-to-day life for humankind has never been quite the same.

Eventually, 2G was rolled out in the United States using CDMA transmission. The year was 1995. The rival CDMA versus GSM communication protocols are the reason that, for many years, not all cell phones were compatible on all cell networks. Fortunately, those days ended with the rollout of 4G LTE service in 2009.

"LTE," by the way, stands for "long-term evolution," a fitting description.

"Those early years at Illinois," Poor says, "is when I began my wireless research work. As I got to know more and more about wireless, I discovered that the problems within the technology were many, but in the beginning I focused mainly on interference mitigation. Today, my work centers on security and latency issues, which are at the forefront now.

"Wireless was pretty much my only research interest for some twenty to twenty-five years," Poor adds. "And it's still my major research interest."

Did Poor realize, when he first started working on CDMA more than forty years ago, that wireless communications would become such a ubiquitous part of global culture not only in first- and second-world countries but also throughout third-world nations?

"No way!" he says. "I had no idea, and I wasn't alone, right?

"You may know the famous story about what the McKenzie company thought the future of cellular was. Reportedly, the company told AT&T, which through its Bell Labs research arm basically invented cellular, not to go into the business. McKenzie did a study and forecast that by the year 2000, there might be a million customers . . . globally. So AT&T sold their wireless interests to McCaw Communications, and they rebranded the offering as Cellular One. By 2000, there were close to a billion customers worldwide. Now there are more cell phone subscriptions, well over eight billion as I recall, than there are people in the world!"

So what were the McKenzie "experts" thinking?

In 1994, AT&T bought Cellular One for a staggering amount of money at the time, $11.5 billion, marking the second-largest merger in US history.

"By the time I got to Princeton, a lot of people were working on matters relating to cellular systems," Poor says. "I mean, for a while, everybody was working on some aspect of wireless. Now of course, there are a lot of other things that have gotten people's attention. But in the nineties, everybody was doing something on wireless. Kinda like today. Everybody seems to be working on AI or machine learning. Same kind of thing."

As for what has piqued Poor's curiosity of late, we'll get to that in a moment.

Regarding his "history" with Texas A&M, Poor says, "I've engaged with colleagues there for at least forty years," which means his connection to the university predates his interest in wireless. When asked if he remembers his first trip to College Station, Poor digs deep looking for an answer.

"I know I was there in the eighties, but I don't really have any record of it," he says, looking through his travel archives while speaking on his cellular phone. "I went there in '92 for a conference, but I know I'd been there before.

"I've done several other things there. I was chair of an academic review committee there. I've given a number of talks over the years. One of my former students became a faculty member there, and I visited him."

Scrolling through the travel files on his cell, Poor continues.

"I see that I gave a talk there in 1998. We could say my experience at Texas A&M dates back at least thirty years ago, but my guess is it's been longer than that. My records don't go back far enough.

"It was in the ancient times," Poor adds, "when 'dinosaurs roamed the earth.' Sometime in the eighties, I'm sure."

In 2017, Poor gave a talk for Texas A&M's Electrical and Computer Engineering Department. During that time, he was approached about a fellowship with the Texas A&M Institute for Advanced Study, soon to be renamed in honor of Jon Hagler. In his nomination of Poor, A&M ECE Professor Xi Zhang offered this ringing endorsement:

> More than 17 ECE department faculty members work in areas closely related to the research work Prof. Poor is conducting. Also, many other faculty members from TAMU's Aerospace, Biomedical, Computer Science & Engineering, Industrial & Systems, and Mechanical Engineering Departments are working on research problems where the expertise and leadership of Prof. Poor will be very valuable. Prof. Poor's extraordinary expertise and leadership in the field of wireless networks, communications, control, smart grids,

> power grids, big-data analysis, information theory, signal processing, etc., will help TAMU faculty to prepare successful research grant proposals in these hot research areas.

"I wasn't aware that I was actually nominated," Poor says, "until I got a call from the dean. At the time, my wife was very ill. So my initial response was, 'Thank you very much. I'm honored, but I can't really do this.' Then my wife passed away not long after that, and I began thinking it might be good to get away, go someplace different, and adopt a different routine.

"I no longer had a reason not to say yes."

Poor met his wife, Connie, while both were attending high school in Phenix City, Alabama, but it would be incorrect to call them "high school sweethearts."

"We knew each other in high school, but we never dated. She actually dated a friend of mine. After they broke up, during the summer before I went off to college, we got reacquainted. I worked at a gas station across the street from where she went to church on Sundays. People would park at the gas station when the church lot got full, and that's where Connie and I got to know each other better."

Four years later, as Poor was preparing to do graduate work at Auburn and shortly after Connie finished nursing school in Birmingham, the two were married.

Poor's move to Princeton to begin work on his PhD uprooted the young couple. While he studied, she worked, her nursing degree enabling her to have what Poor calls "a portable career."

After earning his PhD, Poor and his wife moved to Illinois. Thirteen years later, they moved back to Princeton. "She was very happy about that move," Poor says. "Even as a Southern girl, she always loved the Northeast."

Connie Poor fell ill in the fall of 2017. As a medical professional, she worked closely with doctors, but initially, they could find nothing out of the ordinary. By the time a tumor was located, it had metastasized into other parts of her body. The Poors got the news shortly after Vincent's return from his talk to the Texas A&M ECE group.

"I declined the Hagler Fellowship before we knew just how sick Connie was," Poor says.

She died on March 2, 2018.

By that time, the Poors' two daughters had started lives of their own, one on the West Coast in Los Angeles and the other in New York City. Being home alone got Poor thinking that maybe a change of scenery would be a good thing. He reached back out to John Junkins and asked if he could again be considered for a fellowship.

"It's no longer the Texas A&M Institute for Advanced Study," Junkins told Poor. "It's called the Hagler Institute now, and we would be honored to have you join us."

"It was tough being in the house where Connie and I had lived for so long," Poor says today of his initial time as a widower. "I spent a lot of time thinking about my life and what I had lost. Getting away from that, getting on a plane, going to Texas, and meeting new people turned out to be a good thing for me. Everyone I met at Texas A&M—the people at the Hagler Institute, in the electrical engineering department, and the people in whose home I first stayed—were all so welcoming and understanding."

The accommodations offered by Jim Singleton and his wife, Stephanie Sale, were not merely a spare bedroom. They continue to host Poor and other Hagler Fellows to the home they call "Turkey Creek," which can be best described as a main house surrounded by a quartet of two-story living quarters.

The compound is very "state-of-the-architectural-arts." Not surprising, since Jim is an architect.

"Jim and Stephanie are wonderful hosts," Poor says, "and we have become great friends over the years."

As for Poor's evolving research interests, in addition to those mentioned by Zhang in his Hagler nomination of Poor—control systems, smart grids, power grids, information theory, and signal processing—he's also become an expert on information spread, as it relates to social networks, and viral spread.

"I had collaborated with a colleague at Carnegie Mellon University, Osman Yağan, who is a professor in the Electrical and Computer Engineering Department there, to write a paper on viral spread," Poor says. "Initially, we were working on information spread in social networking, but one of Yağan's colleagues realized that what we were learning was also applicable to the properties of viral spread as it relates to human disease. So we took our research in a more biological direction. It was mostly about mutations and how mutations affected viral spread. We did most of the work on that in 2019, and our paper finally came out in the *Proceedings of the National Academy of Sciences* in March of 2020."

Just as the COVID-19 pandemic was catching the world's attention.

"No kidding!" Poor marvels. "It came out in March of 2020, and so all of a sudden, at least in the academic world, we knew something about the spread of COVID-19."

Poor's scientific expertise was suddenly in great demand.

"Several funding agencies put out what they call 'rapid calls' for grant proposals," Poor says. "We responded to the National Science Foundation, the Army Research Office, and a private foundation. We quickly wrote our proposals and received funding.

"Since then, I've been publishing papers, writing more proposals, and working with people in the life sciences in a variety of places. During the height of the pandemic, there was a lot of funding. A lot of people turned their attention to containing the spread of the coronavirus. And I was one of them. It's really kind of serendipitous how that happened to me at that time."

Looking back, how close to the mark were the hypotheses Poor and his colleagues put forth in their research findings?

"That's a good question," Poor says. "It was a theoretical model we created. We looked at it with some real data, but it was a theoretical model. We are continuing to look at it. I would say what we've done has been more about trying to understand it than trying to match it to real data.

"I did get involved in some other peripheral studies that were using real data. One was in India, and I think our findings matched up to the real data there fairly well.

"It's hard to say to what extent a particular theory impacts something. That might be something I wouldn't mind doing, see how well our models predicted what actually happened. I suspect that with the right parameters, it would be a pretty good model.

"People can cite your work, read your paper, and then go off and use some of your ideas. You hope this happens, but it's not a direct path. You think, 'I did this, I wrote this paper. Now there's this thing that exists because of that.' You don't get that kind of satisfaction very often in my business."

Power grids are another of Poor's recent theoretical research interests. A paper he produced, along with his Princeton colleague Minjie

Chin, for the Institute of Electrical and Electronics Engineers' *Electrification* magazine in 2020 explored "High Frequency Power Electronics at the Grid Edge: A Bottom Up Approach toward the Smart Grid."

In the early months of 2021, the power grid in Texas had the Lone Star State on edge thanks to the collapse of service caused by Winter Storm Uri. Looking back, Texans faced a formidable "one-two punch" from Mother Nature: the coronavirus *and* subfreezing temperatures that KO'ed the electrical system.

"I was still a Hagler Fellow then, but I wasn't in Texas at the time," Poor says. "Some people go for like six months and do their fellowship all at once. I decided to spread it out and come to Texas a couple weeks at a time.

"My understanding of what happened there," Poor continues regarding the Electrical Reliability Council of Texas (ERCOT) debacle, "was a failure of not just the primary electrical grid but also the distribution network for how the electrical grid is powered.

"Texas has the most wind power of any US state, but there's also a lot of fossil fuel used to power the grid. There has to be. It's a big state with a lot of people. Much of the backup generation for plants is powered by natural gas. So it's not just the electric grid that's important. It's also the distribution network for natural gas.

"My understanding is that the cold somehow froze up the natural gas lines, or something caused some problem with the distribution of natural gas. The reserve power plants couldn't get enough fuel to meet the increased demands and failed to generate enough electricity to keep things going."

In one way or another, it seems, Texas offers esteemed academic researchers unprecedented real-world evaluations of their theoretical models; at least it has for Poor. Hagler Institute administrators might do well to make this a selling point for future prospects.

The township of Princeton, not the university there, is the namesake for the original Princeton Institute for Advanced Study, as you will remember from chapter 1 of this book. Having spent more than thirty years in the township, does today's Princeton Institute appear on the radar screen of an accomplished academic like Vincent Poor?

"I think they're symbiotic, really," Poor says of the relationship between the institute and the university. "In the beginning, the institute was established here because of the university. It was actually housed for a time in the math building here. Then they built their own campus. It's actually not far from the university campus. There are people at the university who spend time at the institute, and there are people at the institute who have close connections with academic departments at the university. Some of them even do teaching or collaborative research with people on campus.

"Some faculty, when they go on sabbatical, spend their time at the institute. In many instances, it's a close relationship, but they have their own board of trustees, their own leadership, their own donors, and their own endowment.

"It's not like the Hagler Institute, which has become an integral part of Texas A&M."

When Poor is asked by his Princeton associates to describe Texas A&M's own Institute for Advanced Study, how does he recount his experience?

"I tell them Texas A&M is a great environment," Poor says. "You go there and you meet all these interesting people and they treat you extremely well. It's a great environment for getting research done. Nobody bothers you to be on a committee or teach a course or anything like that. And it's such a friendly place.

"Did I already mention that?" Poor laughs.

"Texas A&M is a pretty high power when it comes to the kind of research you can do. Listen, Texas A&M is well known in academic circles. It has a top-notch electrical engineering department, and so most people in my sphere have been there, visited people there, and some people have even moved there from Princeton. After they retired from here, they moved to Texas A&M."

Indeed, he's right. Included in that group is another former Hagler Fellow and Permanent Fellow, Christodoulos Floudas, as well as the late Ignacio Rodriguez-Iturbe, a former winner of the Stockholm Water Prize for his role in developing the science of hydrology, and Dick Miles, a University Distinguished Professor in the Department of Aerospace Engineering at Texas A&M since 2017 and a world-renowned expert in hypersonics and advanced laser diagnostics.

Might Poor be persuaded to make that move, maybe in retirement? After all, College Station would put him almost exactly halfway between his two daughters.

"That's a good point!" he says. "I really like Princeton, but I really like College Station, too.

"Right now"—and Poor is in his early seventies—"I just can't wrap my head around the idea of retiring. I don't think the people that I just described thought of their move as retirement. I mean, they officially retired from Princeton, but in moving to Texas A&M, they certainly didn't retire. They just went to another place, and they worked there just as hard. That would be me too. I don't think I would just go off and play golf."

Poor pauses for a moment before adding, "Nothing wrong with golf, but that's just not my idea of where I want to be spending my time. At least not right now."

25

Coping with COVID

Rachel F. Moran, 2020–2021

Rachel F. Moran well remembers her March 2020 trip from California to Texas to assume the position of Hagler Institute Fellow at the Texas A&M University School of Law in Fort Worth. At the time, Moran was a professor of law at the University of California, Irvine. She is considered one of the nation's leading scholars on matters of education policy, civil rights, and race and the law. The Hagler Institute appointment was appealing to Moran because she knew Texas's Hispanic citizenry would soon surpass the state's number of non-Hispanic white residents, and she was eager to continue her long-standing work to identify issues facing the nation's Latinx population.

Although excited about the opportunities her Hagler Fellowship would provide, Moran was anxious about the trip. Stories concerning the spread of a so-called coronavirus in Asia were appearing more and more frequently as the lead on television's nightly network newscasts as well as the front pages of the nation's newspapers and news websites.

"I was nervous," Moran says today. "I knew something was happening, and medical experts were beginning to suggest the possibility of a global pandemic. On my way to Texas, I felt compelled to wear a mask at the airport and on the plane but saw few other people doing so."

Over the ensuing months, masks became the norm as humanity's primary defense against the worldwide microbial scourge.

"The Hagler Institute is designed to build community among scholars," Moran says. "I think what makes my story unique is that creating any sense of community became a challenge during the COVID-19 pandemic."

Rachel Moran was born in Kansas City but grew up in the border town of Yuma, Arizona. Yuma sits adjacent to both the state of

California and the nation of Mexico at a bend in the Colorado River. The city got its start in 1851 with the establishment of Fort Yuma at the end of the Mexican–American war. At about the same time, that area of what would become Arizona became a gathering point for those seeking their fortunes during the California Gold Rush.

Today two of Yuma's largest employers are military installations: the Marine Corps Air Station, Yuma; and the US Army's Yuma Proving Grounds.

The Moran family moved to Yuma when Rachel was ten. Her father, Thomas Moran, was a criminal defense attorney of Irish descent; while her mother, Josephine, was a Mexican American court translator and schoolteacher, born in the United States but raised with her extended family in Mexico.

"My father didn't want his marriage to my mother to be seen as out of the ordinary," Moran says. "We moved from Kansas City, Missouri, in the early '60s, when interethnic marriages were looked at a lot differently than they are today."

Her parents' experience inspired Moran to eventually write the book *Interracial Intimacy: The Regulation of Race and Romance*, published by the University of Chicago press in 2003. A synopsis of the book on the publisher's website reads, "As late as the 1960s, states could legally punish minorities who either had sex with or married persons outside of their racial groups. In this first comprehensive study of the legal regulation of interracial relationships, Rachel Moran grapples with the consequences of that history, candidly confronting its profound effects on not only conceptions of race and identity, but on ideas about sex, marriage, and family."

"The book is not just about Latinos," Moran says, "it's about all ethnicities. In it, you really see some of the group differences. It took me about ten years to write it, and I think that comparative approach was relatively new at the time."

During her early years spent in America's heartland, Moran was impacted by racial discrimination herself.

"At a fairly early age, it became apparent to me that race was important to other people," Moran says of her time in Kansas City. "There were certain incidents that happened which made it very clear that for many people, my ethnicity defined me and my worthiness. So if you're a sensitive and intelligent child, as I think I was, you do your best to adapt to those perceptions. It became important to me because I knew it was important to other people, and in ways that might mean I wasn't going to have much of a chance in life.

"When I was in elementary school—this was while we were still in Missouri—I remember walking down the hall one day and one of the teachers said loud enough for me to hear, 'Such a bright girl. Too bad there's no future for her.'

"I was devastated by that remark."

Anticipating these issues, Moran's parents opted not to teach her Spanish as a child. "They feared that things would be even worse for me if I had a Spanish accent," she says. She eventually learned Spanish by taking Spanish-language courses at school.

"I always found refuge in my intellectual pursuits," Moran says. "I loved to read and to do puzzles, and my parents were very supportive of that. I used to like to stay in my room and read and do my puzzles and that became a sort of sanctuary for me. My room was a place where I could escape from the pettiness of other people and think about things that were profound, that made life feel far less confined. That boundlessness of the imagination probably made me more inclined to be bookish and enjoy the life of the mind."

Aided by her mother, Moran honed her abilities in mathematics at home while she was still in grade school. In high school after her family moved to Yuma, Moran continued to excel in the subject. Ultimately, she was chosen for the National Science Foundation Honors Program in mathematics, which led to a summer spent at San Diego State University, where she took college-level math courses while still in high school. It was there that an instructor encouraged her to think big when it came to choosing a college.

"My parents and I just assumed I would go to one of the Arizona public colleges," Moran says. "But my instructor was adamant that I should go somewhere else. He told my parents I was the top person enrolled in the advanced study program and would thrive in a challenging academic environment. My parents didn't want me to go to school too far from home. 'Stanford is where she should go to college,' the instructor told them."

At Stanford, Moran majored in psychology. By her senior year, she pondered continuing her education by either pursuing a PhD in psychology or going to law school. Either way, she would follow in one of her parents' footsteps. While her father practiced law, Moran's mother had taken courses to be certified as a school counselor. Both parents wanted their daughter to do what would make her happy.

The decision Moran reached was impacted by her involvement as a research assistant to a pair of Stanford psychology professors.

"The wife-and-husband duo of Professors Sandra and Daryl Bem were commissioned by the League of Women Voters," Moran remembers. "The League had taken exception with the Department of Labor over the lack of women in the construction industry. This was the late '70s, and at the time there were no guidelines for the hiring of women, in large part because everyone assumed women weren't interested in construction work.

"What the Bems' research study showed was that women weren't applying for construction jobs—which often paid very well—*because* there were no guidelines for their hiring. The lack of guidelines signaled to women that these jobs were not for them and, thus, became a sort of self-fulfilling prophecy. So thanks to the Bems' study, the League of Women Voters was able to put pressure on the Department of Labor to add women as a category to their guidelines for construction hires.

"Shortly after that change had been instituted, my family and I were traveling down a highway and came upon a construction project. Among the workers, I noticed a woman in the midst of the group. She was tall, seemed strong, and was wearing a hard hat. She fit right in. Having seen that with my own eyes, I suddenly realized how law can be powerful and change people's lives for the better."

Thus the accomplished Stanford psychology graduate left the West Coast for the East Coast to attend the Yale Law School.

Moran's curriculum vitae chronicles a distinguished career of teaching, research, and service. Before becoming a Hagler Permanent Fellow in 2020, she spent the bulk of her time at a trio of University of California System schools. From 1982 to 2010, she taught at the University of California, Berkeley School of Law. She took a leave of absence in 2008 to help found the new law school at the University of California, Irvine. In 2010, she was named Dean of the UCLA School of Law, a position she held for five years. After four more years at UCLA as dean *emerita*, Moran rejoined the faculty at the UC Irvine School of Law, the position she jointly held while a Hagler Fellow at Texas A&M.

When Moran first entered law school, she never imagined becoming an academic.

"My mother was a teacher, but no one in my family was an 'academic,'" she says. "I felt destined to practice law, but during my first year of law school at Yale, my property professor called me into his office. Since I had the highest grade in the class, he told me I should think about a career in teaching. I was very appreciative of his interest, but at the time I didn't give his suggestion much thought.

"Then after I graduated, I was clerking for a federal judge in New York, and I got calls, despite the fact I had not applied, from several law schools about interviewing for teaching positions. Nobody I knew—except my law school professors—was teaching as a full-time pursuit. I thought, 'How could these schools possibly be interested in me?' But I interviewed and got an offer from Berkeley. I took the job, but soon after, I realized I would be a more effective teacher if I had some real-world law experience. So I asked for and was granted a leave of absence and went to work at a big firm across the bay in San Francisco."

Once she returned to Berkeley to teach, Moran never again looked back. Instead, she looked for ways to make a difference.

"When I started teaching in 1984, I really wanted to give back because there were hardly any people of color on the faculty at Berkeley. Instructors from minority groups were very rare at the time."

As a trailblazer, Moran felt a strong sense of obligation to address issues she had experienced herself growing up. Among the more than one hundred articles, book chapters, and short commentaries she has authored through the years, many explore such topics as bilingual education, desegregation, and affirmative action. Among her most recent published works are journal articles with titles such as "Latinas in the Legal Academy: Progress and Promise," "Bakke's Lasting Legacy: Redefining the Landscape of Equality and Liberty in Civil Rights Law," "Contested Meanings of Equality: The Unrealized Promise of the Anti-Discrimination Principle and the Uncertain Future of a Right to Education," and "Diversity's Distractions Revisited: The Case of Latinx in Higher Education."

Throughout her career, Moran has been a champion of diversity, equity, and inclusion before those terms were transformed into a divisive ideology. At Texas A&M, she now finds herself as a law school professor in a state that banned diversity offices in public higher education in 2023.

"When I took the full-time position here, some of the people I knew back in California were afraid for me," Moran laughs. "But there was no issue. I have such a long track record on matters relating to civil rights that I'm fairly sure people know what they're getting in terms of my work.

"I think that the impact of eliminating diversity, equity, and inclusion programs is primarily that it renders less visible the work and also ruptures the infrastructure that creates the kind of community in which you can have sustained and important conversations about those issues. And so, you know, it kind of pulls the rug out from under the ability to actually address these concerns. So it's not just that people are less visible, but they don't have the same ability to get together and engage around what should be done.

"Whether or not you call these programs 'diversity, equity, and inclusion,' it seems to me indisputable that universities are obligated to their students to ensure that they have full access to the learning process. And to the extent that those programs were designed to bring down barriers to that access to ensure that people had an equal opportunity to learn, I don't think those responsibilities go away, even if you dismantle the infrastructure that would have been used to achieve those obligations.

"So I think the one question that still must be addressed is, What do you do next to ensure inclusion? And by inclusion, I mean access to the learning process because creating any kind of a hostile environment or a place where people can't learn on equal terms is not only discriminatory; it's bad pedagogy."

Upon her initial arrival in Texas to begin her time as a Hagler Fellow at the Texas A&M School of Law, Moran spent exactly one week in the Lone Star State.

"But it was a wonderful week," she says of time cut short by the coronavirus. "Professor Luz Herrera was my host. She's the associate dean for experiential education at the Texas A&M School of Law. In that short amount of time, I presented to the law school community about my 'Future of Latinos' project underwritten by the American Bar Foundation, and I was able to go to College Station and meet some of the faculty there who work on Latino issues. Professor Herrera and I then went to McAllen in the Rio Grande Valley, and I did a community presentation there on the nature of my work.

"A week later, I was back home in California with everything shut down because of the coronavirus. From that point forward during the rest of my initial Hagler Fellowship, we did everything by Zoom."

"All the on-site gatherings were basically canceled because hosts didn't want people meeting in person. So all the conferences, all the talks, everything became virtual."

Moran smiles as a thought crosses her mind.

"At first, everybody was pretty focused just on learning how to use Zoom."

During that spring semester of 2020, Moran continued to teach her law classes at UC Irvine.

"The one nice thing there was that I had gotten to know the students already. It wasn't like we had never met and didn't know each other at all. So we had some glue from our previous in-person meetings that kind of bound us together. I think everybody, whether at Irvine or in Texas, was pretty tolerant because we were all impacted in the same way. I remember finishing the semester and thinking everything would soon be back to the way it was before. A neighbor in California told me, 'I don't think so. I don't think we're going to be back to normal any time soon. I think this is going to go on a lot longer.'

"And she was right."

As Moran's Hagler Fellowship was extended, all of her engagements continued to be online.

"I presented papers online," she says. "I dealt with my research assistants remotely. I even went to the faculty scholarship retreat online. But I think it was really nice because at least I had had that one in-person visit to Texas where I met people. Because of that, I felt like I had a connection. And as it turned out, I was actually able to be in contact with people more frequently because I didn't have to uproot and travel back and forth.

"Zoom turned out to be a very nice lifeline for a community of intellectual engagement during a time when a lot of people felt pretty isolated and disconnected."

Of the pandemic's aftermath, Moran said during her 2024 interview for this book, "We went through something unprecedented. I think the repercussions are still there, reverberating through our institutions and our society.

"I honestly don't think we've recovered yet. As for the impact on schooling, everyone is focused on learning losses, and those are important. But overarching cognitive issues are going to haunt us for a generation, at least. When these students enter the workplace and tackle job responsibilities, the fact that they have these relational weaknesses caused by the pandemic isolating everyone is going to be important. The social side of the COVID equation still deserves our attention.

"We've seen a pretty significant increase in mental health issues. And some of that has to do with the fact that our personal engagements were limited to online platforms like Zoom. Part of school is learning the cognitive issues—the tasks—but part of it is the socialization of how you interact with other people, how you work cooperatively, and how you behave yourself so others can be involved and contribute. Now that we're all together again, we see students struggle and admit they don't know how to interact with their peers. They're still suffering from isolation and depression.

"It's taking a lot longer for people to learn how to just hang out with each other than anyone probably expected."

Throughout her academic career, Rachel Moran has spiced up her teaching responsibilities with a series of visiting professorships. Similar to being a Hagler Fellow, these roles typically last for a semester and involve relocation. She's taught at law schools affiliated with Stanford, New York University, the University of Miami, and Fordham University.

And then there was her semester at the University of Texas in Austin, when the world faced the prospect of yet another global catastrophe.

"People might recall a little thing called 'Y2K,'" Moran says with a knowing smile. "All the experts were predicting that the world's computers were going to crash when their internal clocks rolled over to the year 2000. Well, I was just getting set to begin my time as a visiting professor at the University of Texas when that happened."

Is Moran's life afflicted with a "Lone Star curse"?

"Because of Y2K, friends discouraged me from flying or driving. They warned me to steer clear of checking into hotels and to carry an ample supply of water when I was away.

"And then . . . nothing happened!"

Which, it turns out, was cause for celebration in a college town like Austin.

"They had a big Y2K event downtown," Moran says, "and the featured musical guest was Lyle Lovett."

A quarter-century later, Rachel Moran is delighted to share an Aggie affiliation with Lovett, Texas A&M Class of '79, thanks to her involvement with the Hagler Institute for Advanced Study. In fact, today she is a Permanent Fellow and a full-time member of the faculty at the Texas A&M School of Law, serving as the director of the educational law program there.

26

"To Be or Not to Be"

Lena Orlin, 2019–2020

A journey into the *New Variorum Shakespeare* website is not for the "faint-hearted" . . . as the character "Woodville" was so described by the Duke of Gloucester in act I, scene 3 of William Shakespeare's play *Henry VI, Part I*:

WOODVILLE. What noise is this? What traitors have we here?
DUKE OF GLOUCESTER. Lieutenant, is it you whose voice I hear?
Open the gates; here's Gloucester that would enter.
WOODVILLE. Have patience, noble duke; I may not open;
The Cardinal of Winchester forbids:
From him I have express commandment
That thou nor none of thine shall be let in.
DUKE OF GLOUCESTER. Faint-hearted Woodville, prizest him 'fore me?
Arrogant Winchester, that haughty prelate,
Whom Henry, our late sovereign, ne'er could
brook? Thou art no friend to God or to the king:
Open the gates, or I'll shut thee out shortly.

Shakespeare lays out Henry VI's story in three plays written in the late-sixteenth century. In part 1, Henry is a mere child, having succeeded his deceased father, Henry V, to the throne at the age of just eight months. In "real life," following the death of Henry V in 1422 and with an infant his successor, the French rebelled against British claims to certain of their territories captured under Henry V's rule.

That rebellion was led by Joan of Arc.

The Duke of Gloucester was Lord Protector of England and Henry V's brother, in obvious rivalry, as Shakespeare wrote, with the Bishop of Winchester, Henry Beaufort, uncle to both Henry V and the Duke of Gloucester. Henry V's death set about family squabbles which ultimately led to the War of the Roses, a series of civil wars fought over control of the English throne from 1455 to 1487. This was a part of English history, Shakespeare believed, worth dramatic exploration.

The New Variorum Edition of Shakespeare, which now resides in its digital form within Texas A&M University's Department of English, dates back to 1871 when American lawyer Horace Howard Furness and his son of the same name sought to collect what had been written by various commentators, critics, and editors about a number of Shakespeare's best-known works. They intentionally called their endeavor "new" because similar efforts had been put forth on several of Shakespeare's works dating back to the late 1700s.

With the "faint-hearted" forewarned, let's dive deeper into the *New Variorum Shakespeare* website, "NVS" for short. A random stop within the play *A Midsommer Nights Dreame* (known today as *A Midsummer Night's Dream*) reveals "textual variants by year" of the work's opening stanza . . . and plenty of examples of "Old-English" spelling:

Now faire Hippolita, our nuptiall hower
Draws on apase: fower happy daies bring in
An other Moone: but oh, me thinks, how slow
This old Moone waues! She lingers my desires,
Like to a Stepdame, or a dowager,
Long withering out a yong mans reuenewe.

While the textual variants focus on single words or phrases, what does one make of this exquisitely crafted refrain as a whole? Well, with the help of the NVS, users are directed to a hyperlinked footnote from the British scholar Leah Scragg and her "interpretative commentary" taken from *Shakespeare's Alternative Tales,* published in 1996: "The imagery here, while conveying the strength of Theseus' desires and his enforced submission to the temporal process, defines authority in the world of the play as female, rather than male. The old moon, explicitly female, is seen as frustrating the happiness of the Duke, and is compared to other authority figures in atypical ascendency over masculine dependents."

For the explanation of but a single word, let's stop on the hyperlink for "happy," from the second line: "Schmidt (1874, 3): 'Favourable, propitious' (*OED*, 3). Cf. 1773. Scott (ed. 1898): 'By the figure of Hypallage, or of Transferred Epithet, the happiness of the expectant lovers is transferred to the time through which they are passing.'"

Such is the detail with which Shakespeare's works are dissected and examined within the NVS website. For nearly 150 years, this information was available only in print form. Imagine the tedium involved in both editing and examining content presented in this manner via the written word.

Beginning in 1933, *The New Variorum Edition of Shakespeare* existed as an official project of the Modern Language Association of America (MLA) under the specific direction of an advisory committee and general editor. In 2019, the NVS executive committee established a "working group" to find a new publisher. Georgetown University English Professor and Shakespearean Scholar Lena Orlin led that effort.

Orlin became a Fellow of the Hagler Institute in 2020.

"For years, the Modern Language Association was the publisher of printed versions of the *Variorum*," Orlin said from London in the summer of 2024 while on a visit to the United Kingdom as part of her duties as a member of the board of the Shakespeare Birthplace Trust. "A few years ago they realized they couldn't keep it up in print form. First of all, it was an extraordinarily expensive project. Scholars spend decades working on this stuff. And secondly, it really needed not to be in print, but to be electronic. And they didn't have the capacity in-house to do a project as technologically sophisticated as this would be.

"In 2019, they convened a working group to figure out where a good home for the NVS might be. And the place that we wound up selecting was Texas A&M. I met with Dr. Laura Mandel, who was the head of the Center of Digital Humanities Research at A&M. She was the one who pitched us to locate the NVS at her school."

At the time, according to Orlin, Texas A&M already had a "distinguished history as a pioneer in important scholarly digital projects," thanks to having migrated the *World Shakespeare Bibliography* (*WSB*) from print to electronic form. This occurred under the leadership of James Harner, a Texas A&M English professor from 1988 until his retirement in 2013. A noted bibliographer himself, Harner had inherited the *WSB* from his A&M colleague Harry Meserole, who began editing the print version of the *Bibliography* in 1976. Soon after the passing of the torch, Harner began imagining a digital future for the *WSB*, which he finally and fully accomplished, to much acclaim, in 2011.

Not long after Texas A&M was awarded the *New Variorum Shakespeare* project, Laura Mandel submitted Lena Orlin's name in nomination to become a Fellow of the Hagler Institute for Advanced Study. Of Orlin, Mandel wrote, "TAMU has been awarded the Digital Publication rights for the *New Variorum Shakespeare*, and Dr. Orlin is working to get TAMU the rights for publishing and cataloguing (via TAMU's *World Shakespeare Bibliography*) images and films from and performances by the Royal Shakespeare Company (UK). Dr. Orlin's presence here could help us turn TAMU into 'The Folger of the West.'"

Located adjacent to the Library of Congress in Washington, DC, the Folger Shakespeare Library houses the world's largest collection of William Shakespeare's printed works. Its namesake, Henry Clay Folger, was a Standard Oil of New York executive whose Shakespearean collection formed the core of what ultimately would be contained within the library, which opened on April 23, 1932, believed to be the anniversary of Shakespeare's birth.

Folger's wife, Emily, also played a key role in the library's birth and growth. According to Orlin, "The future Mrs. Folger held an MA from Vassar College and was ahead of her time in understanding that there were scholarly reasons to consult multiple copies of early books."

Orlin's first job out of college—thanks to her husband landing work as a civil engineer in the Washington, DC, area—was at the Folger Shakespeare Library. She worked there from 1982 to 1996 running the Folger Institute, the library's research center. Thus was deeply instilled within her an appreciation for "The Bard."

"Some people remember the moment they fell in love with dinosaurs," Orlin says bemusingly. "I, on the other hand, do not remember the exact moment I became interested in Shakespeare."

Lena Cowen grew up in Canton, Ohio, south of the Cleveland area, and attended Oberlin College west of Cleveland, where she met her husband, Glenn. She then attended graduate school at the University of North Carolina.

"My freshman year of college," Orlin says, "I took a Shakespeare course and then wound up taking several more. When I went to graduate school, I wasn't actually certain whether I wanted to go into English or art history. I finally decided on English. Several members of the UNC faculty were entrenched in the Renaissance Period, and it turned out those were the courses that interested me the most."

When she was named executive director of the Folger Institute, Orlin had yet to complete her doctoral dissertation. When she decided to cross that bridge, she wound up focusing on "plays that literary scholars call 'domestic tragedies.'"

"The domestic tragedy in literature came of age after Shakespeare's time in the mid-to-late-seventeenth century," Orlin says. "The more of those tragedies I read, the more I got interested in them. They were really radical for their time because they challenged the classical notion of what 'tragedy' actually was. Typically, you had this important and powerful man and he would fall from great height. But in Shakespeare's time, people started writing about ordinary folks, people living with their families who also had tragic experiences."

Orlin's work on her dissertation ultimately led to the other major facet of her career's calling: the study of private domestic life within the Renaissance period.

"Plays in Shakespeare's time were largely performed on an empty stage," Orlin says. "An actor would come out and say to the audience, 'Now we're in the parlor.' And I started wondering, what is the audience supposed to be imagining when somebody refers to a room or refers to even just a piece of furniture in that room? So I started studying that question and traveled extensively around England to look at buildings for the period, to look at domestic furnishings, to think about how Elizabethans understood domestic space, and how they organized their lives within that space. Those inquiries led me to all kinds of historical records.

"In those times," Orlin elaborates, "people wrote wills, often listing all their possessions. By government mandate, those possessions were cataloged or inventoried at time of death. In reading those records, I began to develop an understanding of how people who were not kings were living their lives. I was taking a different approach to history, where people not only pay attention to high politics and international diplomacy, but also to what was happening socially and economically to common people living in the period."

In 2019, as Orlin came aboard to begin a Hagler Institute Fellowship that remains ongoing, she was putting the finishing touches on a book that combined her life's two greatest intellectual interests.

The book, published by Oxford University Press in 2021, is entitled *The Private Life of William Shakespeare*. A *New York Times* review, written by Columbia University Professor James Shapiro, calls the work "a scholarly analysis of Shakespeare's life that reads like a detective story":

> Biographers confronted with the mystery of "How did Shakespeare become Shakespeare?" split into two groups: those who see his early years in his hometown as formative, and those who place greater weight on his experiences in London. Orlin, whose Shakespeare had a "persistent allegiance to his hometown," and whose choices there, she argues, determined the trajectory of his life, belongs to the Stratford camp.
>
> Neither a literary biography nor a full biography, this book looks more narrowly at what surviving documents tell us, and, when their trail runs dry, what documents about his neighbors might reveal about events that defined Shakespeare's Stratford life: his father's financial collapse, his marriage, his homes (including the "Birthplace," likely damaged by fire in the 1590s, then rebuilt), his will and his memorial. Though most of it consists of dense scholarly analysis, it reads like a detective story in which a skilled investigator returns to a cold case.
>
> It amounts to a revisionist portrait of the artist.

As for the matter of what was the biggest "domestic tragedy" in Shakespeare's own private life, Orlin offers this: "Probably that he was not survived by a son. He had three children: two girls and a boy; first, a girl, and then a pair of twins. That boy"—Hamnet was his name—"died when he was eleven. In those times, that was the child who would have carried the Shakespeare name forward.

"I would think of that as his greatest tragedy."

Shakespeare is believed to have written his play *The Tragedy of Hamlet, Prince of Denmark* somewhere between 1599 and 1601. His son, Hamnet, died in 1596. Recent scholarship suggests that Shakespeare may have been interested in this old Scandinavian story on which *Hamlet* is based because, in his continuing grief over the loss of his son, the name "Hamlet" may have caught his attention.

Tragic, too—or nearly so—were efforts on the Texas A&M campus to do away with the *New Variorum Shakespeare* online project.

Lena Orlin never imagined her stint as a Fellow of the Hagler Institute for Advanced Study would include defending a "prized project" both for her personally and for Texas A&M University institutionally.

"It was not a happy time," she says.

A week before Orlin's interview for this book, she wrote about the *Variorum* crisis in a letter to Hagler Institute Director John Junkins and Associate Director Clifford Fry:

> I am writing as we turn a corner with the *New Variorum Shakespeare Editions,* now hosted by the English Department at Texas A&M University. As you know, I serve as chair of the Board of Governors for the *NVS* and represent the *NVS* General Editors and Volume Editors.
>
> We had a difficult time last year, as we considered moving the *NVS* to another location. But we have emerged from it all stronger than any of us could ever have hoped or imagined. For the first time in its history, the *NVS* has a dedicated support staff of research scientists, technical experts, and digital editors. Emily Johansen, Chair of TAMU English, describes the *NVS* as a "flagship project for the department, a groundbreaking venture around which other faculty research in digital humanities and textual editing will coalesce. The scale of work for Shakespeare is so large that if we solve editorial problems for him, we will have solved them for anyone."
>
> In a recent meeting of *NVS* stakeholders—the Board, the General Editors, and the TAMU team—we had the heady experience of looking forward: new ideas, new horizons, new ambitions. Some of us had recently taken part in a conference organized by the English Department on "Folio Futures: Editing Early Modern Plays for Tomorrow's Audiences." This was a brilliant display of twenty-first century innovations in research.
>
> The *NVS* would not be at TAMU without your backing. It is good to have friends in high places! I am so grateful for all the

support you have offered both to me personally and to the *New Variorum* project.

For nine years, M. Katherine Banks served as dean of the Texas A&M College of Engineering. Under her leadership the college grew dramatically: in enrollment, programs, research dollars generated, and number of new faculty hired. Her arrival roughly coincided with the launch of the Texas A&M Institute for Advanced Study, and she was a strong proponent of the program. It was no surprise that when Michael Young announced his resignation as president, Banks quickly emerged as the leading candidate to replace him, ultimately becoming president of the university on June 1, 2021.

"Behind closed doors," as an investigative piece from the *Journal of Higher Education* reported, Banks soon began the process of seeking to implement her new vision for how the university would be run. A management consulting firm was hired, and by the fall of 2021, a new direction began to take shape.

Among the moves garnering the most attention, both inside and outside the university, was her decision to combine the College of Liberal Arts, the College of Science, and the College of Geosciences into a new College of Arts and Sciences. With consolidation came downsizing: in jobs, budgets, and programs. Many members of the Texas A&M Faculty Senate grumbled at the lack of transparency—or desire for input—that came with Banks's decision-making. But she was not to be dissuaded . . .

. . . until what first seemed a triumphant return home by a former A&M student turned college professor brought an ugliness to light that significantly tarnished both Banks's reputation and the school's image.

Kathleen McElroy, a Black woman, graduated from Texas A&M University in 1981 with a degree in journalism. She went on to build a successful career as a reporter, eventually going to work for *The New York Times*. There, over a nearly twenty-year period, McElroy advanced through a number of administrative positions to become assistant managing editor. She left the paper in 2010 to pursue her PhD at the University of Texas. Her doctoral thesis there was entitled "Somewhere Between 'Us' and 'Them': Black Columnists and Their Role in Shaping Racial Discourse."

By 2018, McElroy had become director of the University of Texas School of Journalism and Media. Then in June of 2023, she was selected to return to her alma mater to lead Texas A&M's revamped journalism program. Her hiring was publicly announced on June 13 and celebrated with great fanfare.

But behind "closed doors," the move was questioned by outspoken alumni, including members of the Texas A&M University System Board of Regents. Citing uncertainties regarding the State of Texas's ongoing efforts to dismantle the university's diversity, equity, and inclusion program and policies, McElroy's initial offer of full professor with tenure at A&M was rescinded and replaced with a one-year agreement. At that point, McElroy did what so many other "sources" had done when she was a reporter: She took her story to "the press."

On July 19, President Banks told members of the Texas A&M Faculty Senate that she did not know why the original offer was rescinded. Later that day, the senate passed a resolution to create a fact-finding committee into the mishandling of McElroy's hiring. Two days later, Banks announced her "retirement" as president of the university.

In an internal report on the incident, conducted by the university system's office of general counsel and released less than two weeks after Banks's departure, a trail of emails and text messages painted a very disturbing picture of what had taken place.

According to a *Texas Tribune* story published on August 3, 2023, "The internal inquiry showed that the effort to promote conservative values at A&M extended beyond the journalism program. Texts between [regent Jay] Graham and regent David Baggett about McElroy's hiring referenced Banks' 2022 move to combine several academic programs into a new College of Arts and Sciences as part of a new strategic plan to reshape the university's structure."

"Kathy told us multiple times the reason we were going to combine arts and sciences together was to control the liberal nature that those professors brought to campus," Graham wrote. "This won't happen with this kind of hire," a reference to McElroy's being offered the journalism job.

Banks chose José Luis Bermúdez to be her interim dean of the revamped College of Arts and Sciences. Bermúdez came to Texas A&M in 2010 as dean of the College of Liberal Arts. He subsequently served as associate provost for strategic planning and as acting dean of the College of Dentistry. Prior to the McElroy debacle, Bermúdez was charged with reducing programs within the new College of Arts and Sciences. In those efforts, he set his sights on redefining the nature of the Center of Digital Humanities Research. One way to do that was to eliminate the *New Variorum Shakespeare* project.

Did Bermúdez have specific instructions to do so, or was he just trying to do what ultimately proved to be an impossible job? Whichever the case, he resigned as interim dean four days before Banks's departure. Bermúdez remains on the A&M faculty as a professor of philosophy, although he took an extended leave following his departure as arts and sciences dean.

"I don't think there's any way in which the *Variorum* would still be at Texas A&M if it weren't for the Hagler Institute," Lena Orlin says with conviction. "I was able to talk with John Junkins and Clifford Fry about what was happening. I told them I would have to resign my Hagler Fellowship because I didn't think I could be associated with the school while trying to find an alternative residence for the *NVS*. It would be unethical toward A&M if I were trying to find the *NVS* a safer home someplace else.

"But thanks to John and Clifford, I had the advantage of being able to talk to people in high places within the university. We were not able to save the *World Shakespeare Bibliography*, but we did save the NVS."

From Orlin's vantage point as an outsider to the inner workings of Texas A&M, she is impressed with how the university has turned the corner from its recent "darker days."

"One of the things that has been most interesting to me being 'on the ground' here is how self-healing Texas A&M has been through all of this," she says. "A key element to that was the university's legal division bringing everything out into the open."

The last four years have been a roller coaster ride for Lena Orlin. As one of the world's preeminent experts on Shakespeare—she currently sits on the board of nearly a dozen global Shakespearean organizations—her addition to the Texas A&M faculty as Hagler Permanent Fellow would seem to have been an obvious offer waiting to happen. Unfortunately, not so, as Orlin explains.

"I have to say, one of the things that I admire about the Hagler Institute is that it has been so effective at hiring faculty to come to

Texas A&M full-time. I don't know whether that was John's original intention or whether it's something that he kind of discovered along the way, but it's been a great recruitment tool. I think that's been a really important success story for the Hagler Institute.

"I haven't really thought about that for myself because when I first heard from the Hagler Institute, we were on the cusp of the pandemic. My university, Georgetown, saw that as a good opportunity to encourage older faculty to retire. So I got a buyout offer and retired earlier than I intended to because it was an attractive deal.

"So I kind of feel like I'm retired now," Orlin laughs. "That's why I haven't thought about taking on another full-time job."

Before concluding the Zoom call from London, Orlin asked for the opportunity to add one more thing to her story.

"In comparison to other places where I've had fellowships or residencies, the Hagler Institute offers a very high level of engagement with the university. I think my situation was somewhat unusual because of the crisis that happened last year. I spent much more time trying to save the *New Shakespeare Variorum* than I did doing my own research. But from what I've seen and experienced, people who come here as Hagler Fellows, whether to teach courses or work in labs, engage with students and faculty in a very special way.

"These interactions and engagements are deeply woven into the fabric of Texas A&M in ways that don't happen elsewhere.

"If you had an institute that was located in its own building and all the Fellows just went to that building and did their research, that would be very isolating. But that's not the way it works here. We're integrated into the university community much more than other places.

"I know how much I appreciate the help I received in keeping the *NVS* here. I just hope that Texas A&M recognizes the value that the Hagler Institute brings to the university as a whole."

27

The 5:00 a.m. Phone Call

Donna Strickland, 2021–2022

By the first Tuesday of October 2018, autumnal splendor ran rampant throughout the Canadian province of Ontario. In the far western reaches of the Toronto metropolitan area, morning had yet to be broken.

But for one woman whose life was mostly routine and without great fanfare, this was to become a day unlike any other.

"The 5 a.m. phone call changed Donna Strickland's life in an instant."

So wrote *NBC News* Science reporter Denise Chow in her story about Strickland's landmark achievement.

> It was Tuesday, Oct. 2. A representative from the Royal Swedish Academy of Sciences was on the line from Stockholm calling to inform Strickland, an associate professor at the University of Waterloo in Canada, that she had won the Nobel Prize in Physics.
>
> Strickland, a self-described "laser jock," recalled that it was a surreal moment: "No one can be ready for such an honor."
>
> In winning the coveted award, which she shares with American physicist Arthur Ashkin and Gérard Mourou of France, Strickland became only the third woman ever to win the physics prize, and the first woman to win since 1963. Ashkin will receive half of the $1.4 million prize, and Strickland and Mourou will split the other half.
>
> Strickland was chosen for the award in recognition of her role in developing a technique to generate the shortest, most intense laser pulses ever created. The research, done while Strickland was a graduate student working for Mourou at the University of

Rochester, led to new techniques for corrective eye surgery and other applications in medicine and manufacturing.

A day after the announcement, NBC News MACH called Strickland to ask what it felt like to win a Nobel Prize for work she did in her 20s.

MACH: The research paper that led to your Nobel was published in 1985, when you were only 26. How do you top that?

Strickland: It's hard to copy that, and so I don't know that I've tried. My husband's grandfather was an artist, and he did [a bust of] Alexander Graham Bell, which I have in my foyer. [Bell's] children were very upset with that. They said, "You've made him look like a grumpy old man." And then a newspaper article had Alexander Graham Bell saying, "I am a grumpy old man." He really did talk about how he always thought he could top the telephone. Sometimes you just have to be happy you get one chance in life and not worry too much about whether you get two.

In late 2021, three years and four days after learning she would receive the Nobel Prize, a University of Waterloo press release made this announcement:

University of Waterloo Professor and Nobel Laureate Donna Strickland has been named a Fellow of the Hagler Institute for Advanced Study at Texas A&M University.

The Hagler Institute announced its 10th class of Hagler Fellows on September 28, and Professor Strickland was among the eight honourees. The list of Fellows includes scientists, engineers and scholars who are recognized internationally for their achievements. Each belongs to the National Academies of Sciences, Engineering and Medicine, or hold recognitions of equal stature in their fields.

"Strickland was awarded the 2018 Nobel Prize in Physics for developing chirped pulse amplification, used in corrective eye surgery, industrial machining and medical imaging," says her citation. "She is a member of the National Academy of Sciences and a fellow of the Royal Society of Canada. Strickland will collaborate with faculty and students in the Institute for Quantum Science and Engineering, and the colleges of science and agriculture and life sciences."

Strickland is one of six Nobel Prize winners to be named Hagler Fellows at Texas A&M. The first was Vernon Smith, a member of the inaugural class of 2012–13. Smith won his Nobel Memorial Prize in Economic Sciences in 2002 as a pioneer in the field of experimental economics. Roy Glauber, a member of the TIAS Class of 2013–14, won his Nobel Prize in Physics in 2005 for his work, like Strickland, with laser technology.

Since Strickland, three more Nobel laureates have become Hagler Fellows. Michael Young was a member of the Class of 2022–23 and earned his Nobel Prize for the study of the internal clock contained within cells that helps humans adapt their biological rhythm to the different phases of day and night. Two more Nobel laureates were included in the Class of 2023–24: Eric Maskin, another winner of the Nobel Memorial Prize in Economic Sciences, and Konstantin Novoselov. Maskin's 2007 award stemmed from his work on mechanism-design theory, while Novoselov's Physics Prize in 2010 came as a result of his production of the carbon-based material graphene, which is incredibly thin—just one atom in thickness—yet incredibly strong.

If Strickland is a "laser jock," then her longtime friend Marlan Scully, founder and head of the Texas A&M Institute for Quantum Science and Engineering, is laser physics' "Teddy Roosevelt." The "teddy bear" was named in honor of Roosevelt and one of the few bears he did not attempt to gun down as a big-game hunter. In the same vein, a giant stuffed grizzly bear stands menacingly in the welcome area to Scully's office in the George Mitchell Physics Building on the Texas A&M University campus. It was Scully who shot the bear and Scully, along with his IQSE associate Olga Kocharovskaya, who nominated Strickland for her Hagler Fellowship.

In that nomination they wrote, "[She is a] leading figure in the area of ultrafast laser physics, known for developing high-intensity laser systems for nonlinear optics investigations. She pushes the boundaries of ultrafast optical science to new wavelength ranges such as the mid-infrared and the ultraviolet, using techniques such as two-colour or multi-frequency methods, as well as Raman generation. She is also working on the role of high-power lasers in the microcrystalline lens of the human eye, during the process of micromachining of the eye lens to cure presbyopia."

Of course, their nomination form had the full attention of the Hagler selection committee with the words "2018 Nobel Prize winner."

As for Strickland's work, which became Nobel-worthy more than thirty years after the fact, the *Rochester Review* spelled out the details in the fall 2018 issue of the University of Rochester alumni publication.

> When Strickland entered the University's graduate program in optics, laser physicists were grappling with a thorny problem: how could they create ultrashort, high-intensity laser pulses that wouldn't destroy the very material the laser was used to explore in the first place?
>
> Working with former Rochester engineering professor Gérard Mourou, Strickland developed and made workable a method to overcome the barrier. They called it "chirped pulse amplification."
>
> The problem Strickland and Mourou overcame was one that would have ignited the imagination of any physicist who, like Strickland, experiences scientific discovery as a form of play. When it comes to lasers, she says, "I've always had fun playing with them. I do often think of it as playing, not work."
>
> But their solution had important practical ramifications. In the 30 years since Strickland and Mourou conducted the research, chirped pulse amplification has made it possible to use lasers effectively in a wide range of medical, scientific, and commercial applications.
>
> As laser science grew as a discipline in the 1980s, researchers were incrementally increasing the intensity of laser pulses, resulting in damage to the amplifying material. Mourou had come up with an idea to clear the hurdle by perfecting the technique known as chirped pulse amplification. The technique involved a three-part sequence: stretching a laser pulse thousands of times so that the power was low; amplifying the pulse to higher intensities; and then compressing the pulse in time back to its exact original duration.
>
> Mourou knew the pulse needed to be perfectly compressed, yet still retain its amplification, in order to make the technique work more effectively. "If you can do it exactly, then you can go to much higher power," Mourou says.

He had the idea to put amplification in the middle of the process—a novel concept at the time, according to Strickland. "Different people were trying to get short pulses amplified in different ways, but it was thinking outside the box to stretch first and then amplify," she says.

Chirped pulse amplification, however, was only a theory. Strickland and Mourou still had to make it work. At the Laboratory for Laser Energetics, Strickland tested different laser systems to create laser pulses that were short and high powered, but that wouldn't destroy the amplifying material. In 1985, she succeeded, demonstrating the stunning advance in laser power with the tabletop terawatt laser, or "T-cubed laser."

As for the phone call decades later that would change Strickland's life, it wasn't quite as straightforward as *NBC News* first reported. Let's let Strickland set the record straight.

"I have tried to lead a low-profile life," she says. "I'm sort of an introvert, and so the Nobel committee, as it turned out, couldn't find my phone number. They called Gérard first and after they gave him the news about winning the Nobel Prize, they asked if he knew how to reach me. He had our home phone number, and so the call came in at around five o'clock in the morning."

That's 10 a.m. Swedish time.

"When you hear the phone ring at that time of day, you think it's going to be about bad news," Strickland says. "My husband, Douglas, answered, and the voice at the other end of the line asked for 'Professor Strickland.' Since his last name is Dykaar and he took his PhD in electrical engineering into the private sector, he handed the phone to me. When I said hello, they told me this was an important call from Sweden. 'Please stay on the line.'

"That's when the possibility first hit me. It was October. They're calling from Sweden. Have I just won the Nobel Prize?

"And then the call got disconnected.

"Later they said they thought I hung up on them, assuming it was a prank call, but that wasn't the case. I waited another fifteen minutes or so for someone to call back, and when they didn't, I thought to check my email. And when I did, there was an email saying, 'Please call us,' with a number for the Royal Swedish Academy. So that's when I knew, and technically, I knew before they actually told me. And I immediately thought I knew why I would be getting this phone call because in my career I've only done the one thing that's worthy of the Nobel Prize."

Her PhD thesis at the University of Rochester was the catalyst for the honor, right?

"Well, that's the joke between Gérard and me. It wasn't really my thesis. It was the main tool I helped develop, that 'desktop' laser from which I could do my thesis. Gérard was the idea guy."

And the reason she chose to pursue her graduate work in the United States?

"It was in 1983, a year before I arrived in Rochester, that Gérard got to thinking that one needed to stretch the pulse of a laser in order to amplify it, but no one had the wherewithal to do that. All of that interested me greatly, and Gérard invited me to come and work with him. Then a year later after my arrival, we were at the Ultra-Fast Conference, and the thing that was all the rage there was pulse compression. We learned that if you wanted to compress the pulse, you would actually have to use a very long fiber to allow the dispersion and the nonlinear process that makes the extra colors.

"So the original thinking was to generate a 'linear chirp,' where the red would be way out on the front and the bluest colors way in the back. That was the start, but we needed to compress the colors so that they all came quickly. I remember Gérard gathering our group together at the conference to explain how we would approach that challenge. I was the one in the group already trying to create a high-intensity laser, and I was the only one whose project was about using short pulses to make the power higher to do nonlinear optics. So of all the people in his group, that's why it fell to me to do it."

How long did Strickland actually work on the chirped-pulse-amplitude project?

"We started in the summer of '84 and, and we sent off the publication in the summer of '85."

Strickland continued work toward her PhD for another five years. During that time, she became familiar with work being done by French physicist Anne L'Huillier in the field of high-harmonic generation. "She saw the thirty-third harmonic while I was trying to see the ninth," Strickland says. "No one was expecting that, really. That kind of blew my thesis out of the water."

And in 2023, earned L'Huillier her own Nobel Prize in Physics.

While Strickland was the first woman in more than half a century to win the Nobel Prize in Physics, since then, L'Huillier and Andrea Ghez in 2020 have also broken that gender barrier.

As for how her life has changed since becoming a Nobel laureate, Strickland says, "In ways both good and bad. I don't really like being the female role model. For me, it's too much to wear. It's obviously not something I studied. Frequently I get asked about things I don't know a lot about, and I think, 'I'm not an expert in this field.' You can ask me about short pulse lasers, but don't ask me about what it means to be a female scientist.

"On the other hand, I've had an audience with the pope. I met Rock 'n' Roll Hall of Famer Brian May of Queen when I attended his event, the Starmus Festival. I've sat down with the prime minister of Canada, the president of Spain, and spoken to the leaders of the Mexican government.

"When you receive an honor of this magnitude, there's just things that you end up doing and, all of a sudden, you have this voice, and you have to figure out what you're going to do with that voice."

Out of curiosity, which was the bigger thrill: meeting the pope or Brian May?

"It was the same," she says, "but let me tell you about the time I met another rock star, Steven Tyler of Aerosmith."

Certainly!

"We were getting the same award in 2019 at the St. Regis Hotel in New York. *The Golden Plate of Achievement* is presented each year by the American Academy of Achievement. Tyler was sitting behind my table by himself at the dinner that night. As I remember, he was alone because the person that was going to give him his award hadn't shown up yet."

Later, Tony Fadell, developer of the Apple iPod, did arrive and presented Tyler his prize.

"We were seated in alphabetical order," Strickland continues, "and I happened to be at a table with Kip Thorne, who won the Nobel Prize in Physics in 2017. Nobel Prize winner France Arnold was presenting me my award, and two other Nobel-winning physicists were presenting Thorne his.

"And there was Steven Tyler, behind us, all by himself.

"I don't know why, because Steven Tyler obviously didn't know me from a hole in the ground, but I just turned around and said to him, 'How did a famous guy like you get stuck behind a table full of physicists?' After the words came out of my mouth, I was a little embarrassed because I had yet to become comfortable in dealing with 'fame.' But he was great. In response, he leaned forward, reached out, took my hand, and said, 'I love physicists.'"

Texas A&M University faculty, students, and staff have fallen in love with Donna Strickland during her regular visits as a Hagler Fellow.

And the feeling is mutual.

"Marlan is incredible at gathering us all together at his IQSE and making us all feel like family," Strickland says. "I was on campus in late 2023, and I said to my husband, Doug, who was with me, 'We're going to have to think of something to do for American Thanksgiving.' Well, I soon found out that some seventy people associated with the IQSE all get together and have Thanksgiving dinner at the Institute.

"That dinner was amazing! I remember thinking, 'Oh my God, these people all love each other so much, they're willing to spend time together on their day off!' Many of them brought their families and we all had this big party together. I just think that's amazing!"

Unique to Strickland's Canadian sensibilities regarding the culture surrounding Texas A&M is how Aggies come together for home football games at Kyle Field.

"We don't have the affinity for college sports in Canada that I've seen at Texas A&M," Strickland says. "I was taken to a football game. I got to attend something called the Chancellor's Tailgate at the on-campus hotel and then sit in the Board of Regents suite to watch the game. It is amazing how many alumni come back to campus for this experience.

"This is not something you see in the Canadian educational system. But we also don't have the kind of fundraising prowess that American universities like Texas A&M have. Sports bolsters this incredible school spirit that stays with people. They want to come back and, most importantly, they feel like they should help fund what goes on at their school.

"That's just a whole different ethos of how American and Canadian universities work."

Donna Strickland is an easy person to talk to. As a "laser jock" and mother of two—"I never had time for sabbaticals throughout most of my academic career because I was raising my kids"—she has a down-to-earth sensibility about the twists and turns of life.

So if you could ask a Nobel laureate anything about . . . anything, what would it be?

"Where do you keep the medal?"

The Nobel Prize website, www.nobelprize.org, describes the ceremony at which laureates are honored each December.

"The award ceremony in Stockholm has, almost without exception, taken place at Konserthuset Stockholm (the Stockholm Concert Hall) since 1926. . . . At the [ceremony], speeches present the laureates and their discovery or work, after which His Majesty the King of Sweden hands each laureate a diploma and a medal."

As for that medal, the exact weight varies slightly, as different medals are awarded in the field of Economic Sciences. Each medal is eighteen-karat green gold plated with twenty-four-karat (pure) gold, with an average weight of around 175 grams. Recent Nobel Prize medals are thought to be worth in excess of $10,000.

"The gold medal I received from the King of Sweden is actually in our bank vault," Strickland says, "because my brother was all upset about someone trying to steal it. Otherwise, the fake gold medal, my certificate of achievement, and an original piece of art each honoree is given are on the mantel above our fireplace at home."

In the big scheme of things, at least for someone like Donna Strickland, a Hagler Fellowship pales, at least a little, in comparison to winning the Nobel Prize. However, the announcement the University of Waterloo released regarding her Hagler Fellowship seems to suggest otherwise.

So did it take a lot of convincing before Strickland made the decision to accept her Hagler Fellowship?

"It took some time," Strickland says with a smile on her face. "But it's very hard to say no to Marlan? Have you met him?"

Yes, we have. And we've stood toe-to-toe with his bear.

"I certainly didn't say yes to him while I was out there the first time," Stickland says. "But he had me meet Bill Unruh, who was a Hagler Fellow at the time. Bill is a fellow Canadian. He's at the University of British Columbia, and he's very famous for his work on Hawking radiation.

"Marlan, Bill, and I all went out to dinner together. There, Marlan had Bill tell me what a great thing the Hagler Institute is. So between the two of them, it was hard to say no."

Texas A&M University and its alumni base are glad she said yes.

"Gig 'em, Donna!"

28

Lindau Laureate

Stefan Kaufmann, 2018–2019

For centuries, many of the world's great scientific thinkers spoke German as their native tongue. From Johannes Kepler to Max Planck to Albert Einstein, groundbreaking discoveries and seminal scientific ideas were spawned by those raised in the lands today known as Germany.

In February 1945, on the heels of its defeat in World War II, Germany was "dismembered" geographically by the victorious Allied forces. The country was also punished politically in many important ways. Because of the human atrocities committed under the Nazi regime both before and during the war, Germany's previously distinguished scientific community was ostracized by many of its global peers. It took until 1951 for a pair of medical practitioners from the Bavarian town of Lindau to put forth a concept to reengage German scholars with the rest of the scientific world. They were able to do so with the support of a member of the Swedish royal family who happened to be a nearby neighbor.

Thus were born the Lindau Nobel Laureate Meetings, which for more than seventy years have brought recipients of Sweden's preeminent scientific, economic, and political prize together to engage, at first with German researchers and physicians, and then, ultimately as it does today, with young scientists from around the world for the exchange of knowledge between nations, cultures, disciplines, and generations.

On July 12, 2023, Texas A&M University announced a partnership with the Council for the Lindau Nobel Laureate Meetings to become "the U.S. university provider of students for the annual world gathering." The idea stemmed from a collaboration of minds—both German—on the A&M campus in 2020, a partnership kindled by the Hagler Institute for Advanced Study.

In Germany, some may know the name "Stefan Kaufmann" as a musician who achieved European acclaim with the heavy-metal bands Accept and U.D.O. Back problems forced him to give up playing the drums, but he persevered and became a guitarist.

Others will remember "Stefan Kaufmann" as a German politician and member of the *Bundestag*, the only nationwide representative body that is directly elected by the German people. He held office there from 2009 to 2021 and is openly gay.

Neither of those "Stefan Kaufmanns" played a role in Texas A&M's partnership with the Lindau Nobel Laureate Meetings. That distinction belongs to Hagler Fellow Stefan H. E. Kaufmann, a German immunologist and microbiologist who was a member of the Hagler Class of 2018–2019. The Hagler Institute's Stefan Kaufmann is also a senior professor at the Charité University Clinics in Berlin, a Director Emeritus for the Max Planck Institute for Multidisciplinary Sciences in Göttingen, Germany, and a member of the Council for the Lindau Nobel Laureate Meetings, which governs that organization and, most significantly according to the Lindau website, "ensure[s] that eligible and qualified young scientists get the chance to participate in the Lindau meetings."

In 2017, Ramesh Vemulapalli, a professor and department head in the Texas A&M College of Veterinary Medicine & Biomedical Sciences, put forth Kaufmann's name in nomination to become a Fellow of the Texas A&M Institute for Advanced Study. In that nomination, he wrote of Kaufmann,

> His extraordinary expertise in vaccinology will have a major impact on the discovery, development, and manufacturing of human and animal vaccines in the National Center for Therapeutics Manufacturing, the Texas A&M University Center for Innovation in Advanced Development and Manufacturing, and local and international biopharma companies. His extensive experience with select agents will accelerate the development and testing of vaccines for high-consequence diseases in the new Global Health

Research Center biocontainment facility which will be commissioned for research in mid-2019. Research interests of several Texas A&M faculty are in the areas of Prof. Kaufmann's expertise.

Among those who supported Kaufmann's nomination to become a Hagler Fellow was Garry Adams, a senior professor in the Department of Veterinary Pathobiology and member of the Aggie Class of 1964. According to Kaufmann, his professional relationship with Adams went back many years.

"I met Garry when he came to the University of Ulm on sabbatical in my department sometime in the late 1980s," Kaufmann remembers. "Then a few years later, maybe twenty-five or thirty years ago, Garry invited me to give a talk at Texas A&M.

"The place has grown a lot since then!"

Although Kaufmann has specialized in human immunology for most of his career, his Hagler Fellowship gave him the chance, once again, to work with the eminent scholars and researchers to be found in the Texas A&M College of Veterinary Medicine and Biomedical Sciences. "I learned a lot from my colleagues in veterinary medicine," Kaufmann says. "The study of disease, any disease, whether by veterinary or human doctors and researchers, is important work."

Among those doing important work in veterinary medicine at Texas A&M is Jörg Steiner, a University Distinguished Professor and director of the gastrointestinal laboratory. And like Kepler, Planck, Einstein, and Kaufmann, Steiner is a native German.

"I was born in Germany," Steiner says, "and went to high school there. I then attended the University of Munich and studied veterinary medicine. When I graduated, I did an exchange at Kansas State University. I went there for a year in exchange for a student going to Munich. After that year, it became clear to me that my career would need to be in the US. So after I finished vet school in Germany, I came back and did a residency at Purdue University. I started a PhD there, but then my mentor, David Williams, moved to Texas A&M in 1997 to become a department chair, and I moved with him to help relocate his lab.

"I then finished my PhD here and was supposed to go somewhere else, but I didn't really want to. So David helped me put together a strategy to create a position for me here. Initially, I was a visiting professor, then a clinical assistant professor. Then I got on a tenure track and, you know, the rest is history, so to speak."

As for meeting Kaufmann during his stay on campus as a Hagler Fellow, Steiner says, "You don't always get to meet the Hagler Fellows for a one-on-one conversation, but Stefan was spending a good bit of time with several of my colleagues here at the vet school. And because he's German and I'm German, we became friends, and he eventually told me about the work he's done in Lindau.

"In Germany every scientist knows that meeting and, in fact, I would say that all over Europe everyone in science knows this meeting. But to be honest, I didn't know about Lindau until Stefan first told me about it."

In their discussions about the Lindau Nobel Laureate Meetings, Kaufmann told Steiner very few American students participated. "That surprised me," Steiner says.

"Stefan explained that the University of California System has a partnership agreement with Lindau, but they weren't sending many students to the meetings. It turned out there was no other centralized mechanism for US students to seek involvement."

What Steiner learned about the Lindau meetings from Kaufmann left a lasting impression. Soon Steiner turned his attention to determining a means by which Texas A&M students could gain access to the event. He reached out to Lindau on a personal basis. They told him they weren't interested in establishing new contracts with "isolated universities" because they wanted the "best students from the US," not just students from a particular school.

"If you look at the Royal Academy in London," Steiner says, "each year they announce they are taking applications to attend the Lindau meetings, They get a lot of interest and from that, they can pick the best twenty or thirty students from the UK and that group gets sent to Lindau."

Steiner shared the Royal Academy model with Texas A&M Interim Provost Tim Scott, who liked his idea. Together, Scott and Steiner laid out a plan to get Texas A&M involved as an "institutional clearinghouse." That role, if accepted, would guarantee a certain number of Aggie students be included each year, with the remainder coming from other colleges and universities. But soon, the COVID-19 pandemic hit, and Steiner was forced to shelve his Lindau dream for nearly two years.

The gathering of Nobel laureates in Lindau, located along Germany's southern border with both Austria and Switzerland, was originally the brainchild of local physicians Franz Karl Hein and Gustav Wilhelm Parade. Key to their concept, the pair hoped a local resident, who happened to be the grandson of King Gustav V of Sweden, might help advance their cause.

Count Lennart Bernadotte of Wisborg grew up to be a landscaper, filmmaker, photographer, and activist. As a young Swedish royal, Lennart had fallen in love with the small German island of Mainau in Lake Constance, owned by his grandmother, Queen Viktoria. Mainau is not far from the city of Lindau. In fact, much of Lindau also sits on a Lake Constance island. When Viktoria died in 1930, she left Mainau to Count Lennart, then twenty-one years of age. At that point, he took up management of the "palace island" and eventually moved himself and his family there, becoming active and well known within the Lindau community while doing good works in many other parts of the world.

Since the first Nobel Prizes were awarded in 1901, the Swedish royal family has played a key role in ceremonies recognizing honorees. In fact, Count Gustav, who would later become King Gustav V, handed out the first medallions—in the absence of his father, King Oscar—in the large hall of the Royal Swedish Academy of Music in Stockholm. Because of Count Lennart's connection to the royal family, Drs. Hein and Parade believed he could help influence Nobel laureates to visit Lindau and rekindle relations with Germany's scientific community. Seven laureates attended the first meeting in 1951, devoted to the study of medicine and related fields. They were joined by more than four hundred scientists from across the European continent, including many making a return to their homeland after leaving Germany during the war.

Count Lennart became president of the Lindau council and remained in that position for decades, significantly influencing the direction of the organization. Since 2008, his daughter, Countess Bettina Bernadotte, has served as president of the Council of the Lindau Nobel Laureate Meetings.

As for Jörg Steiner, while pursuing a means to involve Texas A&M and its students in the Lindau Nobel Laureate Meetings, he separately became involved with the Hagler Institute for Advanced Study, first as an elected member of the Institute's Advisory Committee and then as an academic advocate within the veterinary medicine school.

"Every year John Junkins appoints what he calls 'faculty liaisons' to the Hagler Institute," Steiner says. The faculty liaison is a person who connects the Institute to an underrepresented school or department within the university. John looks at where Hagler Fellows' nominations are coming from, and if a school has been underrepresented for a few years, he'll find a liaison within that school to get more Hagler nominations.

"When I became the faculty liaison for veterinary medicine, I was already working to cultivate a relationship with the Lindau meetings," Steiner says. I told John I thought the program fit very well into the portfolio of the Hagler Institute, and John agreed. Ultimately, he offered to pick up 20 percent of A&M's financial requirements to create our Lindau affiliation.

After the worst of the COVID-19 pandemic had subsided, Steiner began discussions with then–Interim Provost Alan Sams about forging a strong relationship with the Lindau Nobel Laureate Meetings. After devising a plan, Sams attended the Lindau meetings in 2023, where he signed the memorandum of understanding that makes Texas A&M a nominating institution.

"The agreement we reached with the Lindau council fit much better under the Hagler Institute umbrella than it did as a product of my lab," Steiner says.

The list of nominating entities for the Seventy-Third Lindau Nobel Laureate Meetings in 2024 included nearly 150 academic institutions from around the world. Only two are American: the University of California System and, for the first time, Texas A&M University.

Along with Steiner and Sams, five Texas A&M graduate and professional students attended the Seventy-Second Lindau Nobel Laureate Meetings in 2023. Included in that group was Evan Lloyd, a fifth year PhD student in the department of biology. The Texas A&M College of Arts and Sciences chronicled his "once-in-a-lifetime experience" at the Lindau convocation:

> This year's prestigious scholarly gathering featur[ed] around 40 Nobel laureates and guests from 90 countries. Lloyd and his fellow students were nominated by faculty and selected based on demonstrated excellence in research documented by publications, conference presentations and/or other research awards and prizes.

Their participation and travel expenses were covered by the Hagler Institute for Advanced Study.

While in Lindau, Lloyd took full advantage of his opportunities to take in Nobel laureate lectures and be exposed to master classes, academic dinners, and both roundtable and panel discussions on a variety of topics, from artificial intelligence in medicine and the effects of climate change on health to the latest in emerging avenues of physiology and medicine-related scientific exploration. As in previous years, the event concluded with an all-day tour of Mainau Island highlighted by a science picnic hosted by the Ministry of Science, Education and the Arts, State of Baden-Württemberg.

"All the speakers gave fascinating and engaging lectures on a wide range of topics," Lloyd said. "Some chose to give an overview of their Nobel Prize–winning work, while others showed us their current work, which proved to be just as groundbreaking."

During the final day of the meeting featuring a boat excursion to Mainau Island and a round table on climate change, Lloyd said he enjoyed the opportunity to strengthen the connections he had made with other young scientists throughout the week.

"All in all, the Lindau meeting was a fantastic, once-in-a-lifetime experience," Lloyd added. "I'm very grateful to have been given the opportunity."

In each year of Texas A&M's four-year memorandum of understanding with the Lindau meetings, Texas A&M will send a total of twenty American students to Germany. At least seven members of that group will be doing postgraduate coursework with the Texas A&M University System. The balance of those twenty students will represent academic institutions from across the United States. In advance of the group's travels to Germany, selected students will meet for two days on the Texas A&M campus to gain an understanding of how to maximize their Lindau experience.

"We'll answer questions and help make travel arrangements," says Steiner, "but we also hope to give them an appetite for coming to Texas A&M one day, perhaps as a postdoc or, possibly, to become a member of our faculty here."

The benefits emanating from the work done through the Hagler Institute for Advanced Study are widespread and far-reaching. For instance, Texas A&M's—and the Hagler Institute's—Lindau Nobel Laureate Meetings story begins with Hagler Fellow Stefan H. E. Kaufmann. His own Lindau experience—as not yet a Nobel Prize winner himself—began as a result of his relationship with another European immunologist.

"I was approached by a Swedish member of the Lindau council, Klaus Kärre," Kaufmann says. "He knew me as a scientist and someone who has a broader view on people and society than a lot of researchers. He asked me if I would like to pursue an affiliation with the Lindau meetings, to which I ultimately said yes. I was there for a few years as a kind of corresponding member to the council before I became a full member. And I have to say, I really like it a lot. It is something which brought back my interest in interdisciplinary activities. And in that respect, it's very similar to my Hagler Institute experience."

On the Lindau meetings website, Kaufmann's bio reads: "Stefan H. E. Kaufmann was Founding Director of the Max Planck Institute for Infection Biology in Berlin (1993). He is now Director Emeritus both there and at the Max Planck Institute for Biophysical Chemistry in Göttingen. He is also Professor of Microbiology and Immunology at the Charité, Berlin, and he is a Faculty Fellow of the Hagler Institute for Advanced Study at Texas A&M University, College Station, USA."

As the Council of the Lindau Nobel Laureate Meetings Scientific co-chair in physiology/medicine, Stefan Kaufmann believes he has played a role in defining the nature of excellence.

"Which is, of course, excellence in science," he says. "Many groups like the Lindau group invite young people only from places like Harvard, Max Planck, and highly regarded institutions like that. But we want to make the Lindau meetings open for the whole world and promote scientific excellence in countries that are less privileged, like on the African continent for example. I think it's a highly interesting endeavor. When I'm at Lindau, I not only go to the talks on physiology and medicine but also go to the chemistry, physics, and even the economics talks, which are actually very interesting because the economics people know how to talk to laypeople like me."

Kaufmann's concern for the "less privileged" permeates much of his research, as Texas A&M faculty and students learned during his 2024 "Eminent Scholar Lecture," a series sponsored by the Hagler Institute. His talk was entitled, "The Fight Against Tuberculosis: Lessons from the Past, Thoughts for the Future." Globally, Kaufmann told his audience, no other infectious disease kills more people.

"It was back in the 1980s," Kaufmann says, "that the World Health Organization realized tuberculosis was truly a big issue. Over the last two hundred years, a billion people have died of tuberculosis, which is an average of five million people a year. When the WHO formed a study group to address the global TB issue, I was asked to participate and that poured me into this."

Over the past half century, TB deaths have been concentrated in third-world countries, but there are recent increases in TB cases in advanced nations as well. Thus, the TB scourge remains a concern to all of humanity.

Following are excerpts from Kaufmann's lecture on the matter:

> One of the most exciting moments in my early days here [as a Hagler Fellow], was when I was invited to a jam session at Clifford Fry's house. That revitalized some memories about my youth and my interest in music. So I thought it would be appropriate to start my talk today about tuberculosis with a kind of historical overview.

(At this early point of his lecture, Kaufmann plays an excerpt from the 1931 Jimmy Rogers song, "TB Blues." Many consider Rogers the father of country music. He contracted tuberculosis in 1924 while working for the railroad, became a full-time musician in 1927, recorded his "TB Blues" song in 1931, and died from the disease in a New York City hotel room in 1933 at the age of thirty-five.)

> *'I've been fightin' like a lion looks like I'm going to lose*
> *I've been fightin' like a lion looks like I'm going to lose*
> *Cause there ain't nobody ever whipped the TB blues*
> *I've got the TB blues'*

(Kaufmann continued his lecture after playing the above refrain from Rogers's song.)

> Despite his "fight like a lion," Jimmy Rogers ultimately died of tuberculosis. And he's not the only famous musician to contract the

disease. Tom Jones developed TB, but recovered. Ringo Starr, the drummer for the Beatles, had TB at age 11. He recovered also. Tom Fogarty, the American guitarist of Credence Clearwater Revival, became infected with HIV and ultimately died of TB in 1990.

This is the "modern" part of the TB story in music. If you look in the past, Frederic Chopin, the famed classical composer, died of TB. Also, Igor Stravinsky. If you look at opera, tuberculosis was a major theme in *La Boheme* and *La Traviata*. Actually, Giuseppe Verdi, the composer of *La Traviata*, developed tuberculosis himself, as did his wife and kids. He survived, but he had to witness the death of his family.

If you look at death caused by infectious diseases, the number-one killer is tuberculosis with about one billion deaths over the last 200 years. That's a lot. That's more than smallpox, malaria, plague, flu, cholera, and AIDS together. And also, you could add COVID-19, because it was overcome so rapidly.

Tuberculosis is, still today, one of the deadliest pathogens. It can be cured, but unfortunately is on the rise. And the multidrug resistant Mycobacterium tuberculosis is the deadliest of all "bad bugs." Bad bugs have been defined as those bugs that will cause big problems over the next years because we don't have new drugs sufficient to control the infectious diseases in the future.

This disease gets very little financial support. The point I want to make here—and we saw this so clearly in the COVID-19 vaccine efforts—a TB vaccine must not only be available, but also it must be accessible at an affordable price. It must be made available and accessible to the low- and middle-income countries. I'm very hopeful that we will see results in clinical trial sites in 11 different places in Sub-Saharan Africa, places like South Africa, Tanzania, Kenya, Uganda, and Gabon.

When COVID occurred, the World Health Organization quickly declared it a global health emergency. And that was very productive because three years later, more or less, the pandemic spread was under control. And WHO then announced the end of COVID-19 as a global health emergency.

In 1993, at a time when 3 million people died of tuberculosis, the WHO declared tuberculosis a global health emergency. Today, the WHO still cannot declare an end to the TB global health

> emergency. Last year, the United Nations convened a meeting promising to solve the issue by putting a lot of money into it. This would not only save lives, but also show a remarkable return of investment. For one dollar invested, seven dollars would be paid back, saved in the cost of human lives if a vaccine could be created effective in controlling TB.

And if such a miracle was to occur, Stefan Kaufmann's name could very well appear among the ranks of those receiving the Nobel Prize, in which case his trips to the island of Lindau on the southern border of Germany would take on a whole new meaning and purpose.

29

An Unequivocal Belief

Rich Milner, 2023–2024

"Every single month since I began in August, I've been on campus for a week."

For Rich Milner, a member of the Hagler Institute for Advanced Study Class of 2023–24, that's a lot of back and forth. Milner's full-time job is at Vanderbilt University in Nashville, Tennessee.

"I think I missed one month," Milner says, "but typically, I come in on a Sunday and fly back on either a Friday or Saturday. So I'm usually there a full five days. I usually fly in early on Sunday so I can schedule lunch and dinner meetings that day. It's important for me to get as much in-person contact as possible. I want to be present to be a part of the culture."

When Rich Milner sends out an email, his signature block is of modest size. It reads,

Rich Milner, Professor of Education
Cornelius Vanderbilt Professor of Education

Meeting Milner in person, one is indeed struck by his modest and humble nature. If he chose to, that signature block on his emails could be a great deal more grandiose.

Rich Milner, Professor of Education
Cornelius Vanderbilt Chair of Education
Immediate Past Joseph A. Johnson Jr. Distinguished Leadership Professor Department of Teaching and Learning
Professor of Sociology (Secondary)
Professor of Education Policy Studies (Secondary)

Founding Director, Initiative for Race Research and Justice
Peabody College of Education and Human Development, Vanderbilt University

No wonder the Texas A&M School of Education and Human Development website announced his arrival on campus with the headline: "Renowned Scholar Named Hagler Institute Fellow."

The news release read in part as follows:

> Dr. H. Richard "Rich" Milner IV was among thirteen distinguished scholars selected by the Hagler Institute last month. He is an elected member of the National Academy of Education and the immediate past president and Fellow of the American Educational Research Association (AERA), the largest research organization in the world. He is the Cornelius Vanderbilt Professor of Education, and by courtesy, Professor of Sociology in the College of Arts and Sciences at Vanderbilt University.
>
> Milner studies people's discourses about race and racism and how school policies influence students' opportunities to learn across urban school contexts. A former high school teacher, community college instructor, journal editor and thought leader, Milner will now bring over 25 years of experience as an educational researcher to work with our students, researchers and faculty.
>
> "I am deeply honored to have this opportunity to work with students, faculty, and staff at Texas A&M University," said Milner. "We have an opportunity to build research agendas that help us know more about ways to support young people and educators across our nation and world, drawing from rigorous research tools to improve human conditions."

Milner's humble nature stems from his small-town roots.

"I grew up in Griffin, Georgia, halfway between Atlanta and Macon. My parents were very, very hardworking. My mother owned a beauty salon, and my dad worked at General Motors for many, many years. Our family had resources, we were solidly middle class, so I didn't grow up in poverty. I had a great upbringing."

Milner attended public schools, and when it was time to go off to college, he had several choices but decided to attend South Carolina

State University, a Historically Black University located in Orangeburg, South Carolina, It was there his perspective on life changed.

"I attended a predominately white high school," Milner says of his time growing up in Griffin. "There were very few Black teachers there. But when I got to South Carolina State, there was something really powerful about all the Black professors with terminal degrees. My physics teacher was Black. My math instructor was a Black instructor. That was a very inspiring time for me."

As was his experience building relationships with peers, who were Black like him.

"I met and made friends there, some of whom I'm still in touch with," Milner says. "Roommates, fraternity brothers, people from different parts of the country. The student body at South Carolina State was geographically diverse. I met people whose parents were medical doctors, engineers, scientists, and college professors.

"That played a role in my belief that I could be successful and make a meaningful contribution as well."

When Milner arrived in Orangeburg in the late summer of 1992, he had thoughts of pursuing a career in the legal profession. As a result, he declared political science as his intended major field of study.

"After a semester," he says, "I realized that was not the area in which I really wanted to spend the next three-and-a-half years."

Milner smiles as he recalls his "fork-in-the-road" moment toward becoming his future self.

"My true passion—back then as it is today—was writing," he says. "In the work I do, what I enjoy doing more than anything else is writing. And I am a really excellent writer!

"I had a mentor in college, Dr. Sarah Favors, who taught graduate and undergraduate courses in English and English education. She told me, 'Rich, you can still do law and major in English.' So I did English for a year and fell in love with reading and immersing myself in the kinds of literature I had never been exposed to before. And then, in my third year of college, I decided to pursue a teaching degree.

"That was the best decision I ever made."

After graduating from South Carolina State with a bachelor's degree in English and a master's degree in English education, Milner taught high school English in Columbia, South Carolina, an hour north of Orangeburg. In 1998, Milner moved to Columbus, Ohio, and became a graduate research associate at The Ohio State University while working toward another master's and his PhD degree in Educational Policy and Leadership.

Rich Milner was a young man on a mission. With his Ohio State graduate degrees in hand, he moved on to become a research assistant professor within the Peabody College of Education at Vanderbilt University, moving to Nashville in 2001. In time, Milner became the first Black person in the college's 125-year history to earn tenure and promotion. He was then recruited to the University of Pittsburgh to become the executive director of the Center for Urban Education and the Helen Faison Professor of Urban Education. He remained at the University of Pittsburgh for five years until he was recruited back to Vanderbilt.

In addition to being a professor at Vanderbilt, he founded the Initiative for Race Research and Justice at the school.

As Milner says, he loves to write. A good example of his writing prowess can be found in an opinion piece he authored in 2019 for *The Hechinger Report*, which, according to its website, covers "inequality and innovation in education with in-depth journalism that uses research, data and stories from classrooms and campuses to show the public how education can be improved and why it matters."

OPINION: Parents, here's why your most important partner may be your child's teacher (and vice versa)

As teachers return to the start of school following summer break, I hope they will take some time to reflect on ways to effectively communicate and partner with the families of the students they teach.

Particularly for educators in early and elementary education, they have a chance to set the course for how families imagine school and family connections. As a parent of young children, I know for sure that I am doing the very best I can to support them for school success. But even as a university professor of education, I still experience a deep level of anxiety and vulnerability about how best to partner with teachers to support my children.

As I've communicated with other families across the United States, I have learned that they too yearn for opportunities to build stronger partnerships with teachers.

Similar to a doctor with a bedside manner that fosters respect and trust when working with patients and their families, teachers can design initial and subsequent interactions with families that demonstrate trust, collaboration and advocacy, giving them a better chance of effectively meeting the needs of students.

I believe teachers can learn something really powerful about how to communicate with families and build sustainable partnerships with them. As teachers prepare to build and cultivate positive, relevant and useful partnerships and collaborations with families, I offer the following seven recommendations:

1. Stress the importance of working together. It is essential for educators to explicitly share that, as a teacher, you are on the same "team" as families and parents. Families need to know that you are working with them to support their children. Some parents have had such negative experiences in schools—as students themselves, and/or with previous teachers of their children—that they need to be reassured that teachers are not working "against" them but rather with them to support their children. In this sense, teachers cannot assume that families know or believe teachers are on their side in the education of their children.
2. Acknowledge families and parents as "knowers." Although teachers can rely on their past experiences as teachers, as former students and/or as parents, it is necessary for them also to acknowledge that families and parents are indeed knowledgeable about their children. Teachers should not act as the main arbiters of knowledge regarding the educational experiences, and consequently the needs, of children when they are communicating with families and working to build trusting and lasting relationships. Allow and encourage families and parents to be "experts" on their children's experiences. Despite what teachers might believe, family members

have deep knowledge about their children—their histories, their preferences, their challenges, their assets—which can be used to bridge and advance learning, curricula, instructional assessment and relational opportunities inside the classroom.

3. Identify and articulate student strengths. Rather than waiting until there's a concern to report, identify assets that you observe among students and articulate those strengths to families. For many families, the only time they communicate with teachers is when something negative has happened or when there is a concern about their children. Focus on, and articulate orally and in writing, assets and the potential that you observe in young people. In education, we tend to focus on what young people do not know or have, and then try to "fill in the gaps." A different approach is to identify assets in young people and build on them. While student assets include cognitive and intellectual strengths, young people bring so many others that go unrecognized—such as kindness and advocacy for friends, strong communication skills and perseverance in the midst of challenges. Recognizing how and when students demonstrate robust knowledge, skills, attitudes, dispositions and behaviors requires expanded lenses among teachers.
4. Express your commitment to development. As teachers are communicating with families, you should express to families that young people develop knowledge, skills and ideal behaviors at different times. Share with families that it is your role as a teacher both to help students develop and to help ease parents' anxieties when their children are not developing a particular desired skill or behavior, for instance, at the same rate as others. Even as you provide recommendations, express to families your belief that students will in fact reach the goals and expectations when you, they (as families) and students work together. Young people, similar to all of us, are and should be viewed as developing beings.

5. Demystify the educational process. Most of the time, families are doing the very best they can to support their young people. Families do not want to feel like they are poorly parenting and supporting their children, just as teachers do not want to feel they are ineffectively teaching. One way to reduce tension and worry among families is for teachers to share personal anecdotes with families about challenges they have had with young people in their own families (sons, daughters, nieces, nephews). Sharing personal challenges in children's learning, development and education provides a space for families to understand that children from different walks of life—even educators'—may have challenges. Sharing these stories can bridge implicit divides between educators and parents because families start to see teachers as real people and consequently may believe teachers are not judging them as parents.
6. Set agreeable forms of communication. As technology becomes more sophisticated every day, you should talk with families about their preferred forms of communication. Knowing whether families are more or less responsive to particular kinds of communication can help facilitate more seamless interactions in the future.
7. Include young people in discourse and decision-making. Although we tend to talk about young people in deciding what is best for them in education, it can be necessary and advantageous to include young people (even elementary students) in the decisions and expectations being designed on their behalf. In this way, teachers, families and students are working collectively to co-construct a learning plan that includes important voices and perspectives. Young people in particular can have powerfully insightful ideas about why challenges may be emerging and how problems can and should be addressed. Talking to young people, rather than talking about them, can prove useful in supporting them for academic, psychological, emotional and social success.

As a new school year gets underway, teachers have an opportunity to reflect on more effective ways to build partnerships with families in order to maximize student learning and development. Unfortunately, parental and family involvement tends to decrease as students get older, although student challenges increase. I hope teachers of all grades, from pre-Kindergarten to high school, will build stronger capacity to support young people. Indeed, our students need and deserve partnerships that are trusting and lasting. Family and teacher connections are essential aspects of student success.

In 2023, Milner released a book entitled *The Race Card*, published by Corwin, a Thousand Oaks, California, company that specializes in creating products that offer educators "resources and services to help every child succeed."

The Corwin website offers this description of Milner's book: "Education leaders are on the frontline in the fight for racial justice and must co-construct practices to disrupt storylines, policies, and practices that perpetuate opportunity gaps. Drawing from established research and the wisdom of teachers, young people, parents, community members, policy advocates, and school leaders, *The Race Card* is a guide for frontline leaders at every level to confront and disrupt racism."

This excerpt from Milner's introduction in the book sets a poignant tone for what follows:

> A few years ago, a group of parents, community members, and policy advocates invited me to share recommendations from my research about ways to support minoritized communities in education. This diverse, but mostly white, group was concerned about racial unrest and potential backlash after the killing of George Floyd at the hand of the state. "Liberal minded" and "forward thinking" were the words used by the group to describe themselves to me prior to and during my visit with them. They prepped me by sharing that they did not want to focus on race or the Floyd murder, per se, but on poverty, economic disparities, drug addiction, and lack of opportunities for many in their community. They

preferred and guided me to prepare my presentation for a much broader conversation about "minority groups," and they wanted specific recommendations about ways to help students "do better" in schools. They pleaded: we must do something in our communities and especially our schools to quell the injustices and social unrest facing the nation. Moreover, they were concerned about deep political divisions and lack of solidarity that seemed to be tearing up the United States of America.

Maintaining my professional obligation and commitment to equity, I prepared and focused my remarks on Black students, students living below the poverty line, students whose first language was not English, recent immigrant students, Muslim students, and LGBTQIA+ students. I also attempted to focus on important and necessary intersections of these students' experiences and identities as I invited those in attendance to imagine equitable policies and practices. In fact, I talked with the group about many of the issues I explore within and throughout the pages of this book.

Drawing from well-substantiated research, I attempted to map and share with the group complexities of the issues we face as a nation, sprinkled with recommendations on what they could (and should!) do in solidarity to make society and schools better. But it appeared that the group wanted simple, fixed, and definitive answers. And they wanted solutions disconnected from the realness of racism. Racism is not a playing matter. Racism is serious, as it hurts and harms the bodies and capacities of those who experience it—young people, educators, families, and communities.

As I presented, those in attendance engaged in sidebar conversations with and among each other reflecting on the data, insights, and recommendations I shared on slides. Although I could not hear their comments between and among themselves during my presentation, their concern became clear to me. After I finished my presentation the very first comment from a participant (with head nods from most of the others in the room) was, "This presentation feels like you are 'playing the race card.'" A bit stunned and disappointed by the comment, my response was firm and resolute and sets the stage for this book: "Yes, I certainly am." Play the race card. School leadership committed to centering racial justice requires that we amplify and intensify focus on race, not

> shy away from it. I reflected: Why should I somehow lie about, move away from, or be afraid to speak truth in this space (or any space) about the lingering and enduring effects of racism in society and education? Moreover, what does it mean to "play the race card," and who decides when such a card is being played?

Milner's *The Race Card* book strikes at the heart of the controversy involving diversity, equity, and inclusion policies—known as "DEI"—on college campuses across the country. Efforts to promote DEI principles have been banned in the state of Texas at public institutions like Texas A&M University.

"That continues to be an area of concern for me," Milner says, "because that is my work. That is my life. I care deeply about the issues of diversity, equity, and inclusion. It's not only aligned with my values, but it's also consistent with my belief that justice, equity, and inclusion are central to what it takes to live in a democracy and to support all students and teachers.

"In fact, as I was considering accepting my Hagler Institute Fellowship, my wife said, 'Rich, I don't want you to get hurt.' So this is a matter that I gave considerable thought to. But it's also an area that I won't back down from.

"I'm a tenured professor, a full professor and endowed chair, a very well-respected scholar at Vanderbilt University. So in a lot of ways that enables me to operate from a place of confidence because I'm not going to negotiate that work and I will not renegotiate what I know to be true and fair in the work I do. Several of my Texas A&M colleagues in the School of Education care deeply about these issues as well. So I've taken on a role of mentoring and encouraging them to continue to think that way.

"From my first visit, I have felt extremely welcomed here," Milner adds. "Admittedly, I've also had some interactions with people who know what I do that haven't been as pleasant. No one has said anything explicit to me about my work and my stand on the DEI matter, but I would say that it would be impossible for me to think about moving permanently to a place where my very being, the very core of what I care about, is called into question and is not honored and valued."

So does it say something about the Hagler Institute that someone like Milner, whose strong personal and professional convictions run contrary to Texas state law, is welcome within its ranks?

"First let me say, I like John Junkins a lot. He and his staff have been extremely kind and gracious to me. Given John's expertise, his pedigree, his level of respect, I think there's a lesson embedded in my selection, right? I think that lesson is probably more implicit than explicit, right?

"But at the same time, I think it's really sad that a diversity of thoughts and perspectives are not welcomed. I will say people have a right to have their views, but those views should be substantiated. And we shouldn't bully either perspective, right? Not in the name of scholarship or academic freedom.

"I think we have to be careful about disrupting the very core of what institutions are supposed to do in order to advance knowledge.

"I don't know if that answers your question, but I don't know how you can write a story about me and my work and not mention this. It's like the 'fly on the wall,' or 'the elephant in the room!'"

Milner laughs at his compound metaphors. And in the "blink of an eye," he returns to his more modest and humble nature.

To know Milner is to like him and to respect him for his unequivocal belief in fairness and justice.

30

An Ode to Alexander

Ram Galindo

Ramiro Galindo was born and raised in Bolivia. He attended Texas A&M University in the early 1960s and was a member of the school's men's gymnastics team. For most of his adult life, Galindo—known as "Ram" to his friends—has lived in the Bryan / College Station area working as a successful entrepreneur and real-estate developer. In 2006, he wrote his autobiography, *The Making of an American*. In 2024, the Ramiro A. Galindo Foundation made a $1 million donation to the Hagler Institute for Advanced Study.

As a writer himself, Galindo is accustomed to telling his own story.

I first learned about the Hagler Institute as a result of a meeting I had with Texas A&M University System Chancellor John Sharp.

The chancellor welcomed me into his office to discuss my plan to build five high-rise residential towers near the Traditions Golf Course in far west Bryan, about six miles from Texas A&M University's College Station campus. This was in 2019.

I'm a longtime entrepreneur, focusing on construction, business development, and real-estate development. The high-rise project was a little out of the ordinary for me, but I intended it to be my crowning achievement in business. I had a master plan. I had drawings. And I had a vision. I had everything ready to begin moving forward with the construction of the first tower.

I knew the university had people visiting here all the time: Sometimes it's for a week, sometimes it's for much longer. I thought administration officials might have an interest in investing in a few units in one of my buildings to take care of important guests.

John Sharp listened to my pitch, and when I was finished, he said, "Ram, the only person I know who might be able to use something like what you're describing is John Junkins. He leads the Hagler Institute for Advanced Study. You should talk with him."

I thanked the chancellor for his hospitality and left the meeting. At the time, I had no idea what the Hagler Institute for Advanced Study was and no idea who John Junkins was. But I called the Hagler Institute office and set up a meeting with its founding director.

From the beginning of our relationship, I've felt a sort of kinship with John Junkins. In our first meeting, I discovered that he and I have good chemistry, and we shared an affection for telling stories. In no time, we struck up a friendship. In learning about his Hagler Institute, I realized that his program was something that I'm naturally interested in. I'm a big proponent of education. I think humanity only improves itself through an emphasis on education, and the Hagler Institute is all about advancing excellence.

Soon after I first met John, the Hagler Institute began a search for a new home. Their offices had been located in the Texas A&M administration building, but a new president was moving in, Kathy Banks, and she needed the space the Hagler Institute had occupied.

I offered to donate my building about five miles west of campus as the new home of the Institute for Advanced Study. I invited John over to let him take a look.

"I really believe in your mission," I told John. "I want to give you this building to use as your offices."

I'm pretty sure he wasn't expecting me to make such an offer. He didn't give me an outright "no," but he said his preference was to remain on campus. I do think he saw a financial benefit in owning the building, whether he moved his offices into it or not. He directed me to contact the Texas A&M Foundation. I did, and discovered that they only accept an asset such as the one I was offering after the prospective donor has died. They told me, "If you sell the building now, you can give us the money," or something like that.

I had no plans to sell the building outright, nor was I in any hurry to meet my maker.

So I turned my attention to other things.

Now that I'm in my eighties, I'd like to think I'm still going strong, both in business and in regard to my health. I've reached a point that most of us strive for in that I can afford to do certain things financially without taking too big a risk. I still wanted to help John and the Hagler Institute. I decided to sell another asset, sold it, and did very well in the deal. So I went back to John and told him that I was ready to make a million-dollar donation to the Institute for Advanced Study.

Needless to say, he was thrilled and eagerly accepted my offer.

By the way, COVID-19 put a hitch in my plans for the high-rise development. I called it "I-Con Towers," which stands for "innovative condominiums." I still have hopes for it, but due to inflation and an increase in interest rates, the expected cost has ballooned into the $400 million range. More on that later.

What I sold to give me the resources to support the Hagler Institute was one of the apartment communities I owned in the Bryan / College Station area.

"But for you to get that money," I told John, "I need you to sit down and let me tell you about the very first 'institute for advanced study.'"

"I know that story," John said. "It was at Princeton, although it was not formally associated with the university."

"No, that's not right," I replied. "The first 'institute' of this kind was built in Egypt, in the great city of Alexandria, three-and-a-half centuries before the Christian era. In other words, about 2,500 years ago."

Here's what I shared with John Junkins about the *Mouseion* of Alexandria.

Today, Alexandria is the second-largest city in Egypt, lying at the western edge of the Nile River on the coast of the Mediterranean Sea. It was founded by its namesake, Alexander III, the Macedonian king more commonly known as Alexander the Great. The city, however, did not take root under him. A few months after establishing Alexandria in approximately 331 BC, Alexander the Great left the area, never to return. He died in Babylonia some seven or eight years later at the age of thirty-two; his cause of death was never truly determined, although by many accounts, he most likely drank himself to death.

The stories of Alexander's travels after death are as compelling as any of the great battles he undertook while alive. His conquests rang in what came to be known as the Hellenistic Age, a fusion of the ancient Greek world with that of the cultures of western Asia, northeastern Africa, and southwestern Europe.

Upon Alexander's death, there was a power grab for his empire among his top-ranking generals. These officers decided to transport Alexander's remains back to Macedonia. To do so, they built a gold sarcophagus and carriage fitting for the king. The generals' plans were to create three independent and democratic states in the kingdom that Alexander ruled. But upon arriving at the Aegean Sea, one of the generals, Ptolemy, had come to realize there was too much hunger for power among his compatriots. So instead of taking the king's body onward to its intended destination, he parted ways with his cohorts and routed Alexander's remains to Alexandria. In so doing, his intent was to establish the city that Alexander had envisioned and to do so "forever, in the great king's memory."

In fact, Alexander's remains were eventually interred in Alexandria and became somewhat of a tourist attraction. Julius Caesar, among many others of note, was known to have paid his respects there.

Rather than wage war against the Egyptians to take control of their land and realize his vision for the city he planned, Ptolemy "compromised." He created a hybrid of the Egyptian religion, injecting his own gods and beliefs. With that, and other like-minded maneuvers, he won over the Egyptian leaders and their people and formed his city, the first city in the history of the world where everyone had the right to vote.

As children of privilege, both Alexander and Ptolemy were educated by the Greek philosopher Aristotle. Thus, knowledge and education were engrained into the young lads' thinking. During the years of Alexander the Great's conquest of much of what today is known as the Middle East, he and Ptolemy took great care to preserve new discoveries, thanks to the inclusion of others within their warfaring inner circle who were also intellectual disciples of Aristotle. Those budding philosophers, naturalists, and scientists preserved all they found that was new to them for transport back to Greece and further study by the learned men of the times there.

The *Mouseion* of Alexandria was founded under Ptolemy I Soter—the general eventually became an Egyptian king—and completed under the watch of his son, Ptolemy II Philadelphus. The term *mouseion,* from which the word "museum" is derived, originally referenced any location dedicated to "the Muses," the inspirational Greek goddesses of literature, science, and art. Ptolemy's plan, aided by his son, was to bring the world's eminent scholars to Alexandria in order to teach others as well as improve life for the Egyptian people. It was with these ideas in mind that Alexandria became the center of knowledge for the known world at that time.

Preeminent thinkers flocked to Alexandria from all over. It was in Alexandria that men of learning postulated with high confidence that the world was not flat and that the earth revolved around the sun. They also mastered the principles of irrigation and navigation and refined the rule of law. Perhaps most importantly, they formed one single written language and recorded their knowledge on tens of thousands of papyrus scrolls, all of which were contained within that part of the *Mouseion* that became known as the Great Library of Alexandria.

It was Germain Bazin, the former curator of the Louvre Museum in Paris, who first suggested some fifty or sixty years ago that the *Mouseion* was "analogous to the modern Institute for Advanced Study in Princeton," and thus, as I concluded in my explanation to John Junkins, an ancient predecessor to the Hagler Institute for Advanced Study at Texas A&M University.

My personal story is not nearly as interesting as that of Alexander the Great. I was born in Bolivia and raised in Cochabamba, which is nestled into a valley—as best a city of more than six hundred thousand people can be "nestled"—of the Andes Mountain Range. The elevation of my home town is 8,392 feet, which, now that I think of it, makes Cochabamba the "mile-and-a-half-high city."

I always tell people I am a "favorite of Fortune." I was born into a home with a good father and a good mother. Growing up, I enjoyed all the attributes of a good life with them. I learned about honesty and hard work and dedication and loyalty, and my parents instilled within me the desire to try to be the best person I could be.

My father had gone to Chile for his advanced schooling and studied civil engineering. When he came back to Bolivia, he did very well. He was a stalwart in the community. As I came of age, my parents wanted me to find my way in Cochabamba. But I decided I needed to see a bit more of the world. This was after the Korean War, and from the news headlines I read about the Asian conflict, I had developed a sense that my destiny could lie elsewhere.

Growing up, I frequented a small store near where I lived that had newspapers and magazines for sale, which included the American publications *Newsweek* and *Time*. At first, I would only browse through those weekly periodicals, but the pictures I found always set my pulse to racing. Soon, I bought the magazines to further explore their content. At the time, I didn't speak or write English, so the words were at first a mystery to me. Still, the pictures—always in black and white back in those days—gave me a visual education and inspired my desire to explore the world beyond Bolivia.

When it came time for me to go to college, my parents wanted me to do so in Chile, where my father had been educated. But I wanted to go to the States, as most of the world, both then and now, calls the United States, that country which I'm proud to have called my home now for many, many years. My parents and I discussed this matter for a while, and then my father finally agreed on my plan. He thought my best option would be to approach the missionaries who ran a Catholic school in Cochabamba. They were Augustinians. My father knew the Augustinians ran an American college located near Philadelphia, Pennsylvania.

That school was Villanova University.

"You can go there," my father told me. Because it was affiliated with the Augustinians and the Catholic Church, he reasoned, "We know you'll be safe there."

The Augustinians in Bolivia assisted me in preregistering at Villanova and did their best to teach me English. But they were mostly Dutch serving as missionaries in a Spanish-speaking country, so I did not become fluent in English under their watch.

Even today, my English is impacted by a heavy dose of my Bolivian accent.

My first time visiting America was when I left for school a month and a half before classes began at Villanova. I flew into New York City, and my plan was to spend a few weeks there immersing myself in the native language before moving on to Philadelphia. But when I stepped off the plane at Idlewild Airport—now called John F. Kennedy Airport—I was immediately overwhelmed by this new frontier. I didn't even make it out of the airport before my plans began to unravel. I remember sitting on a curb at the airport crying. In that moment, I didn't know what to do or where to go. But eventually I composed myself.

No one stopped to offer me any solace for my tears; after all, this was New York. But I eventually found the train that ran from Idlewild into Manhattan. It was there, in the heart of the "Big Apple," that I was hoping to find a place to live to begin acclimating myself to my new life.

More quickly than I could have imagined at the time, I indeed found a foothold in New York City, and I accomplished most of my goals there. I learned English, made a few acquaintances—friendships came as I began classes at college—and had plenty of time to marvel at the most spectacular city I could ever imagine.

I did well my first year at Villanova. I decided to follow in my father's footsteps and declared myself an engineering student. I was a little taken aback at the "costume" I had to wear as a freshman. The beanie and bow tie were the most unsettling part of my regular campus attire. And as autumn turned to winter, I realized I wasn't going to be a big fan of my new East Coast environs, either. The lifestyle was much more "hustle and bustle" than what I was used to back home in Bolivia. It was certainly a lot more crowded, to say the least. And even though I grew up in the Andes Mountains, the dampness of the winter cold in Philadelphia was unlike anything I had experienced to that point of my life.

It was for all those reasons that I decided Villanova wasn't going to be the place for me. So I opted to make good on an adage I had picked up in my indoctrination into American culture: "Go West, young man."

I applied for transfer to several schools and was accepted at Stanford, Colorado School of Mines, the University of Texas, Arizona, and Texas A&M. I went to my dean at Villanova and asked him where I should go. "For the money, and to become a good civil engineer," he said, "your best bet is to go to Texas A&M."

"OK," I told him. "I'm going to Texas A&M."

And when I got there, it was love at first sight.

Since my experience arriving in New York City by myself had started out so dismally, my father came with me as I prepared to start classes at Texas A&M. He also loved the place. He became a big Aggie booster back home. My brothers wound up coming here to attend school, and I think some 250 students from Bolivia eventually followed me to Texas A&M, in large part because of my father.

I got both my bachelor's and master's degrees in civil engineering from Texas A&M. After receiving my undergraduate degree, I went to work for the Texas Highway Department. My job was designing urban bridges in Houston. I enjoyed the challenge, but the experience convinced me I needed to learn more. So I went back to Texas A&M and got my master's. After that, I went to work for Brown and Root, a big, big construction company based in Houston. The Brown family were close friends with then–Texas Senator Lyndon Johnson. In fact, it was his relationship with the Browns that led to locating NASA's Manned Space Flight Center in the Houston area.

Of course, that complex is now known as the Johnson Space Center.

In time, I decided to go into business on my own, and after a failed attempt to establish myself in Bolivia, I returned back to my roots and replanted my flag in Aggieland. The local area has been very good to me, and I've tried to give back in return.

So now you know why the Hagler Institute for Advanced Study matters so much to me. They know how to bring scholars to Texas A&M from all over the world, and they have done it successfully for more than a decade. As you can guess, it costs money to do these kinds of things, so I'm very proud to be a donor.

But my support for the Hagler Institute isn't just financial.

Let's just say I have a lot of friends, many of whom I've made during the "Renaissance Weekend."

Have you heard of the Renaissance Weekend?

Let me explain.

That group meets four times a year. It's a "by-invitation-only" gathering, and it remains a mystery to me how I got invited the first time. One of the big perks in attending these occasions is that you have the opportunity to get acquainted with some of the most successful people in the United States, from a wide range of notable pursuits.

One of the people I've met through the Renaissance Weekend is a University of California at Berkeley astronomer by the name of Alex Filipenko. He has a worldwide reputation in the study of supernovae. When I first heard him speak to our Renaissance group, I was mesmerized with his stories about the universe. Afterward, at the meet and greet that follows each lecture, I went up to him and introduced myself. I said, "We have this institute at Texas A&M, and I think you might be interested in it because I'm certain you could contribute a lot."

I went on to tell him, "They have the money and know how to make it attractive for scholars like you to be involved. It's called the Hagler Institute for Advanced Study."

He told me, "Yeah, I know about Texas A&M. I have a friend that has recently gotten a job there."

Alex and I stayed in touch. A couple months after my return from the Renaissance, I invited Alex to the Texas A&M Physics and Astronomy Festival, an annual event that brings high school students to campus from all over the state. It's become a really big deal and regularly features top-notch guest speakers. I told Alex, "Why don't you come? I'll encourage A&M to pay you a $10,000 speaker's fee to cover your time and expenses."

I got the money together myself, made the donation to underwrite Alex's appearance . . . and he came. This was in 2022, and he was a big hit! So much so that Alex caught the interest of the Hagler Institute and the Texas A&M astronomy faculty and he will be a Fellow in an upcoming class!

As for the I-Con Towers, after the pandemic, I regrouped, hired a new architect, and had every intention of moving forward with the project. The towers were to be thirteen or fourteen floors tall with forty-five to forty-seven units in each tower. I was ready to go to the city again to get building permits when, from out of nowhere, the old architect on the project sued the new architect for "copyright infringement," and I've gotten caught up in the collateral damage and ensuing project delays resulting from that.

I'm now at a point where I could start the project if I really wanted to, but I've gotten distracted with a couple other things. And to be honest, I don't know if I'll be strong enough to resurrect the idea again. It's not in my nature to waffle for long, but, "I'm not getting any younger," as they say.

I may just wind up just selling the land, but the concept for now is still alive.

It would be "Alexandria-worthy," for sure!

Epilogue

31

Student Voices

The Heep Graduate Student Fellowship

From Hagler Institute for Advanced Study
Founding Director John Junkins:

"Around 2010, the Heep family wanted to support graduate research and graduate study at Texas A&M. They gave $30 million to the foundation to fund graduate fellowships. This came in the midst of the post–Robert Gates era, and as a result of the ensuing game of musical chairs among our presidents and provosts, the Heep money was not being spent.

"The interest earnings off that donation were sizable, about $1.2 million a year. That turned out to be a blessing for the Institute for Advanced Study because about the time we were getting started, around the 2011 time frame, I was just beginning to try to round up the moneys for our first year. That was before John Sharp's arrival as chancellor of the Texas A&M University System.

"I went with Ed Davis, who was head of the [Texas A&M] Foundation, to see Pres. Bowen Loftin, and the three of us agreed to use some of the Heep funds for the advanced study institute. That amounted to roughly $400,000 a year, about a third of the interest earnings from the Heep family's original donation. The expressed purpose of those funds was to create graduate student fellowships at $30,000 a year.

"That program continues today.

"The PhD students who receive these fellowship packages are students working under a 'faculty host,' as we call them. These are the individuals who nominate the candidates to become Fellows of the Institute. Those host professors get to designate a student to receive the Heep Graduate Fellowship stipend. The award is $60,000 for two students for one year each, or one student for two years.

"The intent is to create a collaboration between the faculty host, the Hagler Fellow, and the student. The student's PhD research is informed not only by the faculty host, who is usually the student's PhD adviser, but also by the Hagler Fellow as well.

"The Heep Graduate Student Fellowship has dramatically affected many of the recipients' lives and careers. When I speak with them after they are named to be recipients, I try to convey the privilege associated with their designation. In some instances, I have to spar with the nominating faculty entity because they initially send me names of students that turn out to be less exceptional than those we are seeking. Sometimes, I've had to compromise, but I'm still working hard to change the perception of what the Heep Fellowship is about. I want to recruit the best students and match them with the premiere scholars who comprise our roster of Hagler Fellows each year."

John Junkins didn't have to look far to find one of the first Heep Graduate Student Fellows. As a distinguished professor, Junkins served as PhD adviser to Tarek Elgohary, who first arrived on the Texas A&M campus in 2008 to pursue his postgraduate degrees in aerospace engineering.

Elgohary earned a bachelor's degree in mechanical engineering from the American University in Cairo in 2006. He then took a break from his academic pursuits to work as a process engineer for Proctor & Gamble in Dammam, Saudi Arabia. There, he analyzed equipment failures and defects to improve reliability results. He knew at the time that job would be temporary, as he had his sights firmly set on pursuing graduate studies in the aerospace field.

"I always loved flight," Elgohary says today from his office on the campus of the University of Central Florida in Orlando, where he is the director of the Astrodynamics and Space Robotics Laboratory. "Both of my parents were engineers, but as a kid growing up in Egypt, I wanted to be a pilot. As a result of my parents' influence, I became an engineer instead, but even as I took my job at P&G, I knew I would soon be headed to graduate school to study aerospace engineering."

Given Texas A&M's reputation as one of the world's elite aerospace engineering schools, Elgohary made an inquiry directly to one of the department's most noted professors. John Junkins's quick reply and words of encouragement made Elgohary's decision an easy one, and once he got to College Station, he struck up what's turned out to be a lasting relationship with the man who recruited him to become an Aggie.

"Tarek was one of my favorite students," Junkins says, "both before and during his time in the PhD program. We're still in touch a decade after he left here for his 'greener pastures' near the Kennedy Space Center in Florida."

From the beginning of Junkins's time as director of what is now the Hagler Institute for Advanced Study, he's pulled double duty with his responsibilities in the aerospace engineering department. Junkins knew Elgohary had—to employ a term most commonly used to define the uncommon characteristics required to become a space explorer—the "right stuff" to become a Heep Student Fellow. Junkins encouraged him to apply, and he was quickly accepted. Elgohary was paired with Dr. Satya Atluri, a native of India and, at the time, a distinguished member of the University of California, Irvine's aerospace engineering department.

Atluri's page on the Hagler Institute website says, "His work has received support from the National Science Foundation, the U.S. armed forces, the National Aeronautics and Space Administration, the Federal Aviation Administration, the U.S. Department of Energy, and the Nuclear Regulatory Commission, among others."

Given Elgohary's own work toward his PhD in aerospace engineering, his pairing with Atluri seemed an obvious and ideal match.

"Dr. Atluri's first visit to campus as an Institute for Advanced Study Fellow was during the Christmas break in 2013," Elgohary remembers. "We met a couple times then, but there were very few people on campus, so we weren't able to get a lot done. He came back in the spring, but didn't stay long. I had a lot less interaction with him than I expected."

As a result, Atluri made a surprising and seemingly generous offer. He invited Elgohary to come to the University of California, Irvine, and spend the last three months of his student fellowship with him there over the summer. Atluri explained that health issues had prevented him from spending as much time in Texas as his own fellowship required.

Junkins advised Elgohary to accept Atluri's offer, and after arriving in California, the student and his new mentor seemed to hit it off just fine.

"Early on he suggested we put together a paper," Elgohary says. "I was really happy to be able to publish something with him. At the same time, I was also working with one of his postdocs, a very bright person, who's now a full professor in China. I was really enjoying my association with Dr. Atluri. Then one day he called me into his office for a meeting I remember very well to this day. He told me, 'You know what? I don't want to work with you anymore.'

"At that point, I had about a month left in my advanced study fellowship, which I had planned to spend in California with Dr. Atluri.

"I called Dr. Junkins and asked him what I should do. John was very wise. He told me, 'You know, first of all, just calm down. This may be his way of testing you, so just keep doing your work, keep sending him your updates, and eventually he will come around.'"

Elgohary took Dr. Junkins's advice and kept working without the benefit of any feedback from his sponsor. Elgohary dutifully kept sending emails about the progress of his work in Atluri's lab, to which he received no response.

"And then, from out of nowhere, with only a couple of weeks left in my fellowship, Dr. Atluri called me back into his office, started talking research again as if nothing had happened between us, and asked me to stay for the whole year. He told me he would pay for me to remain with him in California."

Elgohary received his PhD from Texas A&M in May 2014, and from that time through June of the following year, he worked as a junior specialist in the Mechanical and Aerospace Engineering Department at UC Irvine. Afterward, he returned to Texas A&M and again worked under Junkins as a postdoctoral research associate.

"Dr. Atluri and I became very good friends," Elgohary says. "Not long after I left California, he took a new job and moved to Texas to become a member of the faculty at Texas Tech. He invited me to spend another summer with him there, which I did. Unfortunately, he died in 2023, but he became an important part of my life. His way of mentorship was a little unorthodox, but he definitely had a huge impact on the way I do things. The way I do research, the way I value research, and the quality of the research I do."

• • •

The following are a sampling of student reports summarizing their experiences as Heep Student Fellows of the Hagler Institute for Advanced Study.

Tarek A. Elgohary
Department of Aerospace Engineering
Hagler Fellow: Dr. Satya Atluri
Year Awarded: 2013

As a PhD student in the dynamics and control group at the aerospace engineering department, having a TIAS Heep Fellowship enabled me to work on a much more diversified spectrum of problems. Through my collaboration with Dr. Atluri, I got introduced to singularity-free methods for solving nonlinear algebraic equations. First, we applied those methods to solve a classical structural mechanics problem, Williams Toggle, that had a numerical singularity and mainly relied on finite elements techniques and complicated arc-length methods for obtaining a valid solution. The techniques we applied were singularity-free and straightforward and proved extremely useful in tackling such a classical problem. We submitted our results to the *CMES* journal, and the work is already published. Next, we imported those techniques from the structural mechanics world to optimal control problems. The formulation and methods proved very effective in solving fixed time optimal control problems with high accuracy and efficiency.

Currently, we are writing our second journal article to document and publish our findings. The work with Dr. Atluri has been extremely rewarding in terms of my publication record and the variety of problems I got introduced to. We agreed to continue our collaboration beyond my fellowship appointment to address more problems and publish more papers.

Today, Tarek Elgohary is an assistant professor at the University of Central Florida's mechanical and aerospace engineering department.

• • •

Naomi A. Sachs
Department of Architecture
Hagler Fellow: Mr. Harold Adams
Year Awarded: 2015

I am extremely grateful for the funding and experience provided by the TIAS Heep Fellowship Award. Working with TIAS Faculty Fellow Harold Adams, FAIA, in the College of Architecture has been very rewarding. Mr. Adams has had a long and distinguished career as an architect—the president (and then CEO, then chairman) of the multinational firm RTKL. His focus on health-care design is also my PhD focus, and our conversations on that topic have been enriching. There is a great deal of talk in the design and academic professions about "interdisciplinary collaboration," yet we see little evidence of it in practice. Mr. Adams has devoted much of his time as a TIAS Fellow to building relationships with practitioners on and off campus, and within and outside of the College of Architecture, to foster a culture of and strategy for true interdisciplinary work. As a TAMU Distinguished Alumnus and a strong supporter of the College of Architecture, Mr. Adams's ongoing commitment to A&M is inspiring. This coming semester, I will be conducting research in California, Connecticut, Oregon, Texas, and Washington. Funds from the TIAS Heep Fellowship Award will enable my travel, a critical part of the dissertation. I also look forward to continuing my work with Mr. Adams this fall and in the future.

Today, Naomi Sachs is an assistant professor in the University of Maryland's Department of Plant Science and Landscape Architecture.

• • •

Anna Salvador
Department of Nutrition and Food Science
Hagler Fellow: Dr. Richard Gibbs
Year Awarded: 2016

As a student in the Threadgill Laboratory, I use mouse models of human health and disease to investigate how interindividual differences impact response to diet. The Hagler Institute for Advanced Study granted me the opportunity to work with Faculty Fellow Dr. Richard Gibbs, founding director of the Baylor College of Medicine Human Genome Center. This fellowship has allowed me to collaborate with his colleagues in the development of four mouse models of gene variants identified by the center. Two independent projects have emerged from my experience as a Fellow. The first project is related to regulation of individual diet response, while the second investigates behavioral, developmental, and molecular concepts.

Early on in the fellowship Dr. Gibbs connected me with Dr. Aniko Sabo to evaluate the role of a gene variant associated with susceptibility to obesity in a human study of lower socioeconomic Hispanic youth. We are using CRISPR/Cas9 technology to engineer the same single amino acid change identified from the study in a mouse model. The mouse strain selected has previously shown to be resistant to the development of metabolic phenotypes associated with obesity. After the mouse colony has been established, we will phenotype the wild-type resistant mouse as compared to the transgenic CRISPRed mouse.

The knowledge I have gained by developing the diet study has provided me with an opportunity to join a second project with Dr. Gibbs and his full laboratory. The focus of this project moves away from diet and centers on the behavioral and developmental phenotypes observed in children that present with a de novo (noninherited) gene variant. We will develop three heterozygous mutant mouse models in order to evaluate whether the condition presents itself by a mechanism of haploinsufficiency or dominant negative protein interaction. Understanding this mechanism will inform future potential drug therapies. In the long term, these potential candidates for therapy will be evaluated in the same mouse model that will have been established to determine the gene's mechanism of action.

The Hagler Institute for Advanced Study engaged me with Dr. Gibbs in the infancy of my academic career, and this has been one of the cornerstones to the development of my position in the laboratory. I anticipate that each of these projects will grow and develop into broader areas of study and I look forward to what the future holds for my work with Dr. Gibbs and the Baylor College of Medicine Human Genome Center. The relationships that have been fostered by my participation in this fellowship will continue to guide my research and primary thesis work in the time to come. I am thoroughly grateful for the experience that I have had with this program.

Today, Anna Salvador is a postdoctoral research associate in the Department of Molecular and Cellular Medicine at the Texas A&M School of Medicine. She also works as a registered dietitian for Aramark Healthcare assigned to Baylor Scott and White Medical Center in College Station.

• • •

Vishala Arya
Department of Aerospace Engineering
Hagler Fellow: Dr. Robert Skelton
Year Awarded: 2017

The Heep Fellowship Award by the Hagler Institute for Advanced Study came as a blessing that allowed me to work not only under one of the finest minds but with an incredibly passionate researcher, Dr. Robert E. Skelton.

Dr. Skelton is TEES distinguished research professor in the aerospace engineering department of Texas A&M University. I started working with him at the beginning of August 2017. He introduced me to the philosophy of how an engineer should build a system thinking, as he stresses on integrating structures with controls. Under him, I worked on modeling and simulating tensegrity structures and learning about the intricacies of controls. Tensegrity structures, made up of bars and strings, are useful in providing minimal mass solutions to many fundamental structural problems due to their high stiffness-to-mass ratio, better shape control without stiffness alteration, and optimization of control energy.

I specifically worked to design wearable active prosthetics for leg amputees. With nearly eighty-five thousand amputations every year, nearly two million people live in the United States suffering from a limb loss. Even with various modern prosthetics available on the market, the discomfort and the struggle that are experienced by these individuals result in a reduced hope for living a completely normal life. Our goal is to transform their disability into a superability by rendering more degrees of freedom and nimbleness in a prosthetic leg than a human leg. This idea was essentially driven by Dr. Manoranjan Majji, assistant professor at Texas A&M in the aerospace engineering department. My research work is an initial step in this direction. It is the first effort at exploring the potential of tensegrity systems in providing a robust and improved artificial leg, making it both very challenging and rewarding. We are currently working on a paper to publish by the end of next month as I am also working on completing my thesis and final defense under the guidance of Dr. Majji.

This fellowship came as a savior at the juncture when I was not sure about being able to fund my studies further. My heartfelt gratitude goes to the Hagler Institute for Advanced Study for providing me the financial security and the mental propulsion to pursue research further (PhD) and treat it as a means to serve humankind. This is indeed a brilliant phenomenon that any student will be fortunate to experience and that provides a student more than he or she can ask for.

I will be graduating in August with a master of science degree with a 4.0 grade-point average and will be continuing at Texas A&M for my doctoral studies.

Today Vishala Arya is a postdoctoral fellow working at NASA's Jet Propulsion Laboratory in Pasadena, California.

• • •

Tasmiah Nuzhath

Department of Health Promotion and Community Health Sciences
Hagler Fellow: Dr. Peter Hotez
Year Awarded: 2019

I received the Hagler Institute for Advanced Study Heep Graduate Fellowship for the second time for a twelve-month period that began on September 1, 2020. The fellowship has afforded me the opportunity to work with Dr. Peter Hotez, an internationally renowned physician and scientist in neglected tropical diseases and vaccine development, as my Faculty Fellow.

Measles vaccination coverage in many nations has dropped below the 95 percent coverage rate necessary to prevent measles outbreaks through herd immunity. We published an editorial, "Combating

Vaccine Hesitancy and Other 21st Century Social Determinants in the Global Fight Against Measles," which focused on identifying key social determinants such as war and conflict, climate change, and the antivaccine movement that are decreasing the measles coverage rate and driving the return of measles. We also created a global metric for vaccine-preventable disease risk that has significant potential to serve as a public health tool for predicting current and future measles outbreaks. I have also published a first-authored paper entitled "Childhood Immunization During the COVID-19 Pandemic in Texas" in *Vaccine*, a Q1 journal, to highlight disruptions to childhood immunization caused by the pandemic. Widespread childhood immunization protects children from highly infectious diseases like measles and helps to maintain herd immunity. Therefore, we also provided recommendations to increase vaccine uptake. I have been interviewed by several small and large market news channels about the findings of the study. We also worked on a second editorial entitled "COVID-19 Vaccine Decisions: Considering the Choices and Opportunities" that provides a decision-making tool kit to empower the general population to increase their confidence in COVID-19 vaccines and make informed choices based on scientific evidence about the available vaccine options.

My fellowship has provided me with a life-changing experience that has enriched both my personal and professional lives in an unparalleled way. This unique experience has afforded me invaluable exposure to research at the highest level and provided me with a life-changing opportunity to learn under the guidance and mentorship of a world-renowned expert in the field of infectious diseases. I have been able to gain new insights and perspectives and acquire data analysis and research skills while also having the unique opportunity to learn about my chosen discipline in a way very few students have the opportunity to do.

This prestigious fellowship has introduced me to the culture of academia, boosted my confidence, and shown me that my goal of having a faculty position at a top research university is possible. This experience has allowed me to begin to create a firm foundation upon which to build a record of accomplishment in obtaining funding for my research, which is an essential skill that hiring committees require in candidates for faculty positions at research universities. The benefits of the fellowship will also enhance the credibility of my research, which

I hope will inspire confidence in governmental and private funders to support my work in future years.

The Heep Fellowship provided me with the financial assistance I required to pursue higher education and allowed me to focus my time and effort on my research pursuits. Even though the duration of the fellowship has been one year, the benefits of this wonderful fellowship will last my lifetime. My anticipated graduation date is July 2023.

Today, Tasmiah Nuzhath is a Postdoctoral Research Fellow in the Department of Global Health and Population at Harvard T. H. Chan School of Public Health.

• • •

Jace Willis
Department of Biomedical Engineering
Hagler Fellow: Dr. Vanderlei Bagnato
Year Awarded: 2020

During my Hagler Heep Fellowship I have been working with Dr. Bagnato in research involving photosensitizers (PS), his area of focus. The fellowship officially began in January 2020; however, I had previously started research with Dr. Bagnato in October 2019, as my adviser, Dr. Yakovlev, had begun a collaboration in the study of photosensitizer feasibility as a treatment for in situ pneumonia-causing bacteria. This early project resulted in one published manuscript in the *Journal of BioPhotonics* and many plans for future research.

Research setbacks due to COVID-19 were successfully mitigated by an agile switch from planned research to performing an in-depth literature review. During the process, interim presentations were given on a tangential topic at an SPIE special forum and SPIE Photonics West in 2021. The full review, “Photodynamic Viral Inactivation: Recent Advances and Potential Applications,” has now been published in the high-impact journal *Applied Physics Reviews* as a featured article. It is our hope that garnering attention of the public and science communities will further improve interest and development of flexible, effective photodynamic treatments for viral and bacterial infections.

Current research efforts are focused on the discovery of synergistic effects between photodynamic inactivation (PDI) and antibiotics for inactivating antibiotic-resistant bacteria. This research theory was presented at the TAMU Biomedical Engineering Annual Research Symposium in 2021. To date, a nonresistant strain has been examined and used for training and development of protocols. In the coming weeks, a wide survey of resistant and nonresistant strains will be tested in batches against combinations of PDI and several antibiotics. Following this broad test, in-depth analysis of key strain-antibiotic pairs will reveal the details of synergistic effects.

Future research efforts comprising my PhD thesis are planned around accelerated aging via photosensitizer-controlled reactive oxygen species (ROS) generation. Though early in planning, controlled ROS generation presents a valuable tool for biological and biomaterial accelerated aging studies where common practice involves nonflexible treatments such as uptake-dependent superoxide-generating molecules in vivo and extreme temperature, pressure, and pH in biomaterials and implants. The proposed studies will provide a simple, controllable mechanism of accelerated aging for studies in biology and implantable materials.

The Hagler Heep Fellowship has provided me not only with financial stability, an elusive thing in these times, but with invaluable relationships and collaborative connections both on campus and internationally. These collaborative research efforts have undoubtedly developed my personal skills and made me capable and eager to lead diverse, interdisciplinary research tasks. Beyond new connections, this experience has helped me in learning to culture these relationships, more easily gain acquaintances, and refine my professional network.

Today, Jace Willis is a postdoctoral research fellow at the University of Texas Health Science Center in Houston.

• • •

Ian Down
Department of Aerospace Engineering
Hagler Fellow: Dr. Kathleen Howell
Year Awarded: 2022

I'm a third-year PhD student in the aerospace engineering department. I began working with Drs. Kathleen Howell (Purdue) and Terry Alfriend (TAMU) in September 2022 through the Worsham Heep Fellowship with the goal of addressing unanswered questions related to cislunar spacecraft formation flying.

Dynamically speaking, formation flying in the cislunar domain extends far beyond the limits of earth two-body gravitational modeling, thus traditional methods of modeling, analysis, and design cannot be applied. To remedy this, we turn to dynamical systems theory (DST). For the last forty years, DST has proved to be an invaluable tool in the numerical exploration of astrodynamics, most notably in the Circular Restricted Three Body Problem (CR3BP) and its transition to full ephemeral models. Center manifold theory, a subfield of DST, is of particular interest in the context of formation flying because the toroidal surfaces associated with the center manifold produce bounded quasiperiodic motion about a periodic orbit, which allows for a great deal of formation design. This work aims to investigate the fundamental structures and dynamics of relative motion on these center manifolds about all relevant families of periodic orbits in the CR3BP, as well as illuminate the utility of linear approximations in predicting these topologies. Specific applications of formation flying in the cislunar domain include both monitoring space traffic on the translunar highway, or positioning a large aperture, multiagent, deep-space telescope in optimal lighting conditions (behind the earth in the sun-earth system). This work will be presented at the American Astronautical Society's Astrodynamics Specialist Conference in Big Sky, Montana, in August of 2023, and will later culminate in an academic journal submission at the end of the calendar year.

To prepare me to tackle the aforementioned research questions, I was granted structured learning in subject areas that are not available at my current institution. To gain insight into a subject area from the subject-matter expert herself has been so invaluable not only for developing meaningful technical contributions but also in evolving my own research path as I navigate the remaining years of my PhD program. On the ground level, Dr. Howell has also connected me to her previous students, whom I've had several meetings with to quickly address numerical implementation issues, undoubtedly alleviating many headaches while pushing my work past these speed bumps. The entire collaborative experience has really been so much fun, and I'm very grateful to have had this unique opportunity while at Texas A&M University!

Today, Ian Down is a PhD candidate working in the Texas A&M Land, Air, and Space Robotics Laboratory.

• • •

Caleb Shackelford
Department of Soil and Crop Sciences
Hagler Fellow: Dr. Donald L. Sparks
Year Awarded: 2023

Our current research efforts have shifted to primarily focus on mineral associated carbon formation and phosphorus sorption via root exudation in drought conditions. Root exudation refers to the phenomenon where plants release various organic compounds through their root tips, thereby modifying the underground environment to better suit their requirements. For instance, plants may exude organic acids to extract nutrients adhered to soil particles or release organic substances that fuel microbial activity, aiding in root-tip hydration during drought. Given this, the objective of our research is to investigate this exudation process during drought and determine its impact on mineral absorption as well as ascertain whether recalcitrant carbon is being generated or simply broken down and converted back into CO_2. Given the significance of climate change and the urgent need for long-term carbon storage, this line of research holds profound importance.

This summer through a collaboration with Dr. Kate Szerlag, I will visit Brookhaven National Lab to use synchrotron-based spectroscopy to identify phosphorus speciation on two mineral types. This will be my first time visiting a national lab, and I am very excited to work alongside some of the world's leading scientists. I will hopefully begin writing a manuscript with the data collected from this visit. Additionally, I have been grateful to mentor an exceptional undergraduate student through this research whom I hope to guide toward research and academia.

I am incredibly grateful for my time with the Hagler Institute, as it has exceeded all my expectations. Dr. Sparks, my mentor, embodies exceptional intelligence, humility, and kindness, and I consider it a true honor to be counted among his students. The generous funding provided by the Institute has enabled me to wholeheartedly pursue my aspirations of becoming an accomplished scientist without the need for external employment, which is an invaluable opportunity. I genuinely believe that this institution has empowered me to strive for greatness, and I hope to be like the Hagler Fellows one day. I am blessed to be a part of the Hagler Institute.

Thank You.

Today, Caleb Shackelford is a graduate Fellow of the Hagler Institute for Advanced Study.

• • •

Miriam Sanders
Department of Teaching, Learning, and Culture
Hagler Fellow: Dr. H. Richard Milner IV
Year Awarded: 2023

Working with Hagler Fellow Dr. Rich Milner, I had the opportunity to complete a program evaluation of urban education and multicultural education doctoral programs in the United States. This research informed the programmatic change of the multicultural education doctoral program in the Department of Teaching, Learning, and Culture. In short, only two universities (i.e., University of Wisconsin–Madison and Texas A&M University) were still providing doctoral programs called *multicultural education* while the field has moved to provide programs in *culturally sustaining* or *urban education*. I gave two presentations on my findings in programmatic meetings to the multicultural education faculty and Dr. Rich Milner. The program is now transitioning to focus on urban education with three potential strands of emphasis for doctoral students (e.g., humanizing pedagogy) in the upcoming year.

Furthermore, Dr. Rich Milner is highly committed to mentoring early career faculty and graduate students. I had the opportunity to coordinate and schedule nine professional development events.

Initiative	Date(s)	Description(s)
Early Career Faculty	09/05/2023	Writing Camp for Early Career Scholars
	12/07/2023	Early Career Faculty Writing Retreat
Doctoral Student Engagement	09/26/2023	Job Search
	10/24/2023	Brown Bag Lunch & Learn: Creating a Research Agenda
	11/14/2023	Networking at Conferences
	12/06/2023	Writing Retreat
	01/31/2024	Writing Retreat
Lunch & Learn Series	03/01/2024	Exploring the Role of Education in the Dismantling of Health Disparities: Lessons Learned from Meharry Medical College
	04/22/2024	History Echoes: Spotlighting the Architects of School Pushout Past and Present

Publications and Conference Presentations

During my fellowship, I have had the opportunity to coedit a special issue for the *Journal of Urban Education & Policy Annuals*. The special issue is a top-down case study of urban education in Texas. My coeditors include Dr. John A. Williams III (i.e., my TAMU graduate research assistant supervisor) and Virginia Redwine Johnson (i.e., upcoming Heep Fellowship recipient in the Department of Teaching, Learning, and Culture). The special issue captures a geospatial analysis of urban schools, administrators in urban schools, recruitment and retention of teachers of color, STEM outcomes in urban school, and urban school policy recommendations. The special issue is set to be released in the fall of 2024.

Value of the Heep Fellowship

I cannot express the value of the Heep Fellowship in words, but I will enumerate the outcomes. This year, I have had the chance to work closely with a leading scholar in education, network with faculty from other universities, and focus on my own research. Since I started the fellowship in fall 2023, I have been able to submit one grant application and eleven research manuscripts for review in journals, seven of which I am the first author of. Furthermore, by focusing on my research, I finished my dissertation and secured a tenure-track assistant professor position at the University of Wyoming starting in fall 2024. I graduate with my PhD in curriculum and instruction August 2024. This opportunity has been integral for advancing my career in terms of networking and honing my skills and creative abilities as a researcher.

Today, Miram Sanders is an assistant professor in the College of Education at the University of Wyoming.

32

Distinguished Lecturer

Julio Ottino, 2023–2024

Since 2012, each class of new Fellows admitted into the Hagler Institute for Advanced Study has included at least one Hagler Institute Distinguished Lecturer. The purpose of the designation, according to John Junkins, "was to choose a stellar communicator whose work was of broad appeal, and therefore increase the impact and internal visibility of the Institute. University leaders are invited, and we typically have the provost and, occasionally, the president of the university in attendance. The audience size varies from 'medium-large' two-digit numbers of faculty, students, and staff to over 300. Nobel Laureate in Economics Vernon Smith drew about 350 people to his lecture and filled the large auditorium at the Annenberg Center annex of Bush Library."

Julio Mario Ottino was named a Hagler Distinguished Lecturer in the Class of 2023–24. His Northwestern University Kellogg School of Management profile offers a sampling of the credentials and accomplishments, which stand him in good stead with the Hagler Institute's core mission of enriching the educational experience at Texas A&M University by fostering collaboration:

> Julio M. Ottino is the Distinguished McCormick Institute Professor and Walter P. Murphy Professor of Chemical and Biological Engineering, and professor (by courtesy) of mechanical engineering at Northwestern University. He is the former dean of the McCormick School of Engineering and Applied Science at Northwestern. He previously held positions at UMass Amherst as well as chaired and senior appointments at Caltech, Stanford, and Minnesota. He is the founder of the Northwestern Institute on Complex Systems

(NICO) and numerous university-wide initiatives, programs, and centers in the areas of design, energy and sustainability, human-computer interaction, and entrepreneurship, with the the Kellogg School of Management, Pritzker School of Law, Feinberg School of Medicine, Medill School of Journalism, and the School of Education and Social Policy within Northwestern, as well as with external partners ranging from the Argonne National Lab to the Art Institute of Chicago.

Ottino, who also serves as a professor of management and organizations at Northwestern, presented his Hagler Institute lecture during the fall semester of 2024. He had been on campus previously for a symposium in late 2023 sponsored by the Hagler Institute and entitled "The Art of Collaboration in Academic Research." That event also featured a pair of Ottino's Northwestern colleagues, John A. Rogers and Yonggang Huang, themselves both Hagler Institute Fellows.

"I'm very proud to tell you I hired both of them and brought them to Chicago," Ottino says of his time as dean of the School of Engineering and Applied Science at Northwestern. "Both men are distinguished scientists, but in the case of Rogers, who is in the National Academy of Science, the National Academy of Engineering, *and* the National Academy of Medicine, one of our trustees told me, 'I'll give you $25 million if you can bring him here.' I wanted to hire him anyway, but the $25 million certainly helped."

Ottino, who is originally from Argentina and had a career as an artist before moving to the United States to pursue his PhD in chemical engineering from the University of Minnesota in 1975, was first introduced to the Hagler Institute by Ned Thomas, another Hagler Fellow who came to work full-time at Texas A&M as a Permanent Fellow in the Materials Science and Engineering Department in 2020.

"Ned was a professor at the University of Minnesota when I was a student there," Ottino says. "He then went to University of Massachusetts at Amherst. My first job was there, so we became colleagues. We wrote five papers together. After that, I moved to Northwestern, and Ned went to MIT and eventually wound up at Rice University in Houston before becoming a Hagler Fellow at Texas A&M.

"He's the one who persuaded me to give the Hagler Institute for Advanced Study serious consideration. He said it was a great opportunity. 'You'll be able to meet a good cross section of people there,' he told me, and that has certainly been the case."

Ottino's talk at the collaboration symposium was entitled "Want Creativity? Learn to Use Tension and Chaos." The subject came from his most recent book called *The Nexus*. In fact, the full title of the book is *The Nexus: Augmented Thinking for a Complex World—the New Convergence of Art, Technology, and Science.*

"I'm very proud of this book," Ottino says. "Let me tell you one thing about the book that no one will notice. Every chapter has a title, and the title is followed by a phrase, and the phrase is followed by a sentence. Every title is like this.

"For example, chapter 3 is called 'Creativity at the Nexus.' The phrase that follows is 'Our Views of Creativity, Shaped by the Renaissance and the Romantic Period.' And then the sentence is 'Myths of the lone creative genius and the burst of inspiration remain with us today.' My hope is that this thread of continuity contributes to a sense of beauty in design that is unconsciously perceived by the reader."

It was that chapter, "Creativity at the Nexus," that Ottino featured in his Hagler lecture at Texas A&M. And with Ottino's permission, we present that chapter from *The Nexus* to provide a taste—albeit without the abundant images that populate the book and were a part of his live presentation—of what his Hagler Institute lecture was all about.

• • •

Creativity at the Nexus

Our Views of Creativity, Shaped by the Renaissance and the Romantic Period

> Myths of the lone creative genius and the burst of inspiration remain with us today.

In what ways are creativity in art, technology, and science different, and in what ways are they similar? A view stemming from philosophy, placing the emphasis on uniqueness, declared that science is ephemeral, and that art is permanent, placing artistic creation on the highest plane. Others have had the same view. Is this actually true? There is more analysis in art than perceived by science and more creativity in science than perceived by art. In fact, somewhat paradoxically, we can often see the evolution of ideas in art, whereas the scaffolding that leads to the final result in science is nearly always removed from sight.

What does it mean to be creative in art, in technology, and in science? An all-encompassing definition of creativity across domains is elusive. Ideally, at some abstract level it may be possible to argue that creativity is the same no matter what the domain. In reality, however, the fields—the collection of arbiters who decide what is creative and what is not—operate under very different rules.

Art is about breaking paradigms and the search for the new, whereas science is about building and exhausting paradigms, and technology is about building and disrupting paradigms. In science, it is good to "stand on the shoulders of giants." In technology, the only reason to stand on older technology is to crush the elder giant. And in art, especially now, it is a bad idea to stand close to anybody. Although combining and meshing what exists to create something new is good in technology, it is frowned upon in art.

Science is about unveiling, revealing, and (sometimes) explaining what is already there. A clear method exists for questioning and deciding what can stay in the edifice of science. The standard view of the scientific method starts with collecting information or data about the phenomenon being studied. On the basis of that information, a preliminary generalization, or hypothesis, is formed, usually by inductive reasoning; this in turn leads by deductive logic to a number of implications that may be tested by further observations and experiment. If the conclusions drawn from the original hypothesis successfully meet all these tests, the hypothesis becomes accepted as a scientific theory or law; if additional facts disagree with the hypothesis, it may be modified or discarded in favor of a new hypothesis, which is then subjected to further tests. This scientific method underlies the growth of science. A few people at the edge of science—a minority in fact, the philosophers of science—worry about how science works. Most scientists do science and do not question its underlying assumptions.

Looking at a few other comparisons across art, technology, and science is also revealing. Start with models of growth. Science builds on the past; progress is wired into the very fabric of science. In a normal mode, science is about methodically building and adding knowledge (though disruption certainly happens, and it is essential as well). Massive revolutions are rare. Technology, on the other hand, is always about building and disrupting—constant disruption is necessary for growth. Art, especially as we move into the twentieth and twenty-first centuries, is all about disruption. Mannerist giants were replaced by baroque, then rococo, then neoclassical painters. And, in the early twentieth century, the modern movement turned the neoclassical academic giants William-Adolphe Bouguereau, Ernest Meissonier, and Jean-Léon Gérôme into forgotten figures in less than twenty-five years. More recently, the modern, postmodern, and post-postmodern may have been the last recognizable periods. Peter Schjeldahl, the art critic for *The New Yorker*, put this in vivid language: "Then those movements, too, disintegrated, and it's been pretty much one damn thing after another ever since."

It would be a stretch to consider science as completely algorithmic, something that can be executed following a series of prescribed steps—though mistakenly, in the popular imagination, the scientific method comes close to this. But when compared with art and technology, it appears to be so. Art is the least algorithmic, unless the algorithm is that there is no algorithm. This is heightened as we approach the present; anything smelling of derivative is bad, and originality is a must.

Ordering Developments in Art, Technology, and Science

Science is inevitable, technology less so.
What then is art, as we approach the present?

Assume that we are given a collection of milestones in art, technology, and science: paintings, examples of working prototypes of various technologies, research papers, all spanning two hundred years but unidentified by specific dates. Would it be possible to put them in the right chronological order?

We would contend that this is possible, in principle, for developments in science. Science often follows a clear path in time. Theories become more sophisticated, with new ones containing older ones as special cases. The same can be said within branches of mathematics; the n-theorem follows logically from a subset of n-1 previous theorems.

Technologies can be ordered as well: the iPad came after the Macintosh and the Macintosh after the IBM 360; CDs came after the record player. But here we must make an important point: these are all *successful* technologies. Many *seemingly* successful technologies do not make it: there are many dead branches in the network of evolution of technologies. Technologies can be forgotten, but more often they are combined and embedded into new technologies. GPS is now embedded in phones, and motion sensors are found in cars and video games.

In aesthetics and design-dominated fields, however, ordering becomes less clear. One would have to be well trained to put in the right chronological order chairs designed in the span of a hundred years. A similar observation can be made about paintings, especially as we get closer to the present. Without some training in art history, it would be nearly impossible to put in the right order a hundred randomly selected paintings done by well-known artists in the twentieth century. There is an intrinsic arrow of time in science; there is no arrow of time in art, especially now. And—as in the previous comparisons and several of the ones to follow—*technology falls between art and science.* The science component within technology can be ordered. The design and aesthetic components less so.

Domains and Fields

How do new ideas get accepted in art, technology, and science?

Creative ideas do not happen in a vacuum. Creativity is not a private enterprise but involves the interaction of three things: (1) a person or a group, (2) a domain—say, any of the branches of mathematics or physics or visual arts or history or sociology—and (3) the field, the set of people who are gatekeepers of the domain.

For an idea to be judged as truly creative, it must affect a domain and become a permanent part of that domain. The character of a domain limits the claims. In some instances—for example, in mathematics—the domain trumps the field; a theorem is either proved or not. In physics, the idea cannot violate the first or second laws of thermodynamics.

Altering the content of a domain in transformative ways is very difficult. People see things through the lens of their times. To accept something new we may have to expand or, more crucially, alter our knowledge base and abandon deeply held beliefs. This is not easy. Even in the fact-driven domains of science, such as physics, people are slow to accept new ideas.

Publications associated with scientific revolutions look momentous to us. A list may include books: Nicolaus Copernicus, *On the Revolutions of the Heavenly Spheres*, published just before his death in 1543; Isaac Newton's *Principia* in 1687; and Charles Darwin's *On the Origin of Species* in 1859. The list may also include papers: Einstein's first paper on relativity in 1905 and Watson and Crick's *Nature* paper in 1953. But Darwin did not triumph overnight. Neither did Copernicus: even though *On the Revolutions of the Heavenly Spheres* was dedicated to the Pope, the Pope took little notice of it, written as it was by a minor Polish functionary; in fact, there were few converts even fifty years after its publication. None of these works received instant acclaim; ideas took time to sink in. Sometimes, a long time: Nicolas Léonard Sadi Carnot wrote *Reflections on the Motive Power of Fire* in 1824, when he was twenty-seven years old. We know now this short 118-page book lay the foundations of an entirely new discipline in physics: thermodynamics. But it was not appreciated for close to a hundred years.

By the time of Newton's death in 1727, Newtonian ideas had yet to conquer continental Europe. But when they did, by the end of the eighteenth century, Newtonian ideas took over the intellectual life of Europe and America. Newton had become the God of the Enlightenment, the period that brought inspiration to the French Revolution and the founders of the United States. To have something taken seriously, it had to be labeled Newtonian. But once acceptance sets in, so does a new normal and the inertia that accompanies it. Newtonian ideas became orthodoxy, and dominated physics for close to two centuries—until relativity and quantum mechanics came into the picture, bringing with them one of the biggest disruptions in physics and the establishment of a new normal.

At the other end of the spectrum, in art, the field trumps the domain. Gatekeepers—critics, museum curators, galleries, art collectors, and the like—collectively decide what gets in. Judging which works of art are worthy of inclusion in the canon of art brings to mind the concept of emergence in complex systems. Here, though, let us say that judgments are not final and are rarely unanimous; ups and downs in artistic reputations occur as a function of time. When we say that van Gogh was unrecognized in his lifetime or that Bouguereau was overrated, we are, in essence, overriding the judgment of the field in their times.

In general, the rules of acceptance by a field are clearest in math and in physical sciences. We could also add theoretical computer science into the mix. There are ways to check and verify results; there is a methodology for moving forward. Today the bar for possible acceptability is having work published in prestigious peer-reviewed journals or presenting work at the right conferences. But even in these fields, things are not always that clear-cut, and anonymity does not preclude bias.

Style also matters. This is important: clean, clear, elegant presentations will typically prevail. There may be two different styles of doing something, and this is true even in mathematics. The word "beauty" often appears in discussions of math. "There is no place in the world for ugly mathematics," G. H. Hardy said. An elegant proof is always better than an inelegant one. But sadly, for purists, mathematics is starting to accept computer-assisted proofs. And few will label such proofs "elegant."

Let us consider here three aspects that cut across the three domains of art, technology, and science: (1) the talent pipeline, (2) the role of groups or communities, and (3) the transparency of the creative processes.

The talent pipeline in science today is a well-oiled machine consisting of a predictable sequence—undergraduate degrees, followed by graduate degrees and, often, postgraduate studies. Research takes place in universities, institutes, and corporate and national labs. Science and engineering faculty positions in research universities almost always require a PhD. There are now virtually no outsiders in science. The pipeline in technology, however, is far less formal, and the pipeline in art, though becoming professionalized, is less clear as well. With a weak support structure, there are still outsiders in art.

Groups and communities are central to the acceptance of ideas in a domain. In terms of the sheer number of components involved, the most "democratic" (and complicated) system is the one underlying the acceptance of a new technology: an invention that results—via knowledge transfer—in an innovation. This is an area where active collaboration takes place between universities and federal labs and organizations with business-sector researchers via joint research and publications. In the United States, both the number and share of joint articles have increased since the 2010s. The innovation system environment is complex. Nowadays, at least in the United States, this involves an inventor, a system of funding (federal funding agencies, venture capitalists, and angel investors, for example), tax and regulatory policies, and a system of laws for intellectual property protection. Added to this are national and international interests, battles about intellectual property protection, issues of public policy, marketing, and social attitudes toward risk, among other concerns. Products now regarded as an integral part of society took years if not decades to develop. In many cases, public acceptance defies the opinion of experts, and so a clearly inferior idea continues to live.

The degree of cooperation within communities can be quantified in science and, to some extent, in technology. This interfaces with issues concerning the transparency of creative processes. There are established ways by which we investigate the output of teams in science (papers published) and in some aspects of technology (patents and licenses). To a large extent, it is easy to know who worked with whom, because listing references to prior scientific work is part of the established practice of science. Some academic disciplines are meticulous in this area; mathematics is at the top of the list.

But despite transparency, the path leading to the final product may not always be clear. In science and technology, we generally see the result and not all the steps in the evolution that may have led to it. The result is important, but the documentation of the evolution is not valued—except a posteriori, by historians of science and technology.

In fact, in science, all dead ends are concealed in the final presentation of results. This happens more so now than in the past. Space in the most prestigious scientific journals, including *Nature, Science, Proceedings of the National Academy of Sciences of the United States of America*, even in this era of digital content, is at a premium, and presentations are highly condensed. Supplementary materials—easy to include in web versions of articles—are often much longer than the publications themselves. First-person description of results, something that was common in the early 1900s and used by many renowned scientists until the 1960s and 1970s, is discouraged in most top journals covering science and engineering. It is unclear if the emergence of a new field could alter norms; in the synthetic biology community, there is an increasing emphasis on the value of publishing positive and negative results (albeit the negative results are often relegated to the online supplement).

On the other hand, the degree of cooperation in art (literature, music, visual arts, for example) is more subtle—co-authorship is rarer—and operates more at the level of influence. Often a movement is seen in retrospect; alliances are informal. Entire careers in art history and criticism are based on dissecting these facts.

The Adjacent Possible—A Recipe for Creative Output

> We can learn to access and extrapolate from existing cutting-edge knowledge.

Can there be a methodology for coming up with new ideas? What ideas and knowledge could emerge if we were to curate connections of individuals at the vanguard of a domain? What new knowledge could be accessed by relatively small steps and extrapolations from existing cutting-edge knowledge? The theoretical biologist Stuart Kauffman, motivated by his work in prebiotic chemistry, coined a suggestive name for what could be called first-order combinations reached from the boundary of this edge: "the adjacent possible."

The adjacent possible can be thought of as a forever-expanding house—opening a door from a room at the boundary of the domain leads to another room, which will have new doors that may lead to other new rooms with yet other doors. In the case of prebiotic chemistry, the adjacent possible defines all the molecular reactions that were directly achievable in the primordial soup. From those chemicals, other molecules formed, and from those, others more complex still, until they hit upon something that could self-replicate. Plants, trilobites, and sponges exist outside the circle of possibility of the primordial soup. The opening of a very large number of doors and the visiting of many rooms would be needed before reaching something resembling a brain.

The adjacent possible, being accessible to the prepared mind, is at the root of so-called multiple discoveries—when different individuals independently come up with significant new ideas at virtually the same time. Though this also happens with teams, the case of individuals is the clearest and provides the best examples. Earlier in the book, for instance, we encountered Lord Kelvin, a towering figure in physics known for many achievements, including his work on the development of the second law of thermodynamics; Kelvin had several ideas that others had as well, almost at the same time: the sociologist Robert K. Merton and his collaborators examined 400 of Kelvin's 661 scientific communications and addresses and found that at least 32 qualified as multiple discoveries (Merton 1961). Kelvin's co-discoverers were an illustrious set—some on par with Kelvin, others a notch below on the prestige scale. This does not diminish Kelvin's greatness. It indicates a considerable number of other top scientists were required to duplicate only a subset of the discoveries that Kelvin made. Many others were with him at the boundary. But Kelvin was a master in opening doors in the adjacent possible.

We can find numerous documented multiple discoveries in both science and technology. The invention of calculus by both Newton and Leibnitz is a famous example. Charles Darwin and Alfred Russell Wallace having similar ideas about the theory of evolution is another. Possibly the clearest example, in terms of documented timing, may be Elisha Gray and Alexander Graham Bell filing patents for the invention of the telephone the very same day. Yet another, less well-known example, but nevertheless central to our understanding of the world, is Brownian motion, named after the botanist Robert Brown. Using a microscope, Brown observed the seemingly random motion of pollen particles suspended in a liquid. Einstein explained this phenomenon in one of his famous string of papers in 1905; the Polish physicist Marian Smoluchowski came up with a similar model in 1906. Einstein moved to other matters; Smoluchowski kept working on these ideas, using them as a basis for his theory of stochastic process (one central equation bears his name). It is important to see the consequences of Brown's discovery.

Why does Brownian motion happen? The pollen particle is hit by molecules of the liquid, but given its small size, the number of hits will not balance out; more hits will come from one side than the other, and the particle will move in what is a seemingly random motion. In fact, random motion and Brownian motion may be considered synonymous. Neither Einstein nor Smoluchowski knew conclusively about the existence of atoms, but their explanation of Brownian motion became the bases of the experiments by Jean Perrin proving conclusively the existence of atoms and molecules, for which he received the 1925 Nobel Prize in Physics. It is no exaggeration to say that we could not possibly understand the world without understanding that the world is made up of atoms.

In a *New Yorker* article titled "In the Air," the popular author Malcolm Gladwell argued that if you put together, in one room, people who were a notch below Kelvin on the prestige scale, you could get a large subset of Kelvin's discoveries without Kelvin being in the room with them. The point being that there are more people a notch below Kevin and much fewer at the level of Kelvin. But Gladwell also argued that this does not apply to artistic genius. "You can't pool the talents of a dozen Salieris and get Mozart's 'Requiem,'" he wrote. "You can't put together a committee of really talented art students and get Matisse's 'La Danse.' . . . A work of artistic genius is singular." This may be true, though some will disagree, especially regarding the tendency to elevate some people to the genius category and separate them from what were somewhat comparable peers. Not that Mozart does not deserve this label, but there seems to be now some motion to diminish the talent gap between Mozart and Salieri.

The Myth of Genius and *Primo Pensieri*

History has taught us to believe in epiphanies and eureka moments.

Retrospective analyses show that artistic masterpieces do not suddenly appear from thin air. A "final" painting can be deceptive when presented without context. Paradoxically, even though bursts of inspiration are strongly associated with visual arts, it is the domain in which the evolution of an idea—the painstaking process that goes on behind what may appear as flashes of genius—is the most transparently and meticulously documented. This is far from uniform across other domains. There are visible fragments of the evolution of creative thought in music and literature, such as surviving book manuscripts or annotated copies of musical scores. And once in a great while we see the editing processes of a great scientific paper (for example, the manuscript of Einstein containing the celebrated $E = mc^2$). By far, visual arts provide the most complete records of creative processes and exploration of ideas. Two examples serve to make the point. Picasso, for instance, did forty-three sketches for *Guernica*. Another notable example of elaboration of an idea is Picasso's *Las Meninas*, a series of fifty-eight paintings done between August and December 1957. The series has Picasso reinterpreting, reconstructing, and recreating the original *Las Meninas*, the 1656 painting by Diego Velázquez, one of the most widely analyzed works in Western painting.

We popularly associate "genius" with the initial burst of an idea—and often we assume that the idea emerges fully formed. There is a historical underpinning associated with equating creative bursts and genius, and this comes largely from art and the Renaissance.

In the Renaissance, the first ideas—the initial sketches, or *primo pensieri*—were highly prized. These sketches, when sufficiently evolved, were called *il modelli*. They served as templates for say, altar-pieces, larger paintings, or frescoes, but they were not intended to be works of art in themselves. Rather they were functional drawings and an integral step in planning compositions for paintings. But they carried a legal force; *primo pensieri* were the Renaissance version of copyrighting an idea, capturing it at the very moment of creation.

The *primo pensieri* concept has a parallel in science, not in reality but in popular imagination. The falling apple hitting Isaac Newton's head? A myth popularized by Voltaire and Disraeli. What about Archimedes running out the bathtub shouting "Eureka!" or Darwin conceiving the theory of evolution after a brief visit to the Galápagos islands in 1835? Probably myths as well. Some eureka moments are more believable than others, but there clearly is a continuum, with the fully formed idea at one extreme. The reality is that epiphanies are an occasional by-product of single-minded attention devoted to difficult problems. The brain is working on an idea without being consciously engaged. The epiphany is then the last piece that solves a puzzle, though the last piece is often no more special than any of the others. The electric light bulb is often used as a metaphor for inspiration or a great idea, but it would be hard to come up with a worse example in the history of technology. Edison worked on thousands of ideas for filaments before hitting on one that worked. Inspired tinkering, yes; sudden burst of creativity, most definitely no. And even more than in art and science, in technology the first idea is just the beginning of a long and often tortuous process. The beginning idea is never close to the end in technology.

We normally get sanitized narratives of creative and scientific outputs; a distilled and cleaned-up view leading smoothly to an end point: perfect buildings with no hints of the scaffold. The reality is that trajectories leading to what is seen as a creative endpoint are messy—and learning about this messiness is good. Knowing some of the history behind creation and discovery may increase creativity rather than diminish it.

The Deception of Final Pictures

Do not be misled by the removal of scaffolds.

"The secret of creativity is knowing how to hide your sources," Albert Einstein reputedly said. Scientists, philosophers, and mathematicians have commented on this, what we call the "deception of final pictures." Many of the highest exponents of creative thought have left no trace of the scaffold. Ludwig Wittgenstein's phrase "kicking the ladder" after climbing it is apt, even if he used the metaphor in a different context. Mathematics provides rich examples. Karl Friedrich Gauss and Bernhard Riemann left no trace of the scaffold that led to their final result. "I did not succeed in compacting the proof as to make it publishable," Riemann said, simply stating four properties of the "Riemann" function. It took thirty years to prove Riemann's first three conjectures.

And even if final results do not look magical but solidly logical, they can hide what may have been myriad tentative steps and dead ends. The physicist Hermann von Helmholtz captured this point when he described the two roads to research: (1) the shaky ladder every researcher must climb to get to the results and (2) the smooth royal path when the results are presented to an audience.

The idea of final pictures also helps perpetuate the myth of lone genius. Exceptions do exist; some people create in relative isolation. But as human myths go, it is a relatively recent one. The Greek embodiment of creative powers resided with the Muses, attendant spirits who inspired creation in literature and the arts. Creativity touched humans from unknowable and distant sources and for unknowable reasons. This went beyond the Greeks. It was not until the 1500s that

Western cultures grew comfortable with recognizing individual accomplishment. The turning point is when *primo pensieri* mattered and the pendulum began to move in the direction of individual human achievement. The idea of genius, as we understand it today, comes from the Renaissance. Giorgio Vassari, history's first art historian, is partially responsible for this. Leonardo, Michelangelo, and Raphael were seen as having talents seemingly beyond human limitations. The idea was perfected in the Romantic period and migrated from art, music, and poetry to science and mathematics.

The genius myth is still in full force today. The label is sometimes assigned by others, but often the "geniuses" themselves help market the myth. Legend has it that Jack Kerouac wrote *On the Road* in three weeks, typing it almost nonstop on a 120-foot roll of paper, which was sold on May 22, 2001, for $2.2 million. It is a nice legend, but the truth is much more complicated.

Critics love to discover unrecognized geniuses. A somewhat recent case is the German physician and writer Hans Keilson. Two of his novels, written more than sixty years ago, were declared to be "the work of genius" in a *New York Times* book review.

Associating one idea to one person is convenient, but it is rarely the whole story. Yet the idea is deeply ingrained in the popular consciousness and now extends even to organizations: Google did not invent the search engine, Nintendo did not invent the video game, and Netflix was not created because of a late fee at Blockbuster.

Creativity in Technology

> The Rule of Three: balance three or more elements to create something unique.

We quoted earlier Jacob Bronowski's definition of creativity: "the scientist or the artist takes two facts or experiences which are separate; . . . finds in them a likeness which had not been seen before, and . . . creates a unity by showing the likeness." The individual ideas may not be breathtaking, but the combination is. We also said that, in technology, derivative is not necessarily a bad word: adaptations and remixes are good. This fact alone makes creativity in technology different from creativity in science and art.

Combining two things to create a third is relatively common. Combining three balanced elements at the same time to produce something unique, something that could not be glimpsed by the components that go in the mixture, is much less so. There is nothing magical about the number three; it could be four, or even five. Nevertheless, it is useful to think of inventions in terms of triads. When seen through this lens, a rule of three seems to underlie much of technology. Consider the following illustrative examples:

- William Morse, credited with inventing the telegraph. This involved combining the electromagnet, battery, and cable making.
- Louis Daguerre, forerunner of photography: *camera obscura*, one chemistry to record images, and another chemistry to fix images.
- The Wright brothers and the airplane: gliders, aluminum, and the wind tunnel.
- Carl Benz and the Benz Patent-Motorwagen, regarded as world's first production automobile: carriage, bicycle, and the internal combustion engine.
- Guglielmo Marconi and the invention of the radio: the telegraph, telephone, and radio frequency vacuum tube.
- Henry Ford and the development of the first affordable automobile: the maturation of the gasoline engine, petroleum refining, and the concept of the assembly line.
- Willis Carrier and the first useful air conditioner system: the centrifugal compressor (which he invented), the electric motor, and the distribution of cheap electricity.
- RCA's David Sarnoff and the development of television: the cathode ray tube, the radio, and the idea of a scanning technique to collect and transmit images.
- IBM's Thomas Watson and the first practical modern computer: silicon transistor, magnetic tape, and the concept of computer logic, a.k.a. software.
- Vinton Cerf, who is closely associated with the invention of the internet: pervasive computing, pervasive telecommunications, and the concept of coded packet routing of information.

Like in much of technology, other names could be associated to many of these inventions. Also, the list could be longer. As a final example, consider the iPhone. The concept could be traced back to George Gilder's prediction of the "teleputer." The iPhone is a triumph of design but involves pieces that were already there: the large-scale integrated (LSI) circuit, the gallium arsenide (GaAs) radio chip, and the lithium battery. And each of them can be traced much further back.

The common denominator in all these? Three enabling technologies, the combination of which yielded something fundamentally different. It is also useful to imagine how entire disciplines—not just technologies—may be combinations of three components. Consider one example out of many. Modern medicine could be imagined as the combination of information in the form of medical science and knowledge (all the underpinning science, beginning with germ theory and moving into molecular biology, genomics), new classes of machines (X-rays, MRI, robotics, diagnostics), and new kinds of materials (pharmaceuticals would fall in this bucket, as well as implantable and surgical materials). Where would AI fit into this picture? Underpinning knowledge? Such questions are what makes this a useful exercise.

> Creativity is essential in art, in science, and in technology. But in what ways are creativity in these three areas different, and in what ways are they similar? The fields determine what is creative, what becomes part of the domains. Science builds on the past, methodically adding knowledge, with radical disruptions happening rarely. Technology is about both building and constant disruptions—disruptions, in either minor or major scales, are essential for growth, and combining things is both necessary and good. Art is all about uniqueness and disruption, and progress has little or no meaning. A common creative signature encompassing all domains is thus elusive. What is permissible in one domain may be frowned upon in another. But there are still lessons that can be extracted across domains. How can we use creativity from another domain to innovate in our own domains?

33

Gala Planning

Howdy! My name is John Junkins. I am the founding director of the Hagler Institute for Advanced Study and will serve as your master of ceremonies this evening.

Thank you for coming tonight! I am pleased to welcome you to our Annual Gala. We have a wonderful program this evening. In fact, several aspects of our program will literally impact the history of Texas A&M University. In addition to inducting our current class of Hagler Institute Fellows, we will hear remarks from Texas A&M President Mark Welsh III. In addition, we will award a very special new honor, the Take Flight Leadership Award, to a truly outstanding individual.

The birth and growth of this institute has been an effort by many people who decided to pull together. The fundamental cornerstone was laid some years ago by something called Vision 2020. *That endeavor, led by President Ray Bowen and the namesake for our Institute for Advanced Study, Jon Hagler, envisioned a Texas A&M where we were constantly working to elevate the faculty. However,* Vision 2020 *was long on ambition and short on means. Inspired by* Vision 2020, *the Hagler Institute today represents those means.*

—Founding Director John Junkins in his welcome remarks at the 2024 Hagler Institute for Advanced Study Gala, March 1, 2024

The annual Hagler Institute for Advanced Study Gala is organized by the Institute's associate director, Amanda Scott. During Scott's undergraduate years as a student at Texas A&M, Ray Bowen served as university president. Scott recalls President Bowen fondly. "He was somebody that people felt very inspired by," she says. "I think we felt like we were in good hands with him as the university's leader because he truly cared about the students."

As for the *Vision 2020* plan Bowen initiated to guide the university to an enhanced academic future, Scott—who was Amanda Finke from nearby Brenham, Texas, during her college years—admits, "I remember it a little bit, but I didn't fully understand all that was going on. I remember hearing about it, but thinking, 'Well, I won't be here then.'

"I didn't know where the road was going to take me."

Scott came to the Texas A&M Institute for Advanced Study in 2016 as a program specialist charged with, among other responsibilities, the planning and execution of the organization's annual black-tie induction ceremony. Today, Scott is an associate director of the Institute, and the "Gala" remains one of her most important and time-consuming responsibilities.

John Junkins says his early vision for welcoming new additions to the Institute was modeled after his own experience being inducted into the National Academy of Engineering.

"There is significant pomp and circumstance associated with election to National Academies," Junkins says, "but with annual classes that round off to one hundred and with a lengthy three-day agenda of other academy business, it is difficult to say more than one sentence about each inductee at the induction event. At Texas A&M, I wanted the Hagler Institute Gala to be a celebration that felt much more extensive and personalized. With just ten to fifteen Fellows inducted here each year, we're able to spend much more time focused on introducing each of our honorees and their spouses, which leaves everyone feeling a lot more special."

Junkins continues, "The Gala provides an opportunity for our distinguished faculty and university leadership to be 100 percent focused on the extraordinary academic talent we bring to the school each year. In addition, the Gala is a great marketing event. Many distinguished friends of the university are in attendance and are frequently inspired to help grow the Institute through major financial endowments."

Thus, for an event planner like Scott—and make no mistake, she does a lot more for the organization in her role as associate director—the stakes are high the night the Institute throws its biggest party.

With two degrees in nutrition from Texas A&M, Scott never imagined she'd be rubbing shoulders with elite academics and getting paid to do so. Her studies in college seemed destined to prepare her for a "career in corporate wellness, health promotion, the food industry, public health, pharmaceutical sales, medical and research laboratories, biotechnology, or government," as the A&M nutrition website spells out today.

After completing her graduate work, Scott took a position with the Texas A&M AgriLife Extension Services.

"I love the land-grant mission of Texas A&M," Scott says. "The extension mission is to take the research that's done on campus and make sure that people benefit from it. My job with Extension Services was program specialist for a federally funded nutritional education program. I had statewide responsibilities, traveled all over Texas, and learned a lot while gaining experience with budgets, personnel issues, trainings, and conference organization."

Putting on educational conferences provided Scott with experience that made her a viable candidate to organize the Hagler Institute Gala. Right?

"Those conferences were not one-night events," Scott laughs. "They would be three or four days for as many as two to three hundred people. The conferences had a different audience and different purpose than the Hagler Institute Gala. I understood that more fully after I arrived at the Institute for Advanced Study, thanks to Susan Wolff, the former director of Research Communications here at Texas A&M. She taught me a lot about how to put on an event like our Gala. She's retired now, but she was fantastic at planning events like ours, and I learned a great deal from her."

Planning each Hagler Institute Gala—which in recent years has taken place during the spring semester—is about a nine-month process, according to Scott. The work doesn't completely consume her, but attention to detail is critical. Each class of Fellows is announced the preceding fall, and one of the first pieces of information they get is the date of the upcoming Gala.

"We've tried several variations on the format through the years," Scott says. "We've tried to enhance the experience through the use of things like different design elements, different kinds of walkways, and altering our seating arrangements. Over the years we've improved our advance communication with the Fellows, because trying to get everyone together the night of the event for photos and to make sure they understand their roles in the program has been more challenging than you might imagine.

"The theme each year is set in collaboration with Research Communications. They help us come up with the color scheme, look of the stage, and the overall event concept. I do help with that and approve the final designs, but it's more organization of the event as a whole that falls onto my plate. That includes communicating with the Fellows, letting them know what to expect, making sure they arrive at the right place and at the right time, making sure the seating is correct for them, and just making sure the program runs smoothly. I also help Dr. Junkins write his script, coordinate the event meal with our caterer, and make sure Greg Tivis and his orchestra are booked for the event."

Greg Tivis, a Hagler Institute Gala mainstay, was born in Palestine, Texas, but grew up in Oklahoma and attended Central State University in Edmond, Oklahoma, on a full piano scholarship. It was there he fell in love with jazz, and before long his studies became secondary to performing in and around the Oklahoma City area, wherever and whenever he could. Four years after graduation, he became a part of the Russ Morgan Orchestra, one of the longest continuously running big bands in America. As a result, Tivis traveled a lot. He says his wife was finally able to get him off the road by landing him a job playing piano five nights a week at the College Station Hilton Hotel.

The couple moved to Aggieland in 1997.

Tivis and Hagler Institute Associate Director Clifford Fry are longtime friends.

"Dr. Fry is a musician as well," Tivis points out. "He and his former band get together for jam sessions at his home several times a year. One of the musicians in one of my bands is a regular attendee at Clifford's events. This mutual friend invited Clifford to one of my monthly jazz jam sessions, and it was there we first hit it off. Before long, we began teaming up and playing together as volunteers at senior facilities around town.

"One of the folks I met at the jam sessions at his home was his neighbor and good friend, John Junkins. Around that same time I had released an album with singer Kelsey Taylor, whom I worked with for several years at Veritas Wine and Bistro, where I played every weekend from 2008 until it closed in 2020. John liked me, but he was especially fond of my jazz collaborations with Kelsey. So when he founded the

Institute for Advanced Study and planned the first Gala, he had Clifford ask me if I could bring my dance orchestra and make sure Kelsey was our vocalist. I was only too happy to comply.

"We've played the Gala every year since the beginning. And John tells me it's a tradition now, so I suspect we will continue to play it as long as he is in charge."

While the big-band sounds of the Greg Tivis Orchestra kick off and conclude the evening's activities at the Hagler Institute Gala, the visual highlight of the event comes with the introduction of each new Fellow and guest lecturer and their grand march through the saber archway formed by the Texas A&M Corps of Cadets' Ross Volunteers.

The Ross Volunteers unit was formed in 1887 and is named in honor of Lawrence Sullivan Ross, former governor of Texas and president of the Agricultural and Mechanical College of Texas from 1891 to 1898. Ross is revered at Texas A&M, and a statue in his honor stands at the center of the main campus. According to the Texas A&M website, "[Ross] was known for his legendary efforts to keep the college open and is often called the embodiment of the Aggie Spirit."

As for the "RVs," as the Ross Volunteers are known in Aggieland, they make frequent appearances both on and off campus. Most notably, they serve as the official honor guard for the governor of Texas and, somewhat surprisingly, the honor guard of King Rex, the King of Mardi Gras as celebrated in the Bayou City of New Orleans. At Texas A&M–sanctioned events honoring the recently deceased, called Silver Taps and Aggie Muster, the drill team regularly performs revered three-volley and twenty-one-gun salutes.

At the Hagler Institute Gala, the sight of the Ross Volunteers at the center of the Bethancourt Ballroom is a majestic one. Clad in their all-white uniforms, the ten cadets march sharply to their positions and hold their sabers aloft, forming a canopy of sorts under which each new Fellow and his or her spouse or guest pass as the cornerstone of their induction.

In many cases, inducted Fellows attend solo, but no one passes beneath the saber arch alone.

"What we do in that case is we recruit student ambassadors from the colleges and schools," Scott says. "When the need arises, I try to pair a student ambassador—usually a PhD student—with a Fellow nominated by their college or department. And the students love it!

"Occasionally, a Fellow may actually know the student ambassador depending on how much time they've already spent on campus and what's been going on in the department. Managing the student ambassadors on an ongoing basis is another big piece of what I do.

"Working with Dr. Junkins, we send out an invitation to all the deans, and they are able to nominate one doctoral student per year to come and serve as an ambassador at the Gala. We do some training with them beforehand. We try to make them feel really welcome because they are spending their own time and money to be there. So we do our best to take care of our student ambassadors."

Scott adds that another student organization has been supporting the Hagler Institute Gala in recent years.

"They are the Maroon Coats," she says. "When a Texas A&M student becomes a sophomore, they are eligible to apply to be a part of the Maroon Coat program through the Texas A&M Foundation. Students who are selected are trained and then become available to assist at events around campus.

"One of the services they can provide, which is really great in the case of an event like ours, is they have golf carts that can help shuttle guests to and from the parking garages near the Memorial Student Center. We've also utilized some of the Maroon Coats to escort Fellows through the saber arch.

"The Maroon Coats, like our student ambassadors, always do a great job. They're just so highly polished and trained. They visit with guests, interact with people, and supplement our staff because we don't have a huge number of people at our disposal."

If it sounds as if good times are had by all, that's almost always the case when it comes to the Hagler Gala. There are the occasional hiccups because the world is not a perfect place. When something unplanned happens on stage, this usually triggers a John Junkins "off-the-cuff" remark in which humor turns the hiccup into a highlight.

"The only real snafus we've had since I've been involved," Scott says, "have really been related to people being unhappy with their seating arrangements, either with the location of the table or perhaps wanting to be seated with someone different.

"Actually," Scott says, "this has only happened twice, so it's not been too much of a problem.

"When you have more than three hundred distinguished individuals in attendance, not every guest can be seated next to the stage. But I hate it when anyone is unhappy. We just have to handle these things when they come up, and try to learn from the experience for next year."

Gen. (Ret.) Mark A. Welsh III served as the keynote speaker for the 2024 Hagler Institute Gala. Welsh became the twenty-seventh president of Texas A&M on December 12, 2023, after serving as the school's interim president for the preceding five months succeeding M. Katherine Banks. Welsh was elevated into the role of interim president after serving as the dean of A&M's Bush School of Government and Public Service.

Welsh, a graduate of the US Air Force Academy and former chief of staff of the Air Force, has been a unifying figure on the Texas A&M campus since assuming the school's presidency, not only for faculty and staff, but also for students and alumni alike. His remarks at the 2024 Hagler Institute Gala offered strong evidence of his conviction and character as well as his devotion to the university.

> Let me tell you what the guy you're looking at isn't, and that's a scholar. So as I was standing up here listening to the introductions of our new Fellows, I felt myself slowly shrinking as I heard their incredible accomplishments. If John would have read off my academic credentials tonight, he would have been finished before I even reached the saber arch!
>
> As you know, John Junkins preceded me in this role as an interim president of Texas A&M, and he was immensely helpful to me when I stepped into that role myself. So John, thank you for that.
>
> It is a thrill for my wife, Betty, and me to be here to welcome you personally to what we believe is the greatest university. It is rare in a room of this size to be able to make a definitive comparison on just about anything, but I'm confident in this one: I am the only person in this room who ever had a semester GPA that started with a "one," which is why I'm completely intimidated right now.
>
> While I'm on the subject of numbers, the number 12 is a pretty special number here in Aggieland. And this 12th class of Hagler

Fellows and Distinguished Lecturers is an incredibly special group of scholars whose career contributions are simply off the charts. And while the group represents a number of different fields, you do have several things in common as far as I'm concerned. The first one is that you're passionate about what you do. The second one is that you're exceptionally good at it. And the third one is that you are not just making a difference, you are literally changing the world. Congratulations again on your remarkable careers.

The best thing from my perspective about this evening as I saw you walk through the saber arch, I know that you'll inspire and enhance the research productivity of our already outstanding faculty, because you've been doing that your entire career. I also know that you'll open the door to amazing learning experiences for our students, and I believe you'll find them to be exceptional. There are smart and talented students everywhere in universities, but Aggie's add a little extra spice to the sauce. I think it's a mix of Texas swagger, country cool, and a ranch work ethic. They go on to become incredible representatives of this university. There are more Aggie CEOs at Fortune 500 companies than from any other university in the country. Our Bush School of Government and Public Service typically puts 80-plus percent of their graduate students into public service jobs each year. They're absolutely committed to serving their fellow citizens and changing our cities, our state, our nation, and our world for the better. Aggie's are leading the charge in the energy business, the banking industry, animation and visualization, robotics, architecture, geosciences engineering, intellectual property law, and many, many other fields of endeavor. And many of the Aggies who are doing that work are in this room tonight. Thank you all for being here.

The engine that powers that success of our Former Students is the educational experience they receive here, and the heart of that experience beats in our remarkable faculty. Nowhere is the success of the Hagler Institute more evident than in its physical impact on our faculty over the last 12 years. Twelve years ago, there were 11 National Academy members teaching at Texas A&M. Today there are 57. Those distinguished scholars and researchers set the pace for over $1.1 billion in research expenditures last year, a number that has steadily grown since the Institute opened its doors. As

John mentioned earlier, a stunning percentage of former Hagler Fellows developed a taste for that spicy sauce that I mentioned, and now serve as full-time members of our faculty and icons of our university. The Hagler Institute is inexorably changing the academic landscape of Texas A&M in an undeniable way. So I would also like to thank the leadership and staff of the Institute for your extraordinary accomplishments. You have built something unique and amazing, and I thank you for it.

Amanda Scott is both proud and grateful to be a part of that Hagler Institute leadership team. And as John Junkins, Clifford Fry, and innumerable Fellows of the Institute both past and present will tell you, she is indeed integral to the organization's success.

"Amanda does so many things to make this Institute work," says Clifford Fry, Hagler Institute for Advanced Study associate director. "Amanda handles all the financial work connected with John Junkins's research grants and support of his PhD students. She assists in the preparation of extensive reports for both the administrative council, which oversees the Institute, and the external advisory board that provides key insights and support for our operation.

"In addition, Amanda helps coordinate our Eminent Scholar Lectures, our symposia, our annual event announcing the newest class of Hagler Fellows, and our well-attended and entertaining annual Christmas party.

"Amanda is also integral to the evaluation process of those nominated to become Fellows of the Institute. She collects data about nominees' research standings in their various fields and handles the input from our faculty advisory board about our nominees. She works closely with John and me in the recruitment process. The Institute is now adding a new program to recruit the nation's top students to work with Hagler Fellows, and she is an integral part of the endeavor.

"It does not matter which event or project is undertaken, the 'devil is in the details,' and Amanda grapples with those details every day. She is key to having made this institute a success."

"I'm very, very, very blessed to be here," Scott says. "I love my job. There are parts that are challenging of course, but I love what I do. You know how some people wake up and say to themselves, 'I don't want to go to work today.' I'm not that person. When I wake up, I think about what I have to get done, and how I'm going to use my time wisely to do that. I'm just very blessed to be here and to be a part of what's going on and the progress that we are making. I'm really proud to be a part of it."

As John Junkins concluded the 2024 Hagler Gala's official program, he announced the recipient of a new award to be handed out to individuals whose contributions to the Institute have, in a real sense, helped the Institute "take flight." And the first recipient of the Take Flight Award was the first and—after thirteen years—the outgoing chairman of the Hagler Institute's external advisory board, Norm Augustine. The following were Augustine's remarks:

> Thank you John, so very much. It is an understatement to say that I am deeply honored. I count as a great privilege to have been able to watch the growth of the Institute over the last thirteen years. I also want to say that I am very proud to be the father of an Aggie. That's my most important contribution of all.
>
> When John first spoke to me, he didn't quite tell the story the way it was. He found me at a meeting in Washington. He asked me if I would serve as chairman of the advisory board of the Institute for Advanced Study. And I, of course, was very flattered. So I said to him, "That sounds wonderful. How many people serve on the advisory board?" And he said, "Well, there are none right now, but when you say yes, we'll have one."
>
> And so I tried to save the day for him and asked, "Well, how many fellows do you have?" "None, yet," was his reply. "What's your budget?" He said, "There's no money yet." About that time, it occurred to me that John was asking me to lead a board that didn't exist for an institute that didn't exist, that had no funds. It just seemed kind of strange to me.
>
> Then it got more confusing. Does anybody here know what pro bono means? John kept using that term in his initial conversation with me. I'm just an engineer. I thought maybe he was trying to speak Spanish.
>
> In any event, when I compare that initial conversation we had with what I see before me tonight with this incredible group of people, I could never have believed what John would be able to accomplish.

I'm reminded of the late great baseball Hall of Famer Yogi Berra, who, as a manager speaking of one of his players, said, "He has exceeded my expectations and done even better." And that's what John's done. I could never have imagined this when I agreed to come on board. Now thirteen years later, I thank God and all of you so much for what you've done here.

Afterword

John Junkins

It is gratifying to look from the bridge of this vessel we call the Hagler Institute for Advanced Study and see that it is now both an important and a permanent part of the Texas A&M University academic fleet. As you now know from the prior pages of this book, a course was charted, and occasionally rough seas were encountered. But in landing more than 150 of the world's most eminent scholars as Fellows and Distinguished Lecturers, our Institute for Advanced Study has navigated a successful course and played a significant role in establishing an elite port of scholarly call at our great and noble institution.

While a favorable wind now blows mostly steady and we are sailing smartly toward a bright and promising horizon, there were many voyages in our early years in which our seas were often turbulent. Had we been unable to stay the course then, our endeavor could easily have been scuttled and submerged.

The Hagler Institute for Advanced Study is not like other entities on campus that focus on a particular area of a particular academic discipline. The Hagler Institute serves the entire university community. To make our organization a reality required not only a vision—drawn in large part from my own experiences and ideals—but also a buy-in from important and key decision-makers. Were it not for the endorsement of the Texas A&M University System Board of Regents, approvals by administrators of the Texas A&M flagship campus, and the wholehearted, full-steam-ahead backing of Chancellor John Sharp, the Institute would not have made it into open waters. But once there, we caught a favorable current and worked hard—both above and below deck—to keep our propeller churning.

These nautical analogies are a tribute to Texas A&M's status as one of only twenty-four institutions in the nation to hold the triple designation of land-grant, space-grant, and sea-grant university.

By the way, Jacques Cousteau, the late great French oceanographer and host of the popular 1960s and '70s television series *The Undersea World of Jacques Cousteau*, would have been an outstanding Fellow of the Hagler Institute for Advanced Study.

"HIAS," as we now refer to it in its abbreviated form.

You've already met a number of the eminent scholars who comprise the ranks of our HIAS Fellowship, but let me pause now and let a trio of their "fellow" Fellows lend you their observations . . .

. . . from the "crow's nest" of the mother ship:

> As a Senior Henry Ford Technical Fellow with extensive industry and academic collaborations, I highly value the Hagler Institute's collaborative model. The Institute's unique platform for partnership and collaboration with external talent goes well beyond the traditional visiting and consulting practices. The Institute excels at transferring innovative ideas and experience and stimulates interactions between TAMU and world level researchers and scientists.
>
> Dimitar Filev
> Ford Research and Innovation Center
> Hagler Institute Class of 2022–23

> I split my time as a Hagler Fellow between College Station and the Texas A&M Law School in Fort Worth. In College Station, I gave talks in the political science department as well as the school of international studies and worked on various projects. In Fort Worth, I met with old colleagues like Charlotte Ku and Nuno Garoupa, as well as many new friends. I found people in both places to be intellectually ambitious, and the range of interesting people at Texas A&M active in my fields was impressive. Looking back, I recognize that several of my subsequent projects were influenced by conversations as a Hagler Fellow.
>
> Tom Ginsburg
> University of Chicago
> Hagler Institute Class of 2017–2018

> One day I was leaving Rudder Tower for lunch across campus. When I got outside, it started raining . . . heavily. As I was wandering around getting absolutely soaked, a young student walked up and invited me to share his umbrella. I'm a big guy and I knew in sharing the umbrella, the student was likely to get soaked, too. So, I passed on the invitation, but that's the kind of thing that happens often at Texas A&M. People are very, very eager and anxious to help you and, and they do so in a very courteous and respectful manner.
>
> William Unruh
> University of British Columbia
> Hagler Institute Class of 2018–2019

Like our Rob Kennicutt, whom you met earlier in this book, Bill Unruh also has strong ties to the greatest theoretical physicist of our time, Stephen Hawking. Unruh and Hawking worked together to refine the foundations of quantum mechanics in relation to black holes.

Once during his Hagler Institute Fellowship, Bill and his wife, Pat, ran into a "black hole" of their own when Bill lost control of the rental car he was driving from Houston to College Station, veering off the road, down an embankment, and into a ditch. It was a serious accident in which Bill suffered a broken leg and other injuries. Unfortunately, Bill spent most of his initial stay with us as a Fellow confined first to a wheelchair and then a walker . . . but not just *any* walker.

When Bill showed up for a dinner event at Chancellor John Sharp's house, he maneuvered himself with what he called the "Maserati of walkers." It was pretty impressive. Lacking foresight, I failed to ask Bill where he got such a formidable contraption. So when I suffered my own broken femur a couple years later, I was relegated to a run-of-the-mill—albeit functional—device, much to my chagrin.

While on the subject of "injuries in the line of duty," I must share a story about one of my favorite Fellows.

Susan Suleiman was a member of the Texas A&M Institute for Advanced Study Class of 2015–2016. She came to us from Harvard, where she was a professor of comparative literature. In addition, Susan was also a scholar-in-residence with the Center for Advanced Holocaust Studies at the Holocaust Memorial Museum in Washington, DC.

Susan is an accomplished author whose works include *Crises of Memory and the Second World War*, *The Némirovsky Question: The Life, Death, and Legacy of a Jewish Writer in Twentieth-Century France*, and *Budapest Diary: In Search of the Motherbook*, about her own childhood growing up in Hungary.

While with us here at Texas A&M, Susan presented an Eminent Scholar Lecture on *The Némirovsky Question*, which at the time she was getting ready to publish. Among the attendees at Susan's lecture were Susan and Mel Glasscock. The Glasscocks are major financial supporters of the university. Mel is a graduate of Texas A&M, Class of 1959. The Center for Humanities Research building on campus is named in his honor, and, in fact, Susan Suleiman's office as a Fellow was located there. It was her personal relationship with the Glasscocks that ultimately led to them also becoming friends of and generous donors to the Hagler Institute.

Susan Suleiman is a diminutive woman, whose height does not exceed five feet. In order to make her visible over the podium behind which she presented her Eminent Scholar Lecture, she stood on a 12-inch-tall wooden box, designed to elevate shorter speakers into clearer view. She was, she said, comfortable with the arrangement, calling it standard procedure for "people my size."

Unfortunately, during her presentation, she forgot about the box, and as she took a step to make a point of emphasis about one of her slides, she fell, hard, onto the stage floor. I was in the front row of the audience and was the first person to get to her while she was still down after her fall.

"Are you all right?" I asked, greatly concerned.

"I'm fine," she replied as I helped her back to her feet. After she "brushed herself off" in the figurative sense, she said, without looking at me, "Let me get back to my lecture." She resumed without missing a beat—or a slide—and delivered a beautiful presentation. One could not tell she was in serious pain from the fall, but afterward, she expressed concern that she may have fractured some bones. She spent the night in a local hospital.

Fortunately, Susan suffered "only" severe bruises. She showed remarkable resilience by completing her inspired talk despite her agonizing pain. As a pole-vaulter in my youth—before the advent of foam mats on which to soften impact with Mother Earth—I knew well the pain one could experience from a fall to the ground. I reasoned Susan's tenacity in the face of physical injury might have been born in her own athletic pursuits as a younger person. Knowing her origins were in Eastern Europe, I thought, "gymnast, perhaps?"

But the real truth was that Susan has a strong "show-must-go-on" mindset, which has served her well in all her endeavors.

In order to bring to College Station esteemed, diverse scholars like Susan Suleiman, Bill Unruh, Tom Ginsburg, Dimitar Filev, and the other Hagler Institute Fellows you've met in this book, enthusiastic participation of leaders from colleges, schools, and departments across the breadth of Texas A&M University is required. Those constituents team with the Hagler Institute staff through the nomination, recruiting, and selection stages as well as in the management of the Fellows' actual time spent on campus. The college and school deans and department heads must cover expenses for housing, travel, and 30 percent of the personal compensation these scholars and industry pioneers receive after agreeing to become Hagler Fellows.

In our early years, we competed with other pressing projects and prospects for the use of discretionary funds by the departments and colleges. For a time we faced a real concern whether academic decision-makers would pay a portion of the high levels of compensation for our visitors, who themselves had no real obligation for their time here other than collaborative research and the sharing of ideas. "Would funds be better spent," we knew local administrators would be asking themselves, "if we provided raises for our own faculty?"

In large part driven by the grassroots ambitions of our faculty to team with top scholars in their field, we were able to convince administrators of the value an Institute for Advanced Study Fellow would bring to their environs.

On the other side of that equation was another obvious and pressing concern: If some of the world's top scholars were approved for doing advanced study at Texas A&M University, would they come for the extended times required to be a Fellow? Top scholars are rarely looking for outside appointments to occupy their time. Our fellowships had to be compelling in nature to generate serious interest among our target candidates.

To maximize our chances, I knew I must assemble a team capable of competently pursuing our goals when the stability of the Institute, and therefore the team members' job security within it, was by no means guaranteed to last. Could I recruit the needed talent, and would the team possess the required skills to be successful in executing that vision? In Clifford Fry and Amanda Scott—whose own stories and achievements you now know—I was blessed to establish a small and highly committed team who have advanced our efforts to "infinity . . . and beyond."

I wonder if Magellan dealt with the same issues as he headed off to parts unknown. Probably not, since he explored the unknown world in a time well before the movie, *Toy Story*, where the tagline "infinity and beyond" was famously birthed.

We all know that it is one thing to propose an idea and quite another to implement it. Like a machine, all the parts must be in place and adequately tested for it to work.

Many of the challenges we faced in our early years were met through sustained efforts to not only make the Institute a reality but also make sure others not directly involved understood it. Particularly noteworthy in this regard were potential donors, whose financial support was critical toward our goal of making the Institute a permanent part of the university. Some outside observers of the Institute might have said in the early days that our long-term success was an unlikely outcome. Those of us who have been here since or near the beginning would suggest that our success has been an important achievement, one that is rapidly elevating Texas A&M to the upper tier of public universities.

Indeed, there is no comparable "excellence engine" in any public university today.

But where to from here?

As I write this, it's the start of a new school year, the fall of 2024. We've just held the event announcing our newest class of Fellows, which I'm pleased to report has reached twenty for the first time in our history. We decided to increase our numbers this year as we had incredibly strong nominations and extra money on hand as a result of downsizing our classes during the COVID-19 pandemic. With this newest class, it becomes crystal clear to me that we have what it takes to attract a world-class group of people from a wide variety of disciplines and that this institute has proven itself to be a great investment for both the present and the future of our university.

How good?

Well, a study covering our first seven years and done at the direction of the provost's office showed that the $17 million invested into covering the costs of bringing Fellows to campus through our 2017–2018 class generated—mostly through new research grants—a total of over $50 million. This money funded the work of Permanent Fellows who joined our faculty full time as well as outstanding faculty hired to team up with them in their pursuits here.

Excellence begets excellence—our research funded has since exploded from $0.8 billion to $1.2 billion.

We are presently engaged in updating this study to cover the Institute's most recent six years of existence. It is virtually certain that the financial impact will be even greater in terms of funds generated and more impressive in terms of fostering academic achievement by our faculty and students.

An important by-product of all these efforts is enhancement of the Texas A&M University brand. And to make that happen, to replicate the numbers of this most recent class of Fellows as we move forward in the future, requires us to continue finding generous people who believe in what we're doing and will help underwrite our efforts in advancing excellence.

Recently, we have been the recipients of two major endowments: $10 million from the Welch Foundation, and another $5 million from our namesake, Jon Hagler. Jon continues to show incredible generosity through his personal support of our future ambitions and initiatives.

How the Welch Foundation came into our fold of benefactors is an interesting story.

Named for the late Houston industrialist and inventor Robert Alonzo Welch, the Welch Foundation is dedicated exclusively to supporting research work done in Texas in the field of chemistry. For virtually every chemist in our state, the Welch Foundation is well known and well regarded for its generous support of their endeavors.

I first came in contact with the Welch Foundation in the second year of the Institute for Advanced Study's existence.

One of our Fellows at the time was renowned chemist Peter Stang, a National Medal of Science winner from the University of Utah. Joseph Newton was then dean of our College of Science, and I arranged for the board chair of the Welch Foundation to meet with Peter, Joe, and me. At that meeting, I took advantage of the opportunity to let the foundation know what we had already accomplished and my vision for where we were headed. Bringing more chemistry scholars like Peter Stang to campus was clearly in the best interests of Texas A&M and was something I believed the Welch Foundation would be interested in supporting. Our host listened attentively to my pitch to fund an endowed chemistry chair within the Institute, and he seemed impressed by my presentation. When I finished though, he squirmed a little in his seat before telling me, "Dr. Junkins, we don't do endowments. We only fund specific research projects."

Which got the attention of both Stang and Newton!

All was not lost.

Rapid rewinding back to some twenty years ago. I received an invitation from Sen. Kay Bailey Hutchison to attend a meeting in Dallas at the offices of philanthropist Peter O'Donnell. In the email she sent me requesting that I join her, she expressed concern that Texas was lagging behind several other states in the funding of science and engineering and STEM technologies. Simply put, she wanted to know why.

"STEM," as I'm sure you know, stands for science, technology, engineering, and mathematics. Interpreted broadly, STEM includes all the science that also enables modern medicine.

At the meeting, which included Nobel laureates Richard Smalley, Robert Curl, and Michael Brown as well as a number of National Academy members who were leaders in research at various Texas colleges and universities, Senator Hutchison informed us that Texas got a much smaller fraction of tax dollars back for research than states like California, New York, and Massachusetts.

"Why are we not doing better?" she asked.

Actually, she already knew part of the answer to her question.

"I'll tell you why," she went on to say. "I see before me more than a dozen leaders in science and engineering in our state and most of you don't really know each other. Right?"

She was correct.

That meeting became the impetus for "TAMEST," the Texas Academy of Medicine, Engineering, Science, and Technology.

It was decided that membership in TAMEST would be automatic. If you were in your respective National Academy, then you were a member if you chose to be. Then we decided TAMEST would put a major focus on identifying, rewarding, and accelerating the careers of the "next generation of rising stars" rather than allowing senior researchers to dominate our meetings. We wanted to provide opportunities where talented young professors could showcase their work. And if you were chosen to make a presentation at one of our meetings or events, you would be rewarded with a check to take home with you.

TAMEST has done important work since its inception in 2004, and in 2013 it was decided to establish the Kay Bailey Hutchison Award in honor of her lighting the fuse that ignited the organization's liftoff. Naturally, Senator Hutchison was the first recipient of that award. And in 2022, I was bestowed the honor, too, in large part as recognition for the creation of the Hagler Institute.

In the audience at the TAMEST awards event that year was a future friend of the Hagler Institute. Gina Luna is a member of the Welch Foundation board of directors and a Texas A&M graduate. Although she was not associated with the foundation when I made my initial visit in 2013, Luna was impressed to hear what I had done with the Hagler Institute during my remarks after receiving the Hutchison Award. Afterward, Luna reached out to Tyson Voelkel of the Texas A&M Foundation and told him the Welch Foundation was interested in supporting the Hagler Institute. Luna explained to Voelkel the Welch Foundation had broadened its scope dramatically in recent years. In fact, in 2020, the foundation made a $100 million endowment gift to Rice University, the largest gift in that school's history and unambiguous evidence that the organization had expanded its benevolent mission.

After Voelkel relayed this information to me, we set up a meeting with Luna and Adam Kuspa, president of the Welch Foundation.

This meandering set of circumstances, I'm pleased to report, has resulted in a $10 million gift to the Hagler Institute from the Welch Foundation, catching the attention of Sen. Kay Bailey Hutchison, who is now a member of the Hagler Institute External Advisory Board. Thus it is fitting and appropriate that Senator Hutchison authored the foreword for this book.

You just never know, do you?

In that same vein, let me offer one final expression of appreciation to former Texas A&M University Chancellor John Sharp.

In addition to our fateful August 2011 lunch meeting at College Station's famed Chicken Oil Company eatery—which effectively cleared the way for the birth of the Texas A&M Institute for Advanced Study—my other most memorable moment with the chancellor came in a phone call with him nine years later.

"John," he said, "you are going to be the interim president of Texas A&M University."

My initial response, I'm a little sheepish to admit, was to reply, "Chancellor Sharp, what have you been drinking?"

Michael Young, one of several great administrative friends to the Institute, had just announced his resignation as Texas A&M president, and as is the protocol most businesses and organizations follow, an interim president would be put into place as the search for Young's permanent successor was orchestrated.

Sharp went on to tell me, "You've given a lot to this university, in your academic research, your involvement with students, and your commitment to creating and sustaining the Institute for Advanced Study. Would you have the time and inclination to head up the entire flagship as we look for a new permanent president?"

I told him I needed some time to reflect on his unanticipated offer and speak to my boss at home, Elouise.

When I got home that evening to share the news with my ever-supportive and beloved wife, Elouise somewhat surprised me by pointing out that I would be the oldest—at seventy-eight—and "least-qualified" president in the history of Texas A&M. Fortunately, she fervently supported me taking the job.

From January 1 through May 31, 2021, I was privileged to experience both the triumphs and the trials of being a university president. During my brief time in office, our campus was rocked, first by the student protests that occurred in the aftermath of the George Floyd murder and then by the COVID-19 pandemic. These proved to be challenges I wouldn't wish on my worst enemy, and thankfully, both the university and the Hagler Institute survived my presidency.

Let me offer here this quick personal message to our greatest chancellor and my good friend, John Sharp.

John, your Chancellor's Research Initiative and its history-making impact on Texas A&M paved the way not only to launching the Institute for Advanced Study but also in positioning us for our enduring success. What we have done together has been both significant and profound. Proof of just how significant can be found in the growth of National Academy members within the Texas A&M faculty community from eleven in 2011 when you first took on your leadership role with the system to more than sixty today. In addition, our externally sponsored research budget now exceeds $1.2 billion, a figure that leads all institutions of higher learning in the Southwest. Our engineering research budget, which represents a large percentage of that total budget, is now number one nationally.

John, your retirement leaves a giant void in Aggieland, but you've departed in the midst of a monumental "winning streak" that makes Aggies everywhere proud.

One of Chancellor Sharp's greatest contributions to the Texas A&M academic brand has been the ongoing investments in facilities and equipment that he directed. There are no institutional peers for such a rapid "excellence advancement" on multiple fronts at any university anywhere.

Which begs the question: What's next?

Aside from continuing to bring the world's finest senior scholars to Texas A&M, perhaps the most important new direction in which the Hagler Institute can head is to attract what I like to call "rising superstars" interested in doing their PhD studies or postdoctoral work at the university. I call this, informally, our "Field of Dreams" initiative.

In that 1989 movie, a mysterious voice encourages an Iowa corn farmer, played by Kevin Coster, to plow down his field of corn and build a baseball field in its place. "If you build it," the voice promises, "they will come." That line has always struck me as an apt description for what we've done at the Hagler Institute.

Each year, we are building a better and stronger class of visiting Fellows, and I'm certain that we can leverage the presence of those highly regarded academics and researchers to attract an elite level of graduate students.

In the fall of 2024, we began soliciting PhD scholars in STEM fields nationwide by offering to provide take-home stipends of $40,000 per year for four years to fund their studies at Texas A&M. In that time, they will team with a rising star rising star faculty and a National Academy member in their field. A comparable program orchestrated by the National Science Foundation provides STEM Fellows $35,000 per year for three years, making our fellowships more financially attractive. As a result, we anticipate generating a significant number of accomplished applicants from which we can fund "the best of the best."

Our goal is to initially award fifteen of these PhD fellowships each year. Growing those numbers thereafter will depend on our ability to attract endowments to sustain and advance the program. Attracting this talent pool to Texas A&M will make a significant impact not only on their future careers and accomplishments and on the academic standing of our university but also on the state of Texas, our country, and the people of the entire world.

As the first Junkins male to graduate from high school and the first resident of my Georgia county to earn a PhD, my life has been spent—as we relocate our use of metaphors—climbing academic and professional mountains. From the point of view of "institutional enhancement initiatives," leading the creation of the Hagler Institute for Advanced Study has been, by far, my most ambitious organization-building undertaking.

Throughout my career, spanning six decades at this point, I have labored to make the investments required to maximize my personal and technical competencies and become the best engineer, professor, researcher, and administrator I can be.

In this way, I hope I've set a positive example for others.

I've taken great pride in the development and accomplishments of my PhD students, like Tarek Elgohary, whom you read about earlier. I recently achieved an important personal milestone, graduating my sixtieth PhD student. Half of that group, like Tarek, have gone on to become professors in their own right. Including the figurative descendants of my former students, I now have well over two hundred PhD branches on my "academic family tree."

I have been extraordinarily blessed throughout my life, with many dominoes falling favorably to see challenging events turn into rewarding experiences. What fraction of this success—and the occasional failures that I have also experienced and learned from—can be attributed to mere good fortune, and what fraction should be attributed to tenacity (and perhaps stubbornness), I will leave for others to decide.

As for the legacy I wish to leave behind, Elouise and I addressed that recently when we arranged an estate gift to the Texas A&M Foundation of our home in College Station and the approximately five acres of land that surround it. This was not done with any intent to suggest "Junkins is retiring"; rather, my wife and I, as we find ourselves "advancing chronologically," realize we soon will be at a point where it will behoove us to downsize. At the same time, we believe our gift can have a positive impact on the university.

Abraham Lincoln is often attributed to having said, "The best way to predict the future is to create it!" Apparently, he never uttered those exact words, but the Abraham Lincoln Presidential Library and Museum offers that "the sentiment matches other things Lincoln is documented to have said."

Regardless, our Hagler Institute for Advanced Study has come a long way, and the seas ahead of us appear both calm and inviting.

You're invited to join Elouise and me and the many other donors and supporters of the Hagler Institute for Advanced Study on the continuation of our cruise. There are many exciting—and perhaps even monumental—adventures to come.

It will be my pleasure to welcome you aboard.

HAGLER INSTITUTE FOR ADVANCED STUDY FELLOWS AND DISTINGUISHED LECTURERS

2012–2013 Class

Hagler Fellow Name	Texas A&M College or School	Texas A&M Department	Home Institution
Dunlap, Jay C.	Science	Biology	Dartmouth College
Liss, Peter S.	Geoscience	Oceanography	University of East Anglia, Norwich, England
Needleman, Alan	Engineering	Materials Science and Engineering	University of North Texas
Roth, Aleda	Mays Business	Information and Operations Management	Clemson University
Smith, Vernon L.	Liberal Arts	Economics	Chapman University
Sreenivasan, Katepalli R.	Engineering	Aerospace Engineering	New York University

2013–2014 Class

Hagler Fellow Name	Texas A&M College or School	Texas A&M Department	Home Institution
Andersson, Leif	Veterinary Medicine and Biomedical Sciences	Veterinary Integrative Biosciences	Uppsala University, Sweden
Atluri, Satya N.	Engineering	Aerospace Engineering	University of California, Irvine
Bouchard, Claude	Education and Human Development	Health and Kinesiology	Louisiana State University

(*continued*)

2013–2014 Class (*continued*)

Hagler Fellow Name	Texas A&M College or School	Texas A&M Department	Home Institution
Floudas, Christodoulos	Engineering	Chemical Engineering	Princeton University
Glauber, Roy	Science	Physics and Astronomy	Harvard University
Howe, Roger E.	Education and Human Development	Teaching, Learning, and Culture	Yale University
Levine, Robert S.	Liberal Arts	English	University of Maryland College Park
Schleich, Wolfgang	Science	Physics and Astronomy	Ulm University, Germany
Stang, Peter J.	Science	Chemistry	The University of Utah

2014–2015 Class

Hagler Fellow Name	Texas A&M College or School	Texas A&M Department	Home Institution
Adams, Harold L.	Architecture	Architecture	RTKL Associates, CEO, Retired
Agrawal, Rakesh	Engineering	Chemical Engineering	Purdue University
Dongarra, Jack	Engineering	Computer Science and Engineering	University of Tennessee & Oak Ridge National Lab
Marras, William S.	Engineering	Industrial and Systems Engineering	The Ohio State University
Moses, Edward	Science	Physics and Astronomy	Giant Magellan Telescope Organization
Oganessian, Yuri	Science	Physics and Astronomy	Joint Institute for Nuclear Research, Dubna, Russia
Skelton, Robert E.	Engineering	Aerospace Engineering	University of California, San Diego

2015–2016 Class

Hagler Fellow Name	Texas A&M College or School	Texas A&M Department	Home Institution
Arnett, W. David	Science	Physics and Astronomy	University of Arizona
Brosnan, John T.	Agriculture and Life Sciences	Animal Science	Memorial University of Newfoundland
Calderbank, Robert A.	Engineering	Electrical and Computer Engineering	Duke University
Delgado, Richard	Law	Law	University of Alabama
Richard A. Gibbs	Medicine	Clinical Science and Translational Research Institute	Baylor College of Medicine
Hedrick, J. Karl	Engineering	Mechanical Engineering	University of California, Berkeley
Holm, Richard H.	Science	Chemistry	Harvard University
King, Michael D.	Geosciences	Atmospheric Sciences	University of Colorado
Polasky, Stephen	Agriculture and Life Sciences	Agricultural Economics	University of Minnesota
Rogers, John A.	Engineering	Mechanical Engineering	University of Illinois Urbana-Champaign
Schartl, Manfred	Science	Biology	University of Wurzburg, Germany
Sinha, Kumares C.	Engineering	Civil Engineering	Purdue University
Suleiman, Susan R.	Liberal Arts	International Studies	Harvard University

Distinguished Lecturer Name	Texas A&M College or School	Texas A&M Department	Home Institution
Starr, Kenneth W.	Science	Institute for Quantum Science and Engineering	Baylor University

2016–2017 Class

Hagler Fellow Name	Texas A&M College or School	Texas A&M Department	Home Institution
Cummins, Christopher C.	Science	Chemistry	Massachusetts Institute of Technology
Daubechies, Ingrid	Science	Mathematics	Duke University
Galloway, Gerald	Texas A&M University at Galveston	Marine Sciences	University of Maryland
Gao, Huajian	Engineering	Mechanical Engineering	Brown University
Giger, Maryellen	Engineering	Biomedical Engineering	University of Chicago
Kennicutt Jr., Robert	Science	Physics and Astronomy	University of Cambridge (England)
Kolb, Charles E.	Geosciences	Atmospheric Sciences	Aerodyne Research Inc.
Kumar, V.	Mays Business	Marketing	Georgia State University
Sage, William M.	Public Health	Environmental and Occupational Health	The University of Texas at Austin
Ulen, Thomas S.	Law	Law	University of Illinois at Urbana-Champaign

2017–2018 Class

Hagler Fellow Name	Texas A&M College or School	Texas A&M Department	Home Institution
Dhir, Vijay	Engineering	Mechanical Engineering	University of California, Los Angeles
Dixon, Richard	Science	Biology	University of North Texas
Epstein, Richard A.	Law	Law	New York University

Hagler Fellow Name	Texas A&M College or School	Texas A&M Department	Home Institution
Ginsburg, Thomas B.	Law	Law	University of Chicago
Hubbard, Jr., James E.	Engineering	Mechanical Engineering	University of Maryland
Stipanowich, Thomas J.	Law / Architecture	Law / Construction Science	Pepperdine University
Tessendorf, Jerry	Architecture	Visualization	Clemson University

Distinguished Lecturer Name	Texas A&M College or School	Texas A&M Department	Home Institution
Chu, Steven	Science	Physics	Stanford University

2018–2019 Class

Hagler Fellow Name	Texas A&M College or School	Texas A&M Department	Home Institution
Bagnato, Vanderlei S.	Engineering	Biomedical Engineering	University of Sao Paulo, Brazil
Duff, Michael	Science	Institute for Quantum Science and Engineering	Imperial College London, England
Huang, Yonggang	Engineering	Mechanical	Northwestern University
Jones, Cameron	Science	Chemistry	Monash University, Australia
Kaufmann, Stefan H.E.	Veterinary Medicine and Biomedical Sciences	Veterinary Pathobiology	Max Planck Institute for Infection Biology, Germany
Poor, H. Vincent	Engineering	Electrical and Computer Engineering	Princeton University
Putnam, Robert D.	Liberal Arts	Political Science	Harvard University

(continued)

2018–2019 Class (*continued*)

Hagler Fellow Name	Texas A&M College or School	Texas A&M Department	Home Institution
Rinaldo, Andrea	Engineering	Ocean Engineering	École Polytechnique Fédérale de Lausanne, Switzerland
Unruh, William G.	Science	Physics and Astronomy	University of British Columbia, Canada
Distinguished Lecturer Name	**Texas A&M College or School**	**Texas A&M Department**	**Home Institution**
Singer, Joseph William	Law	Law	Harvard Law School

2019–2020 Class

Hagler Fellow Name	Texas A&M College or School	Texas A&M Department	Home Institution
Davidovich, Luiz	Science	Institute for Quantum Science and Engineering	Universidade Federal do Rio de Janeiro
Donovan, Sharon	Agriculture and Life Sciences	Nutrition and Food Science	The University of Illinois at Urbana-Champaign
Hamuy, Mario	Science	Physics and Astronomy	Association of Universities for Research in Astronomy (AURA) Observatory, Chile
Hotez, Peter	Public Health / Veterinary Medicine and Biomedical Sciences	Public Health & Veterinary Medicine and Biomedical Sciences	Baylor College of Medicine
Howell, Kathleen	Engineering	Aerospace Engineering	Purdue University

Hagler Fellow Name	Texas A&M College or School	Texas A&M Department	Home Institution
Lyubich, Misha	Science	Mathematics	Stony Brook University
Rousso, Henry	Bush School of Government and Public Service	Bush School of Government and Public Service	The French National Centre for Scientific Research
Shor, Peter	Science	Institute for Quantum Science and Engineering	Massachusetts Institute of Technology
Thomas, Edwin	Engineering	Materials Science and Engineering	Rice University

Distinguished Lecturer Name	Texas A&M College or School	Texas A&M Department	Home Institution
Fineman, Martha A.	Law	Law	Emory University
McCloskey, Deirdre	Innovation	Innovation	University of Illinois at Chicago
Moran, Rachel F.	Law	Law	University of California, Irvine School of Law
Wolf, Susan	Law/Medicine/ Public Health	Law/Medicine/ Public Health	University of Minnesota
Zoghbi, Huda	Medicine	Clinical Science and Translational Research Institute	Baylor College of Medicine

2020–2021 Class

Hagler Fellow Name	Texas A&M College or School	Texas A&M Department	Home Institution
Cooks, Robert Graham	Science	Chemistry	Purdue University
Feinberg, Andrew P.	Medicine	Molecular and Cellular Medicine	Johns Hopkins University
Giovannoni, James J.	Science / Agriculture and Life Sciences	Biology and Horticultural Sciences	USDA-ARS, Boyce Thompson Institute, Cornell

(continued)

2020–2021 Class (*continued*)

Hagler Fellow Name	Texas A&M College or School	Texas A&M Department	Home Institution
Hammond, Paula T.	Engineering	Mechanical Engineering	Massachusetts Institute of Technology (MIT)
Judge, Timothy A.	Mays Business	Management	The Ohio State University
King, Julia	Science/ Engineering/ Architecture	Chemistry	Crossbench Life Peer, House of Lords, London
Ladson-Billings, Gloria	Education and Human Development	Teaching, Learning, and Culture	University of Wisconsin-Madison
Moran, Rachel F.	Law	Law	University of California, Irvine School of Law
Mukamel, Shaul	Institute for Quantum Science and Engineering	Institute for Quantum Science and Engineering	University of California, Irvine
Orlin, Lena Cowen	Liberal Arts	Center of Digital Humanities Research	Georgetown University
Distinguished Lecturer Name	**Texas A&M College or School**	**Texas A&M Department**	**Home Institution**
Lewis, Jennifer A.	Engineering	Biomedical Engineering	Harvard University

2021–2022 Class

Hagler Fellow Name	Texas A&M College or School	Texas A&M Department	Home Institution
Jaffe, Arthur M.	Science	Mathematics	Harvard University
Bowcutt, Kevin G.	Engineering	Aerospace Engineering	Boeing Research and Technology
Chen, Jacqueline H.	Engineering	Mechanical Engineering	Sandia National Laboratories
Elisseeff, Jennifer	Engineering / Medicine	Biomedical Engineering	Johns Hopkins University

Hagler Fellow Name	Texas A&M College or School	Texas A&M Department	Home Institution
Goodson, Theodore	Science	Chemistry	University of Michigan
Sottos, Nancy R.	Engineering	Mechanical Engineering / Materials Science and Engineering	University of Illinois Urbana-Champaign
Strickland, Donna	Institute for Quantum Science and Engineering	Institute for Quantum Science and Engineering	University of Waterloo Ontario, Canada
Zheludev, Nikolay	Science	Physics and Astronomy	University of Southampton, United Kingdom; Nanyang Technological University, Singapore

2022–2023 Class

Hagler Fellow Name	Texas A&M College or School	Texas A&M Department	Home Institution
Bertrand, Guy	Science	Chemistry	Director, UCSD/ CNRS Joint Research Chemistry Laboratory
Cao, Hui	Institute for Quantum Science and Engineering	Institute for Quantum Science and Engineering	Yale University
Cullen, John Michael	Veterinary Medicine and Biomedical Sciences	Small Animal Clinical Sciences	North Carolina State University
Eisenstein, Odile	Science	Chemistry	University of Oslo, Norway
Filev, Dimitar	Engineering	Mechanical Engineering / Engineering Technology and Industrial Distribution	Ford Research and Innovation Center

(continued)

2022–2023 Class (*continued*)

Hagler Fellow Name	Texas A&M College or School	Texas A&M Department	Home Institution
Frumkin, Howard	Public Health	Environmental and Occupational Health	University of Washington
Jonkman, Sebastian	Texas A&M University at Galveston	Marine and Coastal Management	Delft University of Technology, Netherlands
O'Malley, Mark	Engineering	Electrical and Computer Engineering	Imperial College of London
Que, Lawrence	Science	Chemistry	University of Minnesota
Rodrigue, Jean-Paul	Texas A&M University at Galveston	Maritime Business Administration	Hofstra University
Sparks, Donald	Agriculture and Life Sciences	Soil and Crop Sciences	University of Delaware
Sunder, Madhavi	Law	Law	Georgetown Law Center
Young, Michael W.	Science	Biology	The Rockefeller University
Zilberman, David	Agriculture and Life Sciences	Agricultural Economics	University of California, Berkeley
Distinguished Lecturer Name	**Texas A&M College or School**	**Texas A&M Department**	**Home Institution**
Dulac, Catherine	Medicine	Molecular and Cellular Medicine	Harvard University

2023–2024 Class

Hagler Fellow Name	Texas A&M College or School	Texas A&M Department	Home Institution
Calder, Philip	Medicine	Primary Care and Population Health	University of Southampton, United Kingdom
Carriquiry, Alicia	Arts & Sciences	Statistics	Iowa State University

Hagler Fellow Name	Texas A&M College or School	Texas A&M Department	Home Institution
Georgiou, George	Medicine	Medical Education	The University of Texas at Austin
Gookin, Jody Lynn	Veterinary Medicine and Biomedical Sciences	Small Animal Clinical Sciences	North Carolina State University
Jemison, Mae	Engineering Medicine	Engineering Medicine	The Jemison Group, Inc.
Maskin, Eric S.	Arts & Sciences	Economics	Harvard University
Milner, H. Richard	Education and Human Development	Teaching, Learning, and Culture	Vanderbilt University
Myers, Kyle J.	Engineering	Biomedical Engineering	Puente Solutions LLC
Novoselov, Konstantin	Arts & Sciences	Physics and Astronomy	National University of Singapore
Papoutsakis, Eleftherios	Engineering	Chemical Engineering	University of Delaware
Perlman, Radia	Engineering	Computer Science and Engineering	Dell EMC
Sorooshian, Soroosh	Arts & Sciences	Atmospheric Sciences	University of California, Irvine
Tanguay, Robyn	Texas A&M University at Galveston	Marine Biology	Oregon State University
Wagner, Gunter	Agriculture and Life Sciences / Veterinary Medicine and Biomedical Sciences	Animal Science	Yale University

Distinguished Lecturer Name	Texas A&M College or School	Texas A&M Department	Home Institution
Ottino, Julio M.	Engineering	Materials Science and Engineering / Chemical Engineering	Northwestern University
Zoback, Mark	Engineering	Petroleum Engineering	Stanford University

2024–2025 Class

Hagler Fellow Name	Texas A&M College or School	Texas A&M Department	Home Institution
Alberts, Susan	Arts and Sciences	Biology	Duke University
Baecher, Gregory	Engineering	Civil Engineering	University of Maryland College Park
Beatty, Julia	Veterinary Medicine and Biomedical Sciences	Veterinary Small Animal Clinical Sciences	City University of Hong Kong
Berger, James O.	Arts and Sciences	Statistics	Duke University
Chintagunta, Pradeep	Mays Business School	Marketing	University of Chicago
Cryan, John	Medicine	Department of Neuroscience & Experimental Therapeutics	University College Cork, Ireland
Deshpande, Vikram	Engineering	Materials Science and Engineering	Cambridge University, England
Eden, J. Gary	Arts and Sciences	Institute for Quantum Science and Engineering	University of Illinois Urbana-Champaign
Filippenko, Alexei	Arts and Sciences	Physics and Astronomy	University of California, Berkeley
Hashash, Youssef	Engineering and Architecture	Civil Engineering	University of Illinois Urbana-Champaign
Lester, Marsha	Arts and Sciences	Chemistry	University of Pennsylvania
Long, Jeffrey R.	Arts and Sciences	Chemistry	University of California, Berkeley
Malshe, Ajay	Agriculture and Life Sciences	Institute for Advancing Health through Agriculture	Purdue University
Moyer, Mary Pat	Medicine	Medical Physiology	INCELL Corporation LLC
Rappaport, Theodore	Engineering	Electrical and Computer Engineering	New York University

2025–2026 Class (*continued*)

Hagler Fellow Name	Texas A&M College or School	Texas A&M Department	Home Institution
Pollock, Tresa	Engineering	Materials Science and Engineering	University of California, Santa Barbara
Schaufelberger, John	Architecture	Construction Science	University of Washington
Tung, Jenny	Arts & Sciences	Biology	Max Planck Institute for Evolutionary Anthropology

Distinguished Lecturer Name	Texas A&M College or School	Texas A&M Department	Home Institution
Hughes, Thomas J. R.	Engineering and Medicine	Aerospace Engineering	The University of Texas at Austin
Segev, Mordechai (Moti)	Arts & Sciences	Institute for Quantum Science and Engineering	Technion, Israel Institute of Technology

Hagler Fellow Name	Texas A&M College or School	Texas A&M Department	Ho
Sahinidis, Nikolaos	Engineering	Energy Institute	Ge Tec
Simmons, Beth	Bush School	Political Science	Uni Per
Tang, Christopher	Mays Business School	Information and Operations Management	Un Cal An
Wise, Kevin	Engineering	Aerospace Engineering	Th Co

Distinguished Lecturer Name	Texas A&M College or School	Texas A&M Department	H
Goodchild, Michael Frank	Architecture	Landscape Architecture and Urban Planning	Un Ca Ba

2025–2026 Class

Hagler Fellow Name	Texas A&M College or School	Texas A&M Department	H
Broderick, Joan	Arts & Sciences	Chemistry	M U
Butler, Alison	Arts & Sciences	Chemistry	U C B
Coates, Geoffrey W.	Arts & Sciences	Chemistry	C
Grabowski, David C.	Public Health	Health Policy & Management	H S
Kevrekidis, Ioannis (Yannis)	Engineering and Medicine	Chemical Engineering and Cell Biology and Genetics	J U
Pagan, Jose	Medicine	Institute for Healthcare Access, Translational Medical Sciences	N U
Pianta, Robert C.	Education and Human Development	Educational Psychology	U V

SOURCES

CHAPTER 1

Flexner, Abraham. *Abraham Flexner: An Autobiography*. Simon and Schuster, 1960.

Wikipedia Foundation. "Institute for Advanced Study." Last modified April 27, 2025, 16:09 (UTC). https://en.wikipedia.org/wiki/Institute_for_Advanced_Study.

Flexner, Abraham. "The Usefulness of Useless Knowledge." *Harper's Magazine*, no. 179 (June/November 1939). https://www.ias.edu/ideas/usefulness-useless-knowledge.

Altbach, Philip. "Advanced Study Institutes: The Top of the Top." *Inside Higher Ed* (November 18, 2010). https://www.insidehighered.com/blogs/world-view/advanced-study-institutes-top-top.

Wikipedia Foundation. "Philip Altbach." Last modified October 10, 2024, 14:56 (UTC). https://en.wikipedia.org/wiki/Philip_Altbach.

Wikipedia Foundation. "Institute for Advanced Study (Disambiguation)." Last modified December 11, 2024, 15:06 (UTC). https://en.wikipedia.org/wiki/Institute_for_Advanced_Study_(disambiguation).

Zandonella, Catherine. "Einstein's Legacy." Princeton University, January 7, 2016. https://www.princeton.edu/news/2016/01/07/einsteins-legacy.

Achenbach, Joel. "Einstein at Princeton." *Princeton Alumni Weekly* (April 22, 2021). https://paw.princeton.edu/article/einstein-princeton.

Wikipedia Foundation. "Albert Einstein." Last modified June 10, 2025, 22:48 (UTC). https://en.wikipedia.org/wiki/Albert_Einstein.

Institute for Advanced Study. "Establishing the Advanced Study Institute." https://www.ias.edu/sites/default/files/pdfs/publications/establishing-the-ias.pdf.

Dyson, George. "Institute for Advanced Study: The First 100 Years." Institute for Advanced Study, 2016. https://www.ias.edu/ideas/2016/dyson-ias-first-100-years.

Wikipedia Foundation. "Moses Hutzler." Last modified February 18, 2025, 22:27 (UTC). https://en.wikipedia.org/wiki/Moses_Hutzler.

Wikipedia Foundation. "Bamberger's." Last modified May 28, 2025, 11:21 (UTC). https://en.wikipedia.org/wiki/Bamberger%27s.

Wikipedia Foundation. "Louis Bamberger." Last modified March 19, 2024, 08:15 (UTC). https://en.wikipedia.org/wiki/Louis_Bamberger.

Defree, Suzanne. "Einstein Moves to US, October 17, 1933." *EDN*, October 17, 2019. https://www.edn.com/einstein-moves-to-us-october-17-1933/.

Encyclopedia.com. "Institute for Advanced Study." Last updated May 8, 2018, https://www.encyclopedia.com/social-sciences-and-law/education/colleges-us/institute-advanced-study.

Wikipedia Foundation. "Oswald Veblen." Last updated May 25, 2025, 19:49 (UTC). https://en.wikipedia.org/wiki/Oswald_Veblen.

Institute for Advanced Study. "All Scholars." https://www.ias.edu/scholars/all-scholars.

Institute for Advanced Study. "Mission and History." https://www.ias.edu/about/mission-history.

Wikipedia Foundation. "*The World as I See It* (Book)." Last updated March 17, 2023, 14:40 (UTC). https://en.wikipedia.org/wiki/The_World_as_I_See_It_(book).

GoodReads.com "Albert Einstein Quotes—Quotable Quotes." https://www.goodreads.com/quotes/858496-the-pursuit-of-knowledge-for-its-own-sake-an-almost.

Wikipedia Foundation. "Dr. Rer. Nat." Last updated April 29, 2025, 18:42 (UTC). https://en.wikipedia.org/wiki/Dr._rer._nat.

CHAPTER 2

Wikipedia Foundation. "John Junkins." Last updated June 13, 2025, 05:01 (UTC). https://en.wikipedia.org/wiki/John_Junkins.

Spell, Carrie. "Faye Gibbons." Encyclopedia of Alabama, January 19, 2010. https://encyclopediaofalabama.org/article/faye-gibbons/.

Gibbons, Faye. *Some Glad Morning*. Morrow Junior Books, 1982.

Berry College. "Our Rich History." https://www.berry.edu/about/our-rich-history/.

Berry College. "World's Largest Campus." https://www.berry.edu/about/campus.

Wikipedia Foundation. "Marshall Space Flight Center." Last updated June 3, 2025, 23:49 (UTC). https://en.wikipedia.org/wiki/Marshall_Space_Flight_Center.

Georgia Tech University. "About Georgia Tech." https://www.gatech.edu/about.

Kohut, Matthew, and Tracy McMahan. "Rocketing from Past to Future." *ASK Magazine*, no. 46 (Spring 2012). https://appel.nasa.gov/2010/01/01/rocketing-from-past-to-future/.

Wikipedia Foundation. "Samuel Herrick (Astronomer)." Last updated May 6, 2025, 20:09 (UTC). https://en.wikipedia.org/wiki/Samuel_Herrick_(astronomer).

CHAPTER 3

University of Virginia. "About the Shannon Center." https://shannon-center.provost.virginia.edu/about.

Pearce Rotondi, Jessica. "How Nixon's Invasion of Cambodia Triggered a Check on Presidential Power." History.com, April 27, 2020. https://www.history.com/articles/nixon-war-powers-act-vietnam-war-cambodia.

Wikipedia Foundation. "Nationwide Student Anti-War Strike of 1970." Last updated March 12, 2025, 03:37 (UTC). https://en.wikipedia.org/wiki/Nationwide_student_anti-war_strike_of_1970.

Wikipedia Foundation. "Edgar F. Shannon Jr." Last updated April 20, 2025, 02:59 (UTC). https://en.wikipedia.org/wiki/Edgar_F._Shannon_Jr.

Encyclopedia Britannica. "Chicago Seven." Last updated December 20, 2024. https://www.britannica.com/event/Chicago-Seven-law-case.

University of Virginia. "Jefferson's Masterpiece." https://www.virginia.edu/visit/grounds.

Wikipedia Foundation. "USS *Quincy* (CA-39)." Last updated February 4, 2025, 15:52 (UTC). https://en.wikipedia.org/wiki/USS_Quincy_(CA-39).

Friddell, Guy. "Edgar Shannon, Ex-President of U.Va., Dies in His 15 Years Heading the School, He Saw It Through Social Upheaval." *The Virginian-Pilot*, August 25, 1997. https://scholar.lib.vt.edu/VA-news/VA-Pilot/issues/1997/vp970825/08250074.htm.

Wikipedia Foundation. "University of Virginia College of Arts and Sciences." Last updated January 17, 2025, 04:14 (UTC). https://en.wikipedia.org/wiki/University_of_Virginia_College_of_Arts_and_Sciences#Echols_Scholars.

Zechmeister, Gene. "Timeline of the Founding of the University of Virginia." The Jefferson Monticello, July 5, 2011. https://www.monticello.org/research-education/thomas-jefferson-encyclopedia/timeline-founding-university-virginia/.

Wikipedia Foundation. "Doris Kuhlmann-Wilsdorf." Last updated May 25, 2025, 12:52 (UTC). https://en.wikipedia.org/wiki/Doris_Kuhlmann-Wilsdorf.

"Prof. Heinz Wilsdorf." http://www.heinz-wilsdorf.com/.

VMDO. "University of Virginia Wilsdorf Hall (Materials Science Engineering + Nanotechnology)." https://www.vmdo.com/uva-wilsdorf-hall.html.

Pipkin, Josie. "Pioneering Professor Doris Kuhlmann-Wilsdorf Dies; Service Thursday." *UVA Today* (March 30, 2010). https://news.virginia.edu/content/pioneering-professor-doris-kuhlmann-wilsdorf-dies-service-thursday.

CHAPTER 4

Burka, Paul. "Did You Hear the One About the New Aggies?." *Texas Monthly* (April 1997). https://www.texasmonthly.com/news-politics/did-you-hear-the-one-about-the-new-aggies/.

Texas A&M University Corps of Cadets. "Dr. Ray M. Bowen '58." https://corps.tamu.edu/dr-ray-m-bowen-58/.

Association of American Universities. "AAU History." https://www.aau.edu/aau-history.

Burka, Paul. "Storming the Ivory Tower." *Texas Monthly* (October 2012). https://www.texasmonthly.com/news-politics/storming-the-ivory-tower/.

Burka, Paul. "Gig 'em Ray." *Texas Monthly* (May 2002). https://www.texasmonthly.com/articles/gig-em-ray/.

USC Digital Folklore Archives. "Aggie Joke." May 12, 2016. http://folklore.usc.edu/aggie-joke/.

Texas A&M University Office of the President. "Past Presidents." https://president.tamu.edu/administration/past-presidents/index.html.

Wikipedia Foundation. "President of Texas A&M University." Last updated February 24, 2025, 07:48 (UTC). https://en.wikipedia.org/wiki/President_of_Texas_A%26M_University.

Texas A&M Foundation. "Howdy, Mr. President." *Spirit Magazine* (Spring 2016). https://spirit.txamfoundation.com/spring-2016/cover-feature.aspx.

CHAPTER 5

Henton, Lesley. "Vision 2020: Texas A&M's Aspirational Document Revisited." *Texas A&M Stories* (December 31, 2020). https://stories.tamu.edu/news/2020/12/31/vision-2020-texas-ams-aspirational-document-revisited/.

Texas A&M University. "Vision 2020: Behind the Scenes." *Texas A&M Stories* (December 31, 2020). https://stories.tamu.edu/news/2020/12/31/vision-2020-behind-the-scenes/.

Texas A&M University. "Vision 2020 Reflections." Vimeo. Uploaded December 31, 2020. https://vimeo.com/496102300.

Texas A&M University. "Vision 2020: The 12 Imperatives." https://vision2020.tamu.edu/the-twelve-imperatives.

12th Man Foundation. "Eppright Distinguished Donors: Celebrating a Legacy of Giving." https://12thmanfoundation.com/donor-impact/eppright-distinguished-donors/index.html.

Texas A&M University Corps of Cadets. "Colonel George J. Eppright '26." https://www.corps.tamu.edu/colonel-george-j-eppright-26/.

12th Man Foundation. "Col. George Eppright '26." https://www.12thmanfoundation.com/donor-impact/stories/col-george-eppright/index.html.

Texas A&M University Corps of Cadets. "Dr. Ed Davis '67." https://corps.tamu.edu/dr-ed-davis-67/.

12th Man Foundation. "Our History." https://12thmanfoundation.com/about-us/index.html.

Heathman, Claire, comp. "Gen. James Earl Rudder Led Texas A&M Through Immense Change and Growth." *My Aggie Nation* (August 27, 2013). https://myaggienation.com/aggie_profiles/gen-james-earl-rudder-led-texas-a-m

-through-immense-change-and-growth/article_c8ad387e-0f43-11e3-824c-0019bb2963f4.html.

Texas A&M Foundation. "About Us." https://www.txamfoundation.com/about-us.aspx.

Howell, Jack. "The History of the 12th Man." February 4, 2020. https://storymaps.arcgis.com/stories/8682521634724l788efa103a21bccdc1.

CHAPTER 6

Texas A&M University. "Marlan Scully." *Texas A&M Experts.* https://experts.tamu.edu/expert/marlan-scully/.

Nissimov, Ron. "A&M President Says Bonfire Incident Not Reason He's Stepping Down." *Houston Chronicle* (June 7, 2001). https://www.chron.com/news/houston-texas/article/a-m-president-says-bonfire-incident-not-reason-2029264.php.

Ashford-Grooms, Meghan. "Rick Perry Says He Switched to GOP at a Younger Age Than Reagan Did." The Poynter Institute, September 6, 2011. https://www.politifact.com/factchecks/2011/sep/06/rick-perry/rick-perry-says-he-switched-gop-younger-age-reagan/.

Phillips, Don. "'Boll Weevil' Democrat Phil Gramm, Yanked off the Influential House Budget Committee." United Press International (January 4, 1983). https://www.upi.com/Archives/1983/01/04/Boll-Weevil-Democrat-Phil-Gramm-yanked-off-the-influential/6444506541000/.

Texas A&M University College of Arts & Sciences. "Marlan Scully." Physics and Astronomy. https://physics.tamu.edu/directory/scully/.

Abilene Reporter News. "Dr. James E. 'Jim' Womack." August 16, 2023. https://www.reporternews.com/obituaries/arn029753.

Texas A&M University Institute for Quantum Science & Engineering. "Center for Biophotonics." https://iqse.tamu.edu/center-for-biophotonics/.

Texas A&M University Institute for Neuroscience. https://tamin.tamu.edu/.

Texas A&M University College of Arts & Sciences. "Welcome to the Center of Digital Humanities Research." https://liberalarts.tamu.edu/codhr/.

Texas A&M University. "Physicists Develop Effective Anthrax Detection Method." Science Daily, April 13, 2007. https://www.sciencedaily.com/releases/2007/04/070412145523.htm.

Pruitt, Sarah. "When Anthrax-Laced Letters Terrorized the Nation." History.com, October 4, 2018. https://www.history.com/news/anthrax-attacks-terrorism-letters.

Texas A&M University Institute for Quantum Science & Engineering. "Center for Theoretical Physics." https://iqse.tamu.edu/center-for-theoretical-physics/.

Wikipedia Foundation. "Wendy Lee Gramm." Last modified June 2, 2025, 18:28 (UTC). https://en.wikipedia.org/wiki/Wendy_Lee_Gramm.

Wikipedia Foundation. "Elsa Murano." Last modified May 24, 2025, 00:27 (UTC). https://en.wikipedia.org/wiki/Elsa_Murano.

Texas A&M University Department of Sociology. "Howard B. Kaplan." https://artsci.tamu.edu/sociology/people/in-memoriam/howard-b-kaplan.html.

McCord, Joan. "Deviant Behavior in Defense of Self. Howard B. Kaplan." *American Journal of Sociology* 87, no. 5 (March 1982). https://www.journals.uchicago.edu/doi/abs/10.1086/227584.

CHAPTER 7

Hagler Institute for Advanced Study Archives. "TIAS Proposal Draft."

CHAPTER 8

Texas A&M Engineering Experiment Station. "$100 Million Chancellor's Research Initiative Recruits Four Top Scholars to Texas A&M Engineering." February 16, 2015. https://tees.tamu.edu/news/2014/05/100-million-chancellors-research-initiative-recruits-four-top-scholars-to-texas-am-engineering.html.

Hardy, Michael. "Country Revival." *Texas Monthly* (August 2017). https://www.texasmonthly.com/news-politics/country-revival/.

Patel, Vimal. "Former A&M System Chancellor Mike McKinney Ends Turbulent Run." *Bryan-College Station Eagle* (March 12, 2012). https://theeagle.com/news/a_m/former-a-m-system-chancellor-mike-mckinney-ends-turbulent-run/article_109160d2-f7c1-5c45-8c25-d5322a411195.html.

Wikipedia Foundation. "Mike McKinney." Last modified May 7, 2025, 19:06 (UTC). https://en.wikipedia.org/wiki/Mike_McKinney.

Hamilton, Reeve. "A&M Chancellor Mike McKinney Stepping Down." *The Texas Tribune* (May 10, 2011). https://www.texastribune.org/2011/05/10/am-chancellor-mike-mckinney-stepping-down/.

Wikipedia Foundation. "John Sharp (Texas Politician)." Last modified April 17, 2025, 22:24 (UTC). https://en.wikipedia.org/wiki/John_Sharp_(Texas_politician).

Hooks, Chris. "Aggie Years Launched Perry—and a Rivalry." *The Texas Tribune* (August 2, 2011). https://www.texastribune.org/2011/08/02/perry-aggie-years/.

CHAPTER 9

Aitoro, Jill. "30 Years: A Norm Augustine Retrospective." *Defense News* (August 25, 2016). https://www.defensenews.com/30th-annivesary/2016/10/25/30-years-a-norm-augustine-retrospective/.

Reichhardt, Tony. "NASA Needs Direction? Call Norm Augustine!." *Air & Space Magazine* (May 8, 2009). https://www.smithsonianmag.com/air-space-magazine/nasa-needs-direction-call-norm-augustine-58826042/.

Wikipedia Foundation. "Norman R. Augustine." Last updated March 22, 2025, 21:56 (UTC). https://en.wikipedia.org/wiki/Norman_R._Augustine.

CHAPTER 10

Texas A&M University Department of Economics. "Thomas Saving." https://stories.tamu.edu/news/2024/03/14/texas-am-mourns-loss-of-distinguished-professor-emeritus-thomas-saving/.

Strong, Connie. "College Station's Fry Has Had Lifelong Battle with Rett Syndrome." *Bryan-College Station Eagle* (April 24, 2012). https://theeagle.com/brazos_life/college-stations-fry-has-had-lifelong-battle-with-rett-syndrome/article_aebcbafe-5c18-5b43-a834-5629b5a35430.html.

Wikipedia Foundation. "Eric Xu." Last modified on June 13, 2025, 13:05 (UTC). https://en.wikipedia.org/wiki/Eric_Xu.

Texas A&M University Corps of Cadets. "Jerry S. Cox '72." https://corps.tamu.edu/jerry-s-cox-72/.

12th Man Foundation. "2012 E. King Gill Award: Kay and Jerry S. Cox '72." https://www.12thmanfoundation.com/donor-impact/stories/kay-jerry-cox-gill-award/index.html.

Texas A&M University. "Powell Establishes Chairs in Math, Marine Engineering." *Texas A&M Stories* (February 1, 2008). https://today.tamu.edu/2008/02/01/powell-establishes-chairs-in-math-marine-engineering/.

Wikipedia Foundation. "Rett Syndrome." Last modified on Jun 8, 2025, 17:26 (UTC). https://en.wikipedia.org/wiki/Rett_syndrome.

Hutchins, Shana. "Powell Named Texas A&M Distinguished Alumnus." Texas A&M University College of Arts & Science (May 31, 2022). https://artsci.tamu.edu/news/2022/05/powell-named-texas-am-distinguished-alumnus.html.

Fry, Clifford. "Post 8: Ashley and Henry." Clifford Fry Blog. https://www.cliffordlfry.com/blog/rett-syndrome/post-8-ashley-and-henry-2/.

Fry, Clifford. "Special Needs, Special Love." Dr. Fry's Texas Medicine Band Blog. https://mailchi.mp/33b84968d951/greetings-from-clifford-6278774?e=587f3afaa9.

RRC Economics Consulting & Survey Management. "Permanent Staff-Thomas Saving." https://www.rrc-inc.com/our-people/.

Texas A&M University Hagler Institute for Advanced Study. "Huda Zoghbi." https://hias.tamu.edu/fellow/huda-zoghbi/.

Texas A&M University Hagler Institute for Advanced Study Archives. "Hagler Spring Lecture—Huda Zoghbi."

CHAPTER 11

Crosby Ells, Walter. "The Origin and Early History of Sabbatical Leave." *AAUP Bulletin* 48, no. 3 (September 1962). https://www.jstor.org/stable/40222893.

Singletary, G. Page. "History of Sabbaticals (from 'The Gift of a Sabbatical' Series)." LinkedIn, December 18, 2018. https://www.linkedin.com/pulse/gift-sabbatical-part-two-history-sabbaticals-g-page-singletary.

University of North Texas Department of English. "Deborah Armintor Ph.D." https://english.unt.edu/people/deborah-needleman-armintor-phd.

adminSEAS. "Speech of Acceptance of the 2011 Timoshenko Medal by Alan Needleman." iMechanica: Web of Mechanics and Mechanicians, November 16, 2011. https://imechanica.org/node/11423.

Texas A&M University Hagler Institute for Advanced Study. "Nomination Guidelines." April 3, 2023. https://hias.tamu.edu/wp-content/uploads/2023/08/Call-for-nominations-2024-25.pdf.

Georgia Tech University. "Curriculum Vitae of Dimitris C. Lagoudas, Ph.D., P.E." https://president.gatech.edu/sites/default/files/documents/lagoudas_cv2018.pdf.

Texas A&M University Department of Aerospace Engineering. "Dimitris Lagoudas." https://engineering.tamu.edu/aerospace/profiles/lagoudas-dimitris.html.

Wikipedia Foundation. "David Lee (Physicist)." Last modified on March 9, 2025, 09:20 (UTC). https://en.wikipedia.org/wiki/David_Lee_(physicist).

Texas Materials Modeling Network. "Session 1: Materials Chemistry Chair and Moderator: Alan Needleman, Materials Science and Engineering, University of North Texas." December 14, 2012. https://txmmn.net/annual-workshop/2nd-annual-workshop/.

McFarlane, Bruce. "The Academic Sabbatical as a Symbol of Change in Higher Education: From Rest and Recuperation to Hyper-Performativity." *Journal of Higher Education Policy and Management* 45, no. 3 (2023). https://www.tandfonline.com/doi/full/10.1080/1360080X.2022.2140888.

Schabram, Kris, Matt Bloom, and D. J. DiDonna. "The Transformative Power of Sabbaticals." *Harvard Business Review* (February 22, 2023). https://hbr.org/2023/02/research-the-transformative-power-of-sabbaticals.

Harvard University. "The Needleman Lab." https://needleman.seas.harvard.edu/.

Wikipedia Foundation. "*Horton Hatches the Egg*." Last modified on January 3, 2025, 19:17 (UTC). https://en.wikipedia.org/wiki/Horton_Hatches_the_Egg.

Tozke, Deanna. "Three College of Engineering Faculty Among Hagler Institute's First Permanent Members." Texas A&M University Engineering (February 20, 2019). https://engineering.tamu.edu/news/2019/02/three-college-of-engineering-faculty-among-hagler-institutes-first-permanent-members.html.

CHAPTER 12

Visit Sweden. "Uppsala—a Historical Treasure Driving Towards a Net Zero Future." Last updated February 7, 2024. https://visitsweden.com/where-to-go/middle-sweden/uppland/uppsala/.

Texas A&M University. "Prominent Swedish Animal Scientist Currently Serving at Texas A&M Wins Prestigious Wolf Prize." *Texas A&M Stories* (January 24, 2014). https://today.tamu.edu/2014/01/24/prominent-swedish-animal-scientist-currently-serving-at-texas-am-wins-prestigious-wolf-prize/.

Preigh, Margaret. "A Wonder-Full Life: Dr. Leif Andersson." Texas A&M University Veterinary Medicine & Biological Sciences, August 18, 2020. https://vetmed.tamu.edu/news/press-releases/a-wonder-full-life-dr-leif-andersson/.

Andersson, Leif, et al. "Assignment of the Bovine Immunoglobulin Gamma Heavy Chain (IGHG) Gene to Chromosome 21q24 by in situ Hybridization." *Hereditas* 117 (1992): 237–40. https://onlinelibrary.wiley.com/doi/pdf/10.1111/j.1601-5223.1992.tb00020.x.

Wolf Foundation. "Leif Andersson." https://wolffund.org.il/leif-andersson/.

Wolf Foundation. "About the Wolf Foundation." https://wolffund.org.il/about-wolf-foundation/.

Texas A&M University Veterinary Medicine & Biological Sciences. "Texas A&M Professor Andersson Elected into Prestigious American Philosophical Society." July 11, 2017. https://vetmed.tamu.edu/news/press-releases/texas-am-professor-andersson-elected-into-prestigious-american-philosophical-society/.

Texas A&M University Hagler Institute for Advanced Study. "Leif Andersson Curriculum Vitae." https://hias.tamu.edu/wp-content/uploads/2019/11/Andersson_CV.pdf.

Ledbetter, Kay. "Longtime Texas A&M Leader, Nationally Recognized Entomologist Passes." *AgriLife Today* (June 26, 2020). https://agrilifetoday.tamu.edu/2020/06/26/adkisson-longtime-texas-am-leader-nationally-recognized-entomologist-passes/.

Wikipedia Foundation. "Perry Adkisson." Last modified on June 12, 2024, 19:33 (UTC). https://en.wikipedia.org/wiki/Perry_Adkisson.

LinkedIn. "Margaret Preigh."

Wikipedia Foundation. "Qatar Foundation." Last modified on March 30, 2025, 22:41 (UTC). https://en.wikipedia.org/wiki/Qatar_Foundation.

The Bryan-College Station Eagle. "James Womack Obituary." (August 17, 2023). https://www.legacy.com/us/obituaries/theeagle/name/james-womack-obituary?id=52720299.

Texas A&M University Veterinary Medicine & Biological Sciences. "Distinguished Professor Blazes Trail in Genetics Research—a Man with a Plan." June 5, 2017. https://vetmed.tamu.edu/news/press-releases/distinguished-professor-blazes-trail-in-genetics-research-a-man-with-a-plan/.

National Academies. "President Barack Obama's Speech to the National Academy of Sciences [Full Transcript]." April 30, 2013. https://notes.nap.edu/2013/04/30/president-barack-obamas-speech-to-the-national-academy-of-sciences-full-transcript/.

American Philosophical Society. "Leif Andersson." https://diglib.amphilsoc.org/taxonomy/term/16685.

American Philosophical Society. "About the APS." https://www.amphilsoc.org/about.

Swedish University of Agricultural Sciences. "Professor Leif Andersson elected to the American Philosophical Society." (September 13, 2017). https://www.slu.se/en/ew-news/2017/9/leif-andersson-aps-member/.

Uppsala University. "Leif Andersson." https://www.uu.se/en/contact-and-organisation/staff?query=SLU-30.

CHAPTER 13

Levine, Robert S., and Sandra M. Gustafson. *The Norton Anthology of American Literature.* 10th ed. W. W. Norton, 2022. https://wwnorton.com/books/9780393884425/about-the-book/product-details.

Wikipedia Foundation. "*The Norton Anthology of American Literature.*" Last modified on August 21, 2024, 21:12 (UTC). https://en.wikipedia.org/w/index.php?title=The_Norton_Anthology_of_American_Literature&action=history.

Texas A&M University College of Art & Science. "Larry J. Reynolds." Texas A&M University English. https://artsci.tamu.edu/english/contact/profiles/larry-j-reynolds.html.

Donoghue, Steven. "Book Review: *The Lives of Frederick Douglass.*" *Open Letters Monthly.* https://www.openlettersmonthlyarchive.com/olm/book-review-the-lives-of-frederick-douglass.

Texas A&M University Hagler Institute for Advanced Study. "Robert Levine." https://hias.tamu.edu/fellow/dr-robert-levine/.

The University of North Carolina Press. "*Martin Delany, Frederick Douglass, and the Politics of Representative Identity* by Robert S. Levine." https://uncpress.org/book/9780807846339/martin-delany-frederick-douglass-and-the-politics-of-representative-identity/.

The University of North Carolina Press. "*Frederick Douglass and Herman Melville* by Robert S. Levine." https://uncpress.org/book/9780807858721/frederick-douglass-and-herman-melville/.

University of Maryland. "Robert Levine-Other Publications." https://blog.umd.edu/robert-s-levine/other-publications/.

The University of Maryland English. "Curriculum Vitae: Robert Levine." https://english.umd.edu/sites/default/files/2024-03/levinecv2024.pdf.

University of Arkansas-The Wild Blue Yonder: The English MA/PhD Alumni Profile Blog. "From the U of A—to Texas A&M—to California State University, Long Beach: An Interview with Alum (M.A. '11) Dr. Rene H. Treviño." https://englishalumni.uark.edu/from-the-u-of-a-to-texas-am-to-california-state-university-long-beach-an-interview-with-alum-and-professor-rene-trevino-m-a-11/.

CHAPTER 14

Texas A&M University Hagler Institute for Advanced Study. "Wolfgang Schleich." https://hias.tamu.edu/fellow/dr-wolfgang-schleich/.

Austrian Academy of Sciences. "Curriculum Vitae: Wolfgang Peter Schleich." https://www.oeaw.ac.at/fileadmin/mitglieder/cv/CV_Schleich.pdf.

Wikipedia Foundation. "Max Planck." Last modified on June 8, 2025, 10:03 (UTC). https://en.wikipedia.org/wiki/Max_Planck.

Max Plank Institute of Quantum Optics. "Herbert Walther." https://www.mpq.mpg.de/4928685/walther.

CHAPTER 15

Texas A&M University Texas A&M Energy Institute. "Newly Dedicated Giesecke Engineering Research Building to House Energy Institute Faculty and Students." (May 8, 2015). https://energy.tamu.edu/news_item/newly-dedicated-giesecke-engineering-research-building-to-house-energy-institute-faculty-and-students/.

Princeton University Office of the Dean of the Faculty. "Christodoulos Achilleus Floudas." https://dof.princeton.edu/people/christodoulos-achilleus-floudas.

Sullivan, John. "Pioneer of Global Optimization Christodoulos Floudas Dies." Princeton University (August 17, 2016). https://www.princeton.edu/news/2016/08/17/pioneer-global-optimization-christodoulos-floudas-dies.

National Academy of Engineering. "Dr. Ignacio E. Grossmann." https://www.nae.edu/29863.aspx.

National Academy of Engineering. "Professor Christodoulos A. Floudas." https://www.nae.edu/42069/Professor-Christodoulos-A-Floudas.

Thompson, Drew. "Christodoulos Floudas, World-Renowned Scholar, Professor and Director of the Texas A&M Energy Institute Passes Away." Texas A&M University Engineering (August 15, 2016). https://engineering.tamu.edu/news/2016/08/christodoulos-floudas-world-renowned-scholar-professor-and-director-of-the-texas-am-energy-institute-passes-away.html.

Floudas, Christodoulos. "A New Era of Energy." *Spirit Magazine* (Spring 2017). https://spirit.txamfoundation.com/Spring-2016/Opportunity.aspx.

The University of Arizona Chemical & Environmental Engineering. "Paraskevi (Evi) Flouda." https://chee.engineering.arizona.edu/faculty-staff/faculty/paraskevi-evi-flouda.

Google Scholar. "Stratos Pistikopoulos." https://scholar.google.com/citations?user=n2urHL4AAAAJ.

Journal of Petroleum Technology. "In Memoriam: Stephen A. Holditch." (August 8, 2019). https://jpt.spe.org/memoriam-stephen-holditch.

Luedke, Nancy. "Remembering Stephen Holditch." Texas A&M University Engineering (August 13, 2019). https://engineering.tamu.edu/news/2019/08/remembering-stephen-a-holditch.html.

Texas A&M Engineering Experiment Station. "EEI Names Pappas Associate Director, Announces Organizational Changes." (February 16, 2015). https://tees.tamu.edu/news/2011/03/eei-names-pappas-associate-director-announces-organizational-changes.html.

Texas A&M University Texas A&M Energy Institute. "Professor Christodoulos A. Floudas Remembered." (August 15, 2016). https://energy.tamu.edu/news_item/professor-christodoulos-a-floudas-remembered/.

Seaton, Matt. "The Accidental Laureate." *The Guardian* (September 16, 2002). https://www.theguardian.com/books/2002/sep/16/artsfeatures.poetry.

CHAPTER 16

Brown, Bailey. "Stephanie Sale and Jim Singleton Awarded 2021 Downtown Impact Award." *Bryan-College Station Eagle* (December 21, 2021). https://theeagle.com/stephanie-sale-and-jim-singleton-awarded-2021-downtown-impact-award/article_c98e4fde-62b5-11ec-bc38-a3f92387a6b9.html.

Texas A&M University. "Distinguished Alumni 2018—Jim Singleton." *Texas A&M Stories* (June 7, 2018). https://today.tamu.edu/2018/06/07/ten-selected-as-2018-distinguished-alumni-of-texas-am-university/.

Youngkin, Bill. "James M. 'Jim' Singleton, Vietnam War Veteran." *Bryan-College Station Eagle* (May 9, 2016). https://theeagle.com/veterans/james-m-jim-singleton-vietnam-war-veteran/article_0126455a-2c02-11e6-b51a-4fa492d54ebc.html#tncms-source=login.

Wilson, Sarah. "Singleton Tapped as Texas A&M University Distinguished Alumnus." Texas A&M University College of Architecture (March 28, 2018). https://newsarchive.arch.tamu.edu/news/2018/3/28/singleton-distinguished-alumnus-award/.

CHAPTER 17

Texas A&M University. "Distinguished Alumnus Harold Adams to Receive Honorary Texas A&M Ph.D." *Texas A&M Stories* (April 24, 2019). https://today.tamu

.edu/2019/04/24/distinguished-alumni-harold-adams-to-receive-honorary-texas-am-ph-d/.

Rolfing, Phillip. "100 years of architectural education." Texas A&M University College of Architecture. http://archcomm.arch.tamu.edu/archive/news/fall2006/collegeHistory.html.

CRS Center for Leadership and Management in the Design and Construction Industry, in the College of Architecture at Texas A&M University. "LEADERSHIP 1.1 Texas A&M University Harold ADAMS '61." (December 2016). https://www.youtube.com/watch?v=C99sUVxSJZY.

Texas A&M University College of Architecture. "Aggie Leadership C8 Series: Lessons in Leadership." https://www.arch.tamu.edu/impact/centers-institutes-outreach/crs-center/education/special-lectures/aggie-leadership-c8-series/.

American Institute of Architects Baltimore; Baltimore Architecture Foundation. "Press Release: Harold L. Adams to Receive Ted C. Kennedy Award." (July 6, 2021). https://www.aiabaltimore.org/2021-07-06/press-release-harold-l-adams-to-receive-ted-c-kennedy-award/.

New York University Tandon School of Engineering. "Theodore (Ted) S. Rappaport." https://engineering.nyu.edu/faculty/theodore-ted-s-rappaport.

Wikipedia Foundation. "Ronald Reagan Washington National Airport." Last modified on June 12, 2025, 19:41 (UTC). https://en.wikipedia.org/wiki/Ronald_Reagan_Washington_National_Airport.

Wikipedia Foundation. "Eugene Worley." Last updated December 7, 2024, 23:16 (UTC). https://en.wikipedia.org/wiki/Eugene_Worley.

Gunts, Edward. "Harold L. Adams, longtime CEO of RTKL, passes away at 82." *The Architect's Newspaper* (April 13, 2022). https://www.archpaper.com/2022/04/harold-l-adams-longtime-ceo-of-rtkl-obituary/.

RobertRiggs.com. "About." https://robertriggs.com/about/.

University of Maryland School of Architecture, Planning and Preservation. "The School of Architecture, Planning and Preservation Mourns Harold Adams, FAIA." (April 13, 2022). https://arch.umd.edu/news-events/school-architecture-planning-and-preservation-mourns-harold-adams-faia.

Eagle Staff. "Texas A&M Architecture Professor, Distinguished Alumnus Harold Adams Dies." *Bryan-College Station Eagle* (April 20, 2022). https://theeagle.com/news/a_m/texas-a-m-architecture-professor-distinguished-alumnus-harold-adams-dies/article_f02bc264-c0e5-11ec-b9a5-e78e3324fe88.html.

Nira, Richard. "Distinguished Alum Adams '61 Received Honorary Texas A&M Ph.D. at Spring Commencement." Texas A&M University College of Architecture, April 23, 2019. https://newsarchive.arch.tamu.edu/news/2019/4/23/distinguished-alum-adams-61-receive-honorary-texas-m-phd-spring-commencement/.

CHAPTER 18

National Aeronautics and Space Administration. "Marshall Space Flight Center History." https://www.nasa.gov/marshall/marshall-space-flight-center-history/.

Walker, Felysha. "Remembering Dr. Robert 'Bob' Skelton." Texas A&M University College of Engineering, February 22, 2023. https://engineering.tamu.edu/news/2023/02/remembering-dr-robert-bob-skelton.html.

McHarg, Jan. "Building a Habitat for Sustainable Life in Space." *Texas A&M Stories* (February 28, 2019). https://today.tamu.edu/2019/02/28/building-a-habitat-for-sustainable-life-in-space/.

Texas A&M University Hagler Institute for Advanced Study. "Robert Skelton." https://hias.tamu.edu/fellow/dr-robert-skelton/.

UC San Diego Jacobs School of Engineering. "In Memoriam: Professor Robert E. Skelton." https://jacobsschool.ucsd.edu/news/release/3592?id=3592.

Texas A&M University Department of Aerospace Engineering. "Manoranjan Majji." https://engineering.tamu.edu/aerospace/profiles/majji-manoranjan.html.

Wikipedia Foundation. "Purdue University School of Aeronautics and Astronautics." Last modified on May 19, 2025, 16:08 (UTC). https://en.wikipedia.org/wiki/Purdue_University_School_of_Aeronautics_and_Astronautics.

Texas A&M University Land Air and Space Robotics Laboratory. "Dr. Manoranjan Majji." https://lasr.tamu.edu/team/.

Pereira, Matthew. "Tensegrity Structures: What They Are and What They Can Be." *Arch Daily* (June 3, 2018). https://www.archdaily.com/893555/tensegrity-structures-what-they-are-and-what-they-can-be.

Wikipedia Foundation. "Tensegrity." Last modified on May 28, 2025, 06:59 (UTC). https://en.wikipedia.org/wiki/Tensegrity.

McHarg, Jan. "Skelton to receive AIAA Mechanics and Control of Flight Award." Texas A&M Engineering Experiment Station, January 10, 2017. https://tees.tamu.edu/news/2017/01/skelton-to-receive-aiaa-mechanics-and-control-of-flight-award.html.

Brain Corporation. "Tensegrity." Last modified August 31, 2012, 17:51 (UTC). http://www.scholarpedia.org/article/Tensegrity.

Buckminster Fuller Institute. "Tensegrity." https://www.bfi.org/about-fuller/big-ideas/tensegrity/.

TensegrityWiki.com. "Tensegrity/Etymology." Last modified on April 11, 2022, 22:22 (UTC). https://tensegritywiki.com/index.php?title=Tensegrity,_Etymology.

CHAPTER 19

Texas A&M University. "Jon Hagler to Be Presented Honorary Doctorate." *Texas A&M Stories* (May 13, 2015). https://today.tamu.edu/2015/05/13/financierphilanthropist-jon-hagler-to-be-presented-honorary-doctorate/.

Texas A&M Foundation. "Texas A&M Foundation Receives $20 Million Commitment to Name Hagler Institute for Advanced Study." February 20, 2017. https://www.txamfoundation.com/News/Foundation-Receives-20-Million-Commitment-to-Name-Hagler-Institute-for-Advanced-Study.aspx.

Texas A&M University Tyrus Timm Honor Registry. "Jon Hagler." https://tytimm.tamu.edu/hagler-jon-l/.

Brockenbush, Molly. "Great Minds." *Spirit Magazine* (Spring 2017). https://spirit.txamfoundation.com/Spring-2017/Cover-Feature.aspx.

Texas A&M University Corps of Cadets. "Jon L. Hagler '58." https://corps.tamu.edu/jon-l-hagler-58/.

Baylor College of Medicine. "Richard Gibbs." https://www.bcm.edu/people-search/richard-gibbs-22009.

World Biographical Encyclopedia. "Jon Lewis Hagler." https://prabook.com/web/jon_lewis.hagler/1657832.

Wikipedia Foundation. "Savings and Loan Association." Last modified on May 27, 2025, 15:58 (UTC). https://en.wikipedia.org/wiki/Savings_and_loan_association.

Texas A&M University MSC Programs. "J. Wayne Stark." https://mscprograms.tamu.edu/about-us/j-wayne-stark/.

Texas A&M Foundation. "About the President." https://www.txamfoundation.com/contact-us/executive-staff/tyson-voelkel.aspx.

CHAPTER 20

Texas A&M University Department of Physics & Astronomy. "Astronomer Robert Kennicutt Appointed University Distinguished Professor." April 11, 2022. https://physics.tamu.edu/astronomer-robert-kennicutt-appointed-university-distinguished-professor/.

Texas A&M University Hagler Institute for Advanced Study. "Robert C. Kennicutt." https://hias.tamu.edu/fellow/dr-robert-kennicutt-jr/.

U.S. House of Representatives Document Repository. "Curriculum Vitae: Robert C. Kennicutt, Jr." https://docs.house.gov/meetings/SY/SY16/20211201/114271/HHRG-117-SY16-Bio-KennicuttR-20211201.pdf.

Nobel Prize Outreach. "The Nobel Prize in Physics 2011." https://www.nobelprize.org/prizes/physics/2011/summary/.

Wikipedia Foundation. "High-Z Supernova Search Team." Last modified on November 19, 2024, 20:16 (UTC). https://en.wikipedia.org/wiki/High-Z_Supernova_Search_Team.

Texas A&M University Department of Physics & Engineering Festival. "The Greeks, Einstein, and Alien Worlds: The Importance of Eclipses." https://physicsfestival.tamu.edu/nicholas-suntzeff/.

LinkedIn. "Nicholas Suntzeff."

Wikipedia Foundation. "Kennicutt–Schmidt Law." Last modified on March 27, 2024, 21:16 (UTC). https://en.wikipedia.org/wiki/Kennicutt%E2%80%93Schmidt_law.

Kepos, Paula, ed. "Mitchell Energy and Development Corporation." In *International Directory of Company Histories*. St. James Press, 1993. https://www.company-histories.com/Mitchell-Energy-and-Development-Corporation-Company-History.html.

Wikipedia Foundation. "George P. Mitchell." Last modified on April 18, 2025, 03:56 (UTC). https://en.wikipedia.org/wiki/George_P._Mitchell.

Steffy, Loren. "How One Desperate Energy Executive Launched the Fracking Revolution." *Texas Monthly* (October 16, 2019). https://www.texasmonthly.com/news-politics/energy-executive-launched-fracking-revolution/.

Steffy, Loren. *George P. Mitchell: Fracking, Sustainability, and an Unorthodox Quest to Save the Planet*. Texas A&M University Press, 2019.

Eagle Staff. "George Mitchell Loved, Endowed Texas A&M." *Bryan-College Station Eagle* (July 21, 2020). https://theeagle.com/opinion/editorials/george-mitchell-loved-endowed-texas-a-m/article_86a5acae-f864-11e2-a207-0019bb2963f4.html#tncms-source=login.

Wikipedia Foundation. "Washington Huskies Football." Last modified on May 25, 2025, 17:38 (UTC). https://en.wikipedia.org/wiki/Washington_Huskies_football.

Wikipedia Foundation. "Giant Magellan Telescope." Last modified on June 5, 2025, 12:23 (UTC). https://en.wikipedia.org/wiki/Giant_Magellan_Telescope.

Encyclopedia Britannica. "Stephen Hawking." Last modified on May 16, 2025. https://www.britannica.com/biography/Stephen-Hawking.

Texas A&M University Department of Physics & Astronomy. "Star Panel to Meet on Deep Space Observational Astronomy Proposal for Mitchell Institute." October 15, 2003. https://physics.tamu.edu/star-panel-to-meet-on-deep-space-observational-astronomy-proposal-for-mitchell-institute/.

Texas A&M University Mitchell Institute. "A Brief History of the George P. and Cynthia W. Mitchell Institute for Fundamental Physics and Astronomy." https://mitchell.tamu.edu/about/history/.

Berger, Eric. "George Mitchell: A Man Who Quietly Helped Bankroll the Superconducting Super Collider and Saved the Alpha Magnetic Spectrometer." *Houston Chronicle* (August 5, 2013). https://cgmf.org/blog-entry/70/George-Mitchell-A-man-who-quietly-helped-bankroll-the-Superconducting-Super-Collider-and-saved-the-Alpha-Magnetic-Spectrometer.html.

Berger, Eric. "Money Made in Oil, Mitchell Dreamed of the Stars." *Houston Chronicle* (August 2, 2013). http://www.houstonchronicle.com/news/nation-world/article/money-made-in-oil-mitchell-dreamed-of-the-stars-4704432.php?t=e9eed6895e057dc840.

Wikipedia Foundation. "Wendy Freedman." Last modified on May 11, 2005, 17:07 (UTC). https://en.wikipedia.org/wiki/Wendy_Freedman.

Aspen Center for Physics. https://aspenphys.org/.

Texas A&M University Department of Physics & Astronomy. "George Mitchell, Mitchell Foundation Commit $25 Million to Giant Magellan Telescope." May 10, 2011. https://physics.tamu.edu/george-mitchell-mitchell-foundation-commit-25-million-to-giant-magellan-telescope/.

Texas A&M University. "Texas A&M Mourns Legendary Houston Philanthropist and Distinguished Former Student George P. Mitchell." *Texas A&M Stories* (July 26, 2013). https://today.tamu.edu/2013/07/26/texas-am-mourns-legendary-houston-philanthropist-and-distinguished-former-student-george-p-mitchell/.

Kraus, Rachel. "Meet the Man Whose Voice Became Stephen Hawking's." Mashable, March 14, 2018. https://mashable.com/article/stephen-hawking-voice-origin-story.

CHAPTER 21

Pew Research. "What Role Does Caste Play?." August 23, 2022. https://www.pewresearch.org/religion/2022/08/23/what-role-does-caste-play/.

Wikipedia Foundation. "Indian Institutes of Technology." Last modified on June 11, 2025, 07:02 (UTC). https://en.wikipedia.org/wiki/Indian_Institutes_of_Technology.

Wikipedia Foundation. "IRI (Company)." Last modified on May 5, 2025, 07:01 (UTC). https://en.wikipedia.org/wiki/IRI_(company).

Wikipedia Foundation. "*Slumdog Millionaire.*" Last modified on June 15, 2025, 04:40 (UTC). https://en.wikipedia.org/wiki/Slumdog_Millionaire.

Texas A&M Mays Business School. "Amalesh Sharma Curriculum Vitae." https://mays.tamu.edu/wp-content/uploads/2024/02/CV-Amalesh-Sharma-202402.pdf.

DrVKumar.com. "Curriculum Vitae—V Kumar." https://www.drvkumar.com/wp-content/uploads/2021/08/Curriculum-Vita-July-30-2021_VK.pdf.

DrVKumar.com. "Publications." https://www.drvkumar.com/publications/.

DrVKumar.com. "Professor/Thought Leader." https://www.drvkumar.com/.

Kumar, V, and Philip Kotler. *Transformative Marketing: Combining New Age Technologies and Human Insights*. Springer Nature Switzerland, 2024. https://books.google.com/books/about/Transformative_Marketing.html?id=2r20OAEACAAJ&source=kp_book_description.

CHAPTER 22

Wikipedia Foundation. "James E. Hubbard Jr." Last modified on September 29, 2023, 22:23 (UTC). https://en.wikipedia.org/wiki/James_E._Hubbard_Jr.

Texas A&M University Hagler Institute for Advanced Study. "The Mother of Invention: One Researcher's Unorthodox Approach to Human Cognition." https://tamuresearch.foleon.com/hagler-institute/cornerstone-2020/featured-article-3/.

Spacefacts.com. "Biographies of Astronaut and Cosmonaut Candidates—James Hubbard." http://www.spacefacts.de/bios/candidates/nasa10/english/hubbard_james.htm.

Texas A&M Transportation Institute. "Biography of Herbert Richardson." https://static.tti.tamu.edu/swutc.tamu.edu/publications/Bios/Richardson.pdf.

MIT Black History. "NASA Figures: An MIT Constellation." https://www.blackhistory.mit.edu/story/nasa-figures.

Waymarking.com. "Robert Poole Middle School-Hampden Historic District—Baltimore MD." https://www.waymarking.com/waymarks/wm148ZV_.

Texas A&M University Morpheus Lab. "Dr. James E. Hubbard, Jr." https://morpheuslab.engr.tamu.edu/people/dr-james-e-hubbard-jr/.

University of Maryland Morpheus Lab. "Resume of Dr. James Edward Hubbard, Jr." http://www.morpheus.umd.edu/documents/cv-je-hubbard.pdf.

CHAPTER 23

Salazar, Victor. "World-Renowned Researcher to Join Biomedical Engineering." Texas A&M University Engineering, January 6, 2023. https://engineering.tamu.edu/news/2023/01/world-renowned-researcher-to-join-biomedical-engineering.html.

Pontifical Academy of Sciences. "Goals." https://www.pas.va/en/about/goals.html.

Pontifical Academy of Sciences. "Vanderlei Salvador Bagnato." https://www.pas.va/en/academicians/ordinary/bagnato.html.

Texas A&M University Department of Biomedical Engineering. "Vanderlei Bagnato." https://engineering.tamu.edu/biomedical/profiles/bagnato-vanderlei.html.

Wright, Sarah. "George Koster, Physics Professor Emeritus, at 85." Massachusetts Institute of Technology Department of Physics. https://physics.mit.edu/wp-content/uploads/2021/01/physicsatmit_12_inremembrance.pdf.

Texas A&M University. "Texas A&M Hagler Institute Reveals 2018–19 Faculty Fellows, Distinguished Lecturer." Texas A&M Stories (September 21, 2018). https://today.tamu.edu/2018/09/21/texas-am-hagler-institute-reveals-2018-19-faculty-fellows-distinguished-lecturer/.

Pivetta, Marcos. "Vanderlei Bagnato: 'This Isn't a Case of Brain Drain.'" *Revista Pesquisa FAPESP*, no. 329 (July 2023). https://revistapesquisa.fapesp.br/en/vanderlei-bagnato-this-isnt-a-case-of-brain-drain/.

Texas A&M University. "CPRIT Awards $11.3 Million in Cancer-Prevention and Scholar-Recruitment Grants to Texas A&M and Agencies." *Texas A&M Stories* (March 2, 2022). https://today.tamu.edu/2022/03/02/cprit-awards-11-3-million-in-cancer-prevention-and-scholar-recruitment-grants-to-texas-am-and-agencies/.

Reiley, Jennifer. "Researcher Develops New Technique to Sterilize Intubation Tubes." Texas A&M University Engineering, December 17, 2020. https://engineering.tamu.edu/news/2020/12/researcher-develops-new-technique-to-sterilize-intubation-tubes.html.

Optica. "Vanderlei S. Bagnato." Last Updated August 28, 2023. https://www.optica.org/history/biographies/bios/vanderlei_s_bagnato/.

Luedke, Nancy. "Light-Based Therapy Weakens Antibiotic-Resistant Bacteria." *Texas A&M Stories* (October 6, 2022). https://today.tamu.edu/2022/10/06/light-based-therapy-weakens-antibiotic-resistant-bacteria/.

CHAPTER 24

Wikipedia Foundation. "Vincent Poor." Last modified on May 25, 2025, 16:29 (UTC). https://en.wikipedia.org/wiki/Vincent_Poor.

Princeton University Electrical and Computer Engineering. "H. Vincent Poor." https://ece.princeton.edu/people/h-vincent-poor.

Protin, Corey. "The History of Cell Phones and How Drastically They've Changed." *Business Insider* (July 27, 2017). https://www.businessinsider.com/cell-phone-history-cars-mobile-motorola-apple-bell-labs-samsung-google-2017-7.

Wikipedia Foundation. "History of Mobile Phones." Last modified on June 15, 2025, 09:53 (UTC). https://en.wikipedia.org/wiki/History_of_mobile_phones.

Wikipedia Foundation. "Mobile Phone." Last modified on June 3, 2025, 22:36 (UTC). https://en.wikipedia.org/wiki/Mobile_phone.

Ndungu, Solomon. "(CDMA) Code Division Multiple Access." Informa TechTarget, March 12, 2021. https://www.techtarget.com/searchnetworking/definition/CDMA-Code-Division-Multiple-Access.

Sharpened Productions. "CDMA." https://techterms.com/definition/cdma.

Galazzo, Richard. "Timeline from 1G to 5G: A Brief History on Cell Phones." CEGN, September 21, 2020. https://www.cengn.ca/information-centre/innovation/timeline-from-1g-to-5g-a-brief-history-on-cell-phones/.

Wikipedia Foundation. "History of Nokia." Last modified on June 1, 2025, 10:11 (UTC). https://en.wikipedia.org/wiki/History_of_Nokia.

Wikipedia Foundation. "2G." Last modified on June 13, 2025, 17;49 (UTC). https://en.wikipedia.org/wiki/2G.

Fendelman, Adam. "What Is CDMA and How Does It Work?." Lifewire. Last updated on October 22, 2021. https://www.lifewire.com/definition-of-cdma-578666.

Princeton University Electrical and Computer Engineering. "Minjie Chen." https://ece.princeton.edu/people/minjie-chen.

Wikipedia Foundation. "McCaw Cellular Communications." Last modified on August 22, 2024, 07:39 (UTC). https://en.wikipedia.org/wiki/McCaw_Cellular_Communications.

Chen, Minjie, and H. Vincent Poor. "High Frequency Power Electronics at the Grid Edge: A Bottom Up Approach Towards the Smart Grid." *IEEE Electrification Magazine* (2020). https://oar.princeton.edu/bitstream/88435/pr11c1tf9b/1/chen-pesmg-2021.pdf.

The Hertz Foundation. "Dick Miles." https://www.hertzfoundation.org/person/richard-miles/.

NBC News. "Denise Chow." https://www.nbcnews.com/author/denise-chow-ncpn814621.

Princeton University. "Harold Vincent Poor." https://ece.princeton.edu/sites/g/files/toruqf1836/files/documents/cv_poor_march2024.pdf.

Wikipedia Foundation. "AT&T Wireless Services." Last modified on February 26, 2025, 11:49 (CTU). https://en.wikipedia.org/wiki/AT%26T_Wireless_Services.

Wikipedia Foundation. "Ignacio Rodríguez Iturbe." Last modified on January 26, 2025, 18:32 (UTC). https://en.wikipedia.org/wiki/Ignacio_Rodr%C3%ADguez_Iturbe.

Wikipedia Foundation. "Roy J. Glauber." Last modified on May 17, 2025, 11:47 (UTC). https://en.wikipedia.org/wiki/Roy_J._Glauber.

Princeton University Mechanical and Aerospace Engineering. "Richard Miles." https://mae.princeton.edu/people/faculty/miles.

CHAPTER 25

Wikipedia Foundation. "Rachel Moran." Last modified on October 30, 2024, 04:54 (UTC). https://en.wikipedia.org/wiki/Rachel_Moran.

Texas A&M University School of Law. "Rachel Moran." https://www.law.tamu.edu/faculty/faculty-profiles/rachel-f-moran.html.

Moran, Rachel. "*Interracial Intimacy* by Rachel Moran." University of Chicago Press, 2003. https://press.uchicago.edu/ucp/books/book/chicago/I/bo3683147.html.

University of California–Irvine School of Law. "UCI Law Distinguished Professor of Law Rachel F. Moran Named 2020–2021 Hagler Fellow." https://www.law.uci.edu/news/in-the-news/2020/moran-hagler-fellow.html.

CHAPTER 26

Texas A&M University College of Arts & Sciences. "Digital Future for New Variorum Edition of Shakespeare." November 19, 2019. https://liberalarts.tamu.edu/blog/2019/11/19/digital-future-for-new-variorum-edition-of-shakespeare/.

Ramirez, Reyes. "Much Ado About Shakespeare: Department of English Inherits Project 150 Years in the Making." Texas A&M University College of Arts & Sciences, April 8, 2024. https://artsci.tamu.edu/news/2024/04/much-ado-about-shakespeare-department-of-english-inherits-project-150-years-in-the-making.html.

Dabbs, Thomas. "SoS #28 | Lena Cowen Orlin | The Private Life of William Shakespeare." *Speaking of Shakespeare* (January 28, 2022). https://speakingofshakespeare.buzzsprout.com/1732460/episodes/9967792.

McGee, Kate. "Texas A&M President Katherine Banks Resigns amid Fallout from Failed Hiring of Journalism Professor." *Texas Tribune* (July 21, 2023). https://www.texastribune.org/2023/07/21/tamu-president-resign-journalism/.

Georgetown University. "Lena Cowen Orlin: Bio and Feature Works." https://gufaculty360.georgetown.edu/s/contact/00336000014RctmAAC/lena-orlin.

Johns Hopkins University Press. "In Memoriam: James L. Harner." March 15, 2017. https://www.press.jhu.edu/newsroom/memoriam-james-l-harner.

Daniels, Tiffany. "15 Everyday Words and Phrases Created by Shakespeare." Associated Newspapers Limited, April 22, 2016. Last updated on December 11, 2019. https://metro.co.uk/2016/04/22/15-day-to-day-quips-created-by-shakespeare-5731805/.

Encyclopedia Britannica. "Humphrey, Duke of Gloucester Fictional Character." https://www.britannica.com/topic/Humphrey-duke-of-Gloucester.

Wikipedia Foundation. "*Henry VI, Part 1*." Last updated on January 13, 2025, 18:28 (UTC). https://en.wikipedia.org/wiki/Henry_VI,_Part_1.

Open Source Shakespeare. "History of *Henry VI, Part I*." https://www.opensourceshakespeare.org/views/plays/play_view.php?WorkID=henry6p1&Act=1&Scene=3&Scope=scene.

LeGare, Hannay. "Love's Labor Is Not Lost with Shakespeare." Texas A&M College of Arts & Sciences, February 14, 2020. https://liberalarts.tamu.edu/blog/2020/02/14/loves-labor-is-not-lost-with-shakespeare/.

Werstein, Paul. "About." New Variorum Shakespeare. https://newvariorumshakespeare.org/about/.

New Variorum Shakespeare. "A Midsummer Night's Dream: Variorum Reader." https://newvariorumshakespeare.org/edition/mnd/.

Folger Shakespeare Library. "About." *World Shakespeare Bibliography*. https://www.worldshakesbib.org/about.

Wikipedia Foundation. "Folger Shakespeare Library." Last modified on June 7, 2025, 22:25 (UTC). https://en.wikipedia.org/wiki/Folger_Shakespeare_Library.

Shapiro, James. "A Scholarly Analysis of Shakespeare That Reads like a Detective Story." *New York Times* (November 23, 2021). https://www.nytimes.com/2021/11/23/books/review/private-life-of-william-shakespeare-lena-cowen-orlin.html.

Texas A&M University College of Arts & Sciences. "Jose Bermúdez." https://artsci.tamu.edu/contact/profiles/bermudez-jose.html.

Wikipedia Foundation. "Kathleen McElroy." Last modified on May 25, 2025, 13:06 (UTC). https://en.wikipedia.org/wiki/Kathleen_McElroy.

Moody, Josh. "The Slow, Then Sudden Downfall of a University Leader." *Inside Higher Ed* (August 10, 2023). https://www.insidehighered.com/news/governance/executive-leadership/2023/08/10/slow-then-sudden-downfall-university-leader.

McGee, Kate. "Top Texas A&M Officials Were Involved in Botched Recruiting of Journalism Professor, Who Will Receive $1 Million Settlement." *Texas Tribune* (August 3, 2023). https://www.texastribune.org/2023/08/03/texas-am-regents-kathleen-mcelroy/.

Texas A&M University Department of English. "Kris May." https://artsci.tamu.edu/english/contact/profiles/kris-may.html.

CHAPTER 27

Lindau Nobel Laureate Meetings. "Donna Strickland on Her First Lindau Meeting." Uploaded on July 3, 2019. https://www.youtube.com/watch?v=-KPaod-GoUo.

Fortin, Jacey. "For Just the Third Time in 117 Years, a Woman Wins the Nobel Prize in Physics." *New York Times* (October 2, 2018). https://www.nytimes.com/2018/10/02/science/donna-strickland-nobel-prize-physics.html.

Nobel Prize. "Do What You Love." Uploaded on December 6, 2018. https://www.youtube.com/shorts/gIDUYiOnZQM.

Chow, Denise. "Donna Strickland's Long Journey from Laser Jock to Third Woman Ever to Win Physics Nobel." *NBC News* (October 4, 2018). https://www.nbcnews.com/mach/science/donna-strickland-s-long-journey-laser-jock-third-woman-ever-ncna916451.

Nobel Prize Outreach. "Anne L'Huillier." https://www.nobelprize.org/prizes/physics/2023/lhuillier/facts/.

Wikipedia Foundation. "Anne L'Huillier." Last modified on May21, 2025, 17:58 (UTC). https://en.wikipedia.org/wiki/Anne_L%27Huillier.

Chapman University. "Dr. Vernon Smith." https://www.chapman.edu/our-faculty/vernon-smith.

Berkowitz, Rachel. "The Women Who Win." American Physical Society.(September 30, 2021). https://physics.aps.org/articles/v14/134.

Nobel Prize Outreach. "The Nobel Prize Award Ceremonies and Banquets." June 16, 2025. https://www.nobelprize.org/ceremonies/the-nobel-prize-award-ceremonies-and-banquets/.

American Academy of Achievement. "2019 Summit." https://achievement.org/summit/2019/.

Valich, Lindsey. "Love at First Light." *Rochester Review* 81, no. 1 (Fall 2018). https://www.rochester.edu/pr/Review/V81N1/0501_nobel.html.

University of Waterloo. "Donna Strickland." https://uwaterloo.ca/physics-astronomy/profile/strickla.

Brubacher, Ben. "How Bell's Theorem Proved 'Spooky Action at a Distance' Is Real." *Quantum Magazine* (July 20, 2021). https://www.quantamagazine.org/how-bells-theorem-proved-spooky-action-at-a-distance-is-real-20210720/.

McKevitt, Greg. "John Bell: The Belfast Scientist Who Proved Einstein Wrong." *BBC News* (November 4, 2014). https://www.bbc.com/news/uk-northern-ireland-29904682.

CHAPTER 28

Hutchins, Shane. "Biology Grad Student Enjoys 'Once-in-a-Lifetime Experience' at Lindau Nobel Laureate Meeting." Texas A&M College of Arts & Sciences, July 21, 2023. https://artsci.tamu.edu/news/2023/07/biology-grad-student-enjoys-once-in-a-lifetime-experience-at-lindau-nobel-laureate-meeting.html.

Wikipedia Foundation. "History of East Germany." Last modified on June 10, 2025, 11:23 (UTC). https://en.wikipedia.org/wiki/History_of_East_Germany.

Have Fun with History. "Famous German Scientists." (January 25, 2024). https://www.havefunwithhistory.com/famous-german-scientists/.

Texas A&M University Hagler Institute for Advanced Study. "A Conversation with Dr. John Cullen." August 24, 2023. https://www.youtube.com/watch?v=Ew5XG4ykHiU.

Lindau Nobel Laureate Meetings. "Council & Foundation." https://www.lindau-nobel.org/council-foundation/.

Webb, Brandon. "Texas A&M Signs Partnership Agreement with Nobel Laureate Organization." *Texas A&M Stories* (July 12, 2023). https://today.tamu.edu/2023/07/12/texas-am-signs-partnership-agreement-with-nobel-laureate-organization/.

Henschen, Folke. "From the First Nobel Prize Award Ceremony, 1901." *Nobel Prize Outreach*. https://www.nobelprize.org/ceremony/from-the-first-nobel-prize-award-ceremony-1901/.

Swedish Royal Court. "Count Lennart Bernadotte." https://www.kungahuset.se/english/royal-house/the-bernadotte-dynasty/count-lennart-bernadotte.

Lindau Nobel Laureate Meetings. "The History of the Lindau Nobel Laureate Meetings." https://www.lindau-nobel.org/history/.

Lindau Nobel Laureate Meetings Annual Report 2017. "Countess Bettina Bernadotte." https://www.mediatheque.lindau-nobel.org/files/37387.

Lindau Nobel Laureate Meetings. "About the Lindau Meetings." https://www.lindau-nobel.org/about-the-meetings/.

Texas A&M University. "Graduate Students Selected to Attend Prestigious Nobel Laureate Meeting." *Texas A&M Stories* (June 23, 2023). https://today.tamu.edu/2023/06/23/graduate-students-selected-to-attend-prestigious-nobel-laureate-meeting/?_gl=1*5yl9so*_ga*MzkoNDEyMjYxLjE3MDcoOTUyMjY.*_ga_SJ5GMNoZQL*MTcxMzI4MTcoNS4xNC4wLjE3MTMyODE3NDYuNTkuMC4w.

Wikipedia Foundation. "Lennart Bernadotte." Last updated on March 23, 2025, 03:43 (UTC). https://en.wikipedia.org/wiki/Lennart_Bernadotte.

Lindau Nobel Laureate Meetings. "Nominating Institutions for the 73rd Lindau Nobel Laureate Meeting (Physics)." https://www.lindau-nobel.org/nominating-institutions-lin024/.

Texas A&M University Hagler Institute for Advanced Study. "Stefan H. E. Kaufmann." https://hias.tamu.edu/fellow/stefan-h-e-kaufmann/.

Wikipedia Foundation. "Stefan Kaufmann (Politician)." Last modified on June 11, 2025, 16:34 (UTC). https://en.wikipedia.org/wiki/Stefan_Kaufmann_(politician).

Wikipedia Foundation. "Stefan Kaufmann (Musician)." Last modified on April 1, 2024, 12:23 (UTC). https://en.wikipedia.org/wiki/Stefan_Kaufmann_(musician).

Encyclopaedia Metallum. "Stefan Kaufmann." Last modified on June 6, 2025, 13:13 (UTC). https://www.metal-archives.com/artists/Stefan_Kaufmann/21621.

Myers, Megan. "Texas A&M University Spotlights Immunology Research at Inaugural Symposium." Texas A&M University Veterinary Medicine & Biomedical Sciences, February 28, 2023. https://vetmed.tamu.edu/news/internal-news/inaugural-immunology-symposium/?highlight=stefan+kaufmann.

Wikipedia Foundation. "Stefan H. E. Kaufmann." Last modified on May 25, 2025, 01:59 (UTC). https://en.wikipedia.org/wiki/Stefan_H._E._Kaufmann#cite_note-2.

Texas A&M University Hagler Institute for Advanced Study. "Eminent Scholar Lecture Stefan H. E. Kaufmann—the Fight Against Tuberculosis: Lessons from the Past, Thoughts for the Future." https://hias.tamu.edu/wp-content/uploads/2024/02/Eminent-Scholar-Lecture-Stefan-Kaufmann-2024.pdf.

Max Planck Institute. "Stefan Kaufmann Becomes Emeritus Director at the Max Planck Institute for Biophysical Chemistry." May 25, 2020. https://www.mpinat.mpg.de/649228/pr_2011.

Texas A&M Veterinary Medicine & Biomedical Sciences. "Texas A&M Professor Garry Adams Receives AVMA Award." August 14, 2017. https://vetmed.tamu.edu/news/press-releases/texas-am-professor-garry-adams-receives-avma-award/.

Rodgers, Jimmie. "T.B. Blues." Genius Holdings, LLC. https://genius.com/Jimmie-rodgers-tb-blues-lyrics.

Wikipedia Foundation. "Jimmie Rodgers." Last modified on May 16, 2025, 23:06 (UTC). https://en.wikipedia.org/wiki/Jimmie_Rodgers.

Wikipedia Foundation. "Mycobacterial Cervical Lymphadenitis." Last modified on May 18, 2025, 17:25 (UTC). https://en.wikipedia.org/wiki/Mycobacterial_cervical_lymphadenitis.

Wikipedia Foundation. "Girolamo Fracastoro." Last modified on May 26, 2025, 15:07 (UTC). https://en.wikipedia.org/wiki/Girolamo_Fracastoro.

World Health Organization. "Tuberculosis." (March 14, 2025). https://www.who.int/news-room/fact-sheets/detail/tuberculosis.

Barberis, I., N. L. Bragazzi, L. Galluzzo, and M. Martini. "The History of Tuberculosis: From the First Historical Records to the Isolation of Koch's Bacillus." *Journal of Preventative Medicine and Hygiene* 1, no. 58 (March 2017). https://pmc.ncbi.nlm.nih.gov/articles/PMC5432783/.

CHAPTER 29

Milner, Richard. "OPINION: Parents, Here's Why Your Most Important Partner May Be Your Child's Teacher (and Vice Versa)." *Hechinger Report* (August 29, 2019). https://hechingerreport.org/opinion-parent-teacher-partnerships/.

Vanderbilt University. "Rich Milner." https://news.vanderbilt.edu/experts/expertprofile/?expert_username=rich.milner.

Vanderbilt University. "Curriculum Vitae: H. Richard Milner IV." (February 2023). https://cdn.vanderbilt.edu/vu-web/people-manager/files/people-255-PDF.MilnerCV2023-20230620095618.pdf.

Sage Publications. "*The Race Card* by H. Richard Milner IV." Corwin, 2023. https://www.corwin.com/books/the-race-card-285430.

DrLuciaHodgson.org. "Dr. Lucia Hodgson." https://drluciakhodgson.org/about/.

Linnaeus University. "New Sweden Texas: Swedish Settlers and Slavery in Nineteenth-Century Texas." November 17, 2023. https://lnu.se/en/meet-linnaeus-university/current/news/2023/lnucc-news---project-grant-by-the-swedish-research-council/.

Wikipedia Foundation. "Jerome Loving." Last modified on April 1, 2023, 16:12 (UTC). https://en.wikipedia.org/wiki/Jerome_Loving.

Lowcountry Panorama. "SC State Awards Distinguished Emeritus/Emerita Status to 11 Retirees." (May 26, 2022). https://www.lowcountrypanorama.com/lifestyles/sc-state-awards-distinguished-emeritus-emerita-status-to-11-retirees/article_8743071e-dd6c-11ec-a384-d7eb55a7dbc8.html.

Texas A&M University College of Education. "Renowned Scholar Named Hagler Institute Fellow." (October 19, 2023). https://education.tamu.edu/renowned-scholar-named-hagler-institute-fellow/.

CHAPTER 30

Wikipedia Foundation "Alexander the Great." Last modified on June 16, 2025, 11:35 (UTC). https://en.wikipedia.org/wiki/Alexander_the_Great.

Wikipedia Foundation. "History of Alexandria." Last modified on May 29, 2025, 05:41 (UTC). https://en.wikipedia.org/wiki/History_of_Alexandria.

Wikipedia Foundation. "Library of Alexandria." Last modified on June 14, 2025, 07:42 (UTC). https://en.wikipedia.org/wiki/Library_of_Alexandria.

Wikipedia Foundation. "Alexandria." Last modified on June 11, 2025, 02:27 (UTC). https://en.wikipedia.org/wiki/Alexandria.

Wikipedia Foundation "Alex Filippenko." Last modified on June 7, 2025, 21:23 (UTC). https://en.wikipedia.org/wiki/Alex_Filippenko.

CHAPTER 31

University of Central Florida. "Tarek Elgohary." https://mae.ucf.edu/person/tarek-elgohary/.

Texas A&M University Hagler Institute for Advanced Study. "Satya Atluri." https://hias.tamu.edu/fellow/dr-satya-atluri/.

LinkedIn. "Tarek Elgohary."

Texas Tech University Edward E. Whitacre Jr. College of Engineering. "In Memorium: Dr. Satya Atluri." August 16, 2023. https://www.depts.ttu.edu/coe/wcoenews/posts/2023/08/satya-alturi-memoriam.php.

CHAPTER 32

Texas A&M University Hagler Institute for Advanced Study. "The Art of Collaboration in Research." https://hias.tamu.edu/fall-symposium-2023/.

Texas A&M University Hagler Institute for Advanced Study. "John A. Rogers." https://hias.tamu.edu/fellow/dr-john-rogers/.

Texas A&M University. "Scholars Will Discuss Academic Collaboration at Hagler Institute Symposium." *Texas A&M Stories* (October 31, 2023). https://today.tamu.edu/2023/10/31/scholars-will-discuss-academic-collaboration-at-hagler-institute-symposium/.

JulioMarioOttino.com. "About JMO." https://www.juliomarioottino.com/about/.

YUMPU Media Publisher. "Julio Ottino Biography." https://www.yumpu.com/en/document/read/9230044/julio-m-ottinos-resume-mccormick-school-of-engineering-#google_vignette.

Northwestern University Kellogg School of Management. "Julio Ottino." https://www.kellogg.northwestern.edu/faculty/directory/ottino_julio.aspx.

Ottino, Julio Mario, with Bruce Mau. *The Nexus: Augmented Thinking for a Complex World—the New Convergence of Art, Technology, and Science*. MIT Press, 2022.

CHAPTER 33

Texas A&M University Nutrition. "B.S. Nutrition." https://nutrition.tamu.edu/bs-nutrition/.

Texas A&M University AgriLife. "Mission and History." Texas A&M AgriLife Extension Services. https://agrilifeextension.tamu.edu/about-2/mission-history/.

Texas A&M University. "Pennies on Sully." https://www.tamu.edu/campus-community/traditions/aggie-culture/pennies-on-sully.html.

Texas A&M University Corps of Cadets. "Ross Volunteers." https://www.corpsofcadets.org/the-corps/rv.aspx.

Texas A&M University College of Arts & Sciences. "Hagler Institute Inducts 2023–24 Fellows Class." March 4, 2024. https://artsci.tamu.edu/news/2024/03/hagler-institute-inducts-2023-24-fellows-class.html.

Texas A&M University Office of the President. "General (Ret.) Mark A. Welsh III." https://president.tamu.edu/about/index.html.

Texas A&M University Hagler Institute for Advanced Study. "2013 Gala." https://www.flickr.com/photos/tamu_tias/albums/72157708302223445/.

EPILOGUE

Lee, Jack. "Norman E. Borlaug: The Aggie Who Fed the World." *The Battalion* (November 29, 2022). https://thebatt.com/life-arts/norman-e-borlaug-the-aggie-who-fed-the-world/.

Texas A&M University AgriLife. "Continuing Dr. Norman Borlaug's Mission." Norman Borlaug Institute. https://borlaug.tamu.edu/.

Texas A&M University AgriLife. "Tribute to Dr. Norman Borlaug." Norman Borlaug Institute. https://borlaug.tamu.edu/dr-norman-borlaug/.

"Arid Zone Research." *Nature* 169 (January 5, 1952). https://www.nature.com/articles/169011a0.pdf.

Wikipedia Foundation. "Norman Borlaug." Last modified on May 23, 2025, 09:43 (UTC). https://en.wikipedia.org/wiki/Norman_Borlaug.

National Aeronautics and Space Administration. "Aaron Cohen." https://www.nasa.gov/people/aaron-cohen/.

Hamer, James L. "Harrison T. Meserole." Texas A&M University OAKtrust Digital Repository. https://oaktrust.library.tamu.edu/server/api/core/bitstreams/bdf18a5f-bfb2-44f1-b299-263e3eb6f613/content.

National Aeronautics and Space Administration Jet Propulsion Laboratory. "Dr. Frank J. Malina." https://www.jpl.nasa.gov/who-we-are/faces-of-leadership-the-directors-of-jpl/dr-frank-j-malina-1912-1981/.

Texas A&M University. "Hagler Institute Presents Malina Medallion Award to Ray Rothrock '77." *Texas A&M Stories* (March 2, 2023). https://today.tamu

.edu/2023/03/02/hagler-institute-presents-malina-medallion-award-to-ray-rothrock-77/.

Wikipedia Foundation. "Frank Malina." Last modified on May 26, 2025, 01:53 (UTC). https://en.wikipedia.org/wiki/Frank_Malina.

Chapman, Justin. "JPL Co-Founder and Rocketry Pioneer Frank Malina Finally Gets His Due (Part 2)." *Pasadena Now* (January 29, 2023). https://pasadenanow.com/main/jpl-co-founder-and-rocketry-pioneer-frank-malina-finally-gets-his-due-part-2-of-2.

Wikipedia Foundation. "Paul Revere." Last modified on May 31, 2025, 21:20 (UTC). https://en.wikipedia.org/wiki/Paul_Revere.

Urban Dictionary. "Tea-Sip." https://www.urbandictionary.com/define.php?term=tea-sip.

National Wrestling Hall of Fame. "Dr. Peter Likens." https://nwhof.org/national-wrestling-hall-of-fame/bio/136.

Ralston, Jeannie. "The Triple Crown." *Spirit Magazine* (Summer 2019). https://spirit.txamfoundation.com/summer-2019/cover-feature.aspx.

Texas A&M University Hagler Institute for Advanced Study. "Ned Thomas." https://hias.tamu.edu/fellow/ned-thomas/.

American Institute of Physics Publishing. "Unruh, William on 2021 May 20." May 20, 2021. https://www.aip.org/history-programs/niels-bohr-library/oral-histories/47038.

Texas Academy of Medicine, Engineering, Science & Technology. "Kay Bailey Hutchison Distinguished Service Award." https://tamest.org/kay-bailey-hutchison-award/.

The Welch Foundation. "Gina Luna." https://welch1.org/about/board-of-directors/gina-a-luna.

Williams, Mike. "Largest Gift in Rice History Establishes the Welch Institute." Rice University News and Media Relations (September 2, 2020). https://news.rice.edu/news/2020/largest-gift-rice-history-establishes-welch-institute.

The Welch Foundation. "The Welch Foundation Announces Retirement of Dr. Peter B. Dervan." (November 18, 2021). https://welch1.org/news-reports/news/the-welch-foundation-announces-retirement-of-dr-peter-b-dervan.

Junkins, John L. "From the Founding Director." *Hagler Institute for Advanced Study Newsletter*, no. 27 (September 2022). https://myemail.constantcontact.com/Hagler-Institute-for-Advanced-Study-Newsletter--Issue-27--Special-Edition.html?soid=1129174453048&aid=uHrQwhl-JD8.

Variety Testing. "Texas Corn Variety Trials." https://varietytesting.tamu.edu/corn/.

McKinney, Dave. "Governor Goofs Honest Abe Quote." *National Public Radio Illinois* (February 2, 2018). https://www.nprillinois.org/lincoln/2018-02-02/governor-goofs-honest-abe-quote.

INDEX

Photo insert images are indicated by *g1*, *g2*, *g3*, etc.